D1246780

CAMBRIDGE TRACTS IN MATHEMATICS

71. *Finite Free Resolutions*

D. G. NORTHCOTT

Town Trust Professor of Mathematics, University of Sheffield

71. Finite free resolutions

CAMBRIDGE UNIVERSITY PRESS

CAMBRIDGE

LONDON·NEW YORK·MELBOURNE

Published by the Syndics of the Cambridge University Press
The Pitt Building, Trumpington Street, Cambridge CB2 1RP
Bentley House, 200 Euston Road, London NW1 2DB
32 East 57th Street, New York, NY 10022, USA
296 Beaconsfield Parade, Middle Park, Melbourne 3206, Australia

Library of Congress Cataloging in Publication Data
Northcott, Douglas Geoffrey.
Finite free resolutions.
(Cambridge tracts in mathematics; no. 71)
Bibliography: p.
Includes Index.
1. Algebra, Homological.   2. Modules (Algebra)
I. Title.   II. Series
QA169.N66   512'.55   75–31397

ISBN: 0 521 21155 7

First published 1976

Printed in Great Britain at the
University Printing House, Cambridge
(Euan Phillips, University Printer)

Contents

Contents

Contents

Preface

This Cambridge Tract originated in a seminar given by J. A. Eagon while he was visiting Sheffield University during the session 1972/3. The aim of the seminar was to report on some recent discoveries of D. A. Buchsbaum and D. Eisenbud concerning finite free resolutions. I found myself fascinated by the subject and Eagon and I had many discussions on different aspects of it. In the end we were able to construct what we considered to be a simplified treatment of certain parts of the theory, and our ideas appeared subsequently in a joint paper.

I continued to think about these matters after Eagon had left Sheffield, and during 1973/4 and 1974/5 gave seminars covering an enlarged range of topics, but still using what I regard as elementary methods. The elementary approach was based on the belief that, for those sections of the theory I was considering, Noetherian conditions were never really necessary; consequently I was committed to showing that, where such considerations had previously been used, a way of getting rid of them could be found. Now the parts of the theory in which Noetherian properties had originally played an apparently vital rôle were, to a considerable extent, concerned with applications of the concept of grade; and I had, for some time, known of M. Hochster's approach to a theory of grade in which it was not necessary to restrict oneself to Noetherian modules. Thus the programme I had set myself hinged on adapting Hochster's ideas in order to produce a simplified theory of grade applicable to general modules over an arbitrary commutative ring. At the same time the theory had to be rich enough for the applications that I had in mind.

The results of the attempt to construct an appropriate theory of grade will be found in Chapter 5, and with the aid of what is contained there the original programme was carried through. One advantage that has emerged is that the account in the following pages demands remarkably little in the way of prerequisites. All that is required is a knowledge of the basic properties of modules and linear mappings, and, in a few places, a facility in working with tensor products is presupposed. Otherwise, with the exception of the Appendices, the account is quite self-contained. In particular I have resisted any temptation to use the theory of exterior algebras in the main text because, at this level, it is not particularly helpful and there are some key places where its use seems to have definite disadvantages. Of course it is to be expected that, as the theory grows, general structural considerations will come to play a more dominant rôle and that *ad hoc* computational arguments will be increasingly hard to find. I have tried, at appropriate places, to indicate how the subject of finite free resolutions has grown and to mention the names of those who have contributed to its development. Here I would like to thank those who came to my lectures and who encouraged me by their continued interest. I must add a special word of thanks to P. Vámos whose wide knowledge proved most helpful and whose comments on particular points led to many improvements; and to D. W. Sharpe for discussions on the rôle of grade in the theory of linear equations, for bringing various errors to my attention, and for help with reading the proofs. Also I am much indebted to C. J. Knight who invariably assists me with queries that have to do with Topology. But above all I wish to place on record my appreciation of the work done by my secretary, Mrs E. Benson, who, by producing the whole typescript with her characteristic skill and good judgement, has again enabled me to turn the notes for a seminar into a book.

Sheffield   D. G. NORTHCOTT
July 1975

1. *Matrices and determinants*

General remarks

Throughout Chapter 1, R will denote a commutative ring with an identity element. Usually the identity element of R will be denoted by 1, but 1_R will be used if we wish to draw attention to the ring in which we are working. This chapter is devoted to reviewing certain well known basic facts concerning matrices and determinants and organizing them in a way which will be useful later.

1.1 Matrices

By a *matrix with entries in R*, or an *R-matrix*, it is customary to understand a rectangular array of elements taken from the ring R. However there are many situations where the order of the rows and the order of the columns is not important. Again, in the theory of vector spaces, a linear mapping of one finite dimensional vector space into another can be described by means of a matrix as soon as each of the two spaces involved has been provided with a base. Now it can happen that one or possibly both of the spaces in question has dimension zero. To deal with such situations we need, for example, the notion of a matrix which has p rows (say) and zero columns. It is because of considerations such as these that we make a fresh start and approach the idea of a matrix from a slightly more general standpoint.

Let M and N be finite sets. By an $M \times N$ *matrix* with entries in R we shall mean a mapping

$$A : M \times N \to R \tag{1.1.1}$$

of the cartesian product $M \times N$ into the ring R. Suppose that we have such a matrix. Let $m \in M$ and $n \in N$. Then (m, n) has an

[1]

image in R which we may denote by a_{mn}. As an alternative notation for the matrix A we shall use

$$A = \|a_{mn}\| \quad (m \in M, n \in N). \tag{1.1.2}$$

This will be abbreviated to $A = \|a_{mn}\|$ if there is no risk of confusion.

Given an $M \times N$ matrix $A = \|a_{mn}\|$ we can associate with it an $N \times M$ matrix $C = \|c_{nm}\|$ by requiring that $c_{nm} = a_{mn}$ for all $m \in M$ and $n \in N$. The matrix C is called the *transpose* of A and we shall indicate the connection between the two by writing $C = A^{\mathrm{T}}$. Evidently if C is the transpose of A, then A is the transpose of C.

Now let M, N and Q be finite sets. Suppose that $A = \|a_{mn}\|$ is an $M \times N$ matrix and $B = \|b_{nq}\|$ is an $N \times Q$ matrix. This time we define an $M \times Q$ matrix $C = \|c_{mq}\|$ by putting

$$c_{mq} = \sum_{n \in N} a_{mn} b_{nq} \tag{1.1.3}$$

when $m \in M$ and $q \in Q$. The matrix C is called the *product* of A and B and we write $C = AB$. Clearly

$$(AB)^{\mathrm{T}} = B^{\mathrm{T}} A^{\mathrm{T}}. \tag{1.1.4}$$

Note that the possibility that N may be the empty set is not excluded. In such a situation the product AB will be the zero $M \times Q$ matrix, that is to say each of its entries will be the zero element of R.

Let p and q be positive integers and take $M = \{1, 2, ..., p\}$, $N = \{1, 2, ..., q\}$. For these particular choices of M and N it is customary to refer to an $M \times N$ matrix as a $p \times q$ matrix and we shall observe this custom. Also if $A = \|a_{jk}\|$, where $1 \leqslant j \leqslant p$ and $1 \leqslant k \leqslant q$, is a $p \times q$ matrix we shall, when convenient, exhibit the relation between A and its individual entries by writing

$$A = \begin{Vmatrix} a_{11} & a_{12} & a_{13} & \cdots & a_{1q} \\ a_{21} & a_{22} & a_{23} & \cdots & a_{2q} \\ \vdots & \vdots & \vdots & \cdots & \vdots \\ a_{p1} & a_{p2} & a_{p3} & \cdots & a_{pq} \end{Vmatrix}.$$

In fact most of our matrices will be presented in this way and it is only in a few (but important) situations where the extra generality will prove useful.

1.2 Determinants

Let M be a finite set. An $M \times M$ matrix will be called a *square matrix of type* M. The product of two such matrices will again be a square matrix of type M. Note that this multiplication is associative but not in general commutative. Furthermore the *identity matrix*, that is the matrix

$$I_M = \|\delta_{\mu m}\| \quad (\mu, m \in M), \qquad (1.2.1)$$

where $\delta_{\mu m}$ is zero if $\mu \neq m$ and is 1_R otherwise, is neutral with respect to multiplication.

Let $A = \|a_{\mu m}\|$ be a square matrix of type M and let π be a permutation of the set M. The *determinant* of A, which we denote by $\det(A)$ or $|A|$, is defined by

$$\det(A) = \sum_{\pi} \operatorname{sgn}(\pi) \left(\prod_{\mu \in M} a_{\mu \pi(\mu)} \right). \qquad (1.2.2)$$

Here $\operatorname{sgn}(\pi) = +1$ if π is an even permutation and $\operatorname{sgn}(\pi) = -1$ if it is an odd permutation, i.e. $\operatorname{sgn}(\pi)$ is the *signature* or *parity* of π. In this connection it is convenient to have a convention to cover the case where M is the empty set, that is where we have to do with a square matrix with zero rows and zero columns. In fact we shall define the determinant in this case to be the identity element of R. Thus symbolically

$$\det(\| \cdot \|) = 1_R. \qquad (1.2.3)$$

It has already been remarked that the square matrices of type M form a system which is closed under multiplication and which has the matrix $I_M = \|\delta_{\mu m}\|$ as identity element. The system is also closed with respect to the process of replacing a matrix by its transpose. This structure has a familiar connection with the theory of determinants. Thus if A, B are square matrices of type M, then

$$\det(AB) = \det(A)\det(B), \qquad (1.2.4)$$

$$\det(I_M) = 1_R, \qquad (1.2.5)$$

and

$$\det(A^{\mathrm{T}}) = \det(A). \qquad (1.2.6)$$

An $M \times M$ matrix A is called *invertible* if there exists an $M \times M$ matrix C such that $AC = I_M = CA$. (Of course if C exists, then it is unique.) C is called the *inverse* of A and it is denoted by A^{-1}. It is well known that A is invertible if and only if det (A) is a unit of the ring R. Invertible matrices are also known as *unimodular* matrices.

In the three exercises which follow p and q denote positive integers.

Exercise 1.† *Let B be a $p \times p$ matrix (with entries in R) and $c_1, c_2, ..., c_p$ elements of R such that $(c_1, c_2, ..., c_p) B = 0$. Show that $c_i \det (B) = 0$ for $i = 1, 2, ..., p$.*

Exercise 2. *Suppose that $AB = A$, where $A = \|a_{jk}\|$ is a $p \times q$ and $B = \|b_{km}\|$ a $q \times q$ matrix with entries in R. Let $\mathfrak{A}$ be the ideal generated by the a_{jk} and $\mathfrak{B}$ the ideal generated by the b_{km}. Show that there exists $\beta \in \mathfrak{B}$ such that $(1 - \beta) \mathfrak{A} = 0$.*

We recall that an element e, of R, is called an *idempotent* if $e^2 = e$. Both the zero element and the identity element are idempotents. An idempotent which is different from these is called a *non-trivial* idempotent. An integral domain, for example, has no non-trivial idempotents.

Exercise 3. *Suppose that $A\Omega A = A$, where A is a $p \times q$ matrix and Ω a $q \times p$ matrix. Let $\mathfrak{A}$ be the ideal generated by the entries in A. Show that there exists an idempotent α such that $\mathfrak{A} = R\alpha$. (Hence $(1 - \alpha) \mathfrak{A} = 0$ and if R has no non-trivial idempotents, then either $\mathfrak{A} = 0$ or $\mathfrak{A} = R$.)*

1.3 The exterior powers of a matrix

Throughout section (1.3) the letters p, q and t will denote positive integers. Suppose that $\nu \geqslant 0$ is an integer. We shall denote by S_ν^p the set of all sequences $J = \{j_1, j_2, ..., j_\nu\}$, where $1 \leqslant j_1 < j_2 < ... < j_\nu \leqslant p$. Evidently S_ν^p contains $\binom{p}{\nu}$ members. Thus S_ν^p is empty when $\nu > p$ and is non-empty in all other cases. In particular it contains a single member when $\nu = 0$.

† Solutions to the exercises will be found at the end of the chapter.

Let $A = \|a_{jk}\|$ be a $p \times q$ matrix with entries in R. Suppose, for the moment, that $1 \leqslant \nu \leqslant \min(p, q)$. If now $J = \{j_1, j_2, ..., j_\nu\}$ belongs to S_ν^p and $K = \{k_1, k_2, ..., k_\nu\}$ to S_ν^q, then we put

$$A_{JK}^{(\nu)} = \begin{Vmatrix} a_{j_1 k_1} & a_{j_1 k_2} & \cdots & a_{j_1 k_\nu} \\ a_{j_2 k_1} & a_{j_2 k_2} & \cdots & a_{j_2 k_\nu} \\ \vdots & \vdots & \cdots & \vdots \\ a_{j_\nu k_1} & a_{j_\nu k_2} & \cdots & a_{j_\nu k_\nu} \end{Vmatrix} \tag{1.3.1}$$

so that $A_{JK}^{(\nu)}$ is a typical $\nu \times \nu$ minor of A. Let us keep ν fixed. Then the $A_{JK}^{(\nu)}$, where $J \in S_\nu^p$ and $K \in S_\nu^q$, may be regarded as the entries in an $S_\nu^p \times S_\nu^q$ matrix $A^{(\nu)}$. This matrix $A^{(\nu)}$ is called the νth *exterior power* of A. At this point we relax the restriction that was placed on ν and regard $A^{(\nu)}$ as being defined for all $\nu \geqslant 0$. Note that, by (1.2.3), $A^{(0)}$ is the 1×1 matrix $\|1_R\|$ and that $A^{(1)} = A$. Note also that *if C is the transpose of A, then $C^{(\nu)}$ is the transpose of $A^{(\nu)}$*. Again, *the exterior powers of an identity matrix are themselves identity matrices.* In this context a matrix with zero rows and zero columns counts as an identity matrix.†

Let $A = \|a_{jk}\|$ be a $p \times q$ matrix and $B = \|b_{km}\|$ a $q \times t$ matrix. Put $C = AB$, say $C = \|c_{jm}\|$, and suppose for the moment that $1 \leqslant \nu \leqslant \min(p, t)$. If now $J = \{j_1, j_2, ..., j_\nu\}$ belongs to S_ν^p and $M = \{m_1, m_2, ..., m_\nu\}$ to S_ν^t, then

$$C_{JM}^{(\nu)} = \begin{vmatrix} c_{j_1 m_1} & c_{j_1 m_2} & \cdots & c_{j_1 m_\nu} \\ c_{j_2 m_1} & c_{j_2 m_2} & \cdots & c_{j_2 m_\nu} \\ \vdots & \vdots & \cdots & \vdots \\ c_{j_\nu m_1} & c_{j_\nu m_2} & \cdots & c_{j_\nu m_\nu} \end{vmatrix}$$

$$= \begin{vmatrix} \sum_\alpha a_{j_1 \alpha} b_{\alpha m_1} & \sum_\beta a_{j_1 \beta} b_{\beta m_2} & \cdots & \sum_\gamma a_{j_1 \gamma} b_{\gamma m_\nu} \\ \sum_\alpha a_{j_2 \alpha} b_{\alpha m_1} & \sum_\beta a_{j_2 \beta} b_{\beta m_2} & \cdots & \sum_\gamma a_{j_2 \gamma} b_{\gamma m_\nu} \\ \vdots & \vdots & \cdots & \vdots \\ \sum_\alpha a_{j_\nu \alpha} b_{\alpha m_1} & \sum_\beta a_{j_\nu \beta} b_{\beta m_2} & \cdots & \sum_\gamma a_{j_\nu \gamma} b_{\gamma m_\nu} \end{vmatrix},$$

whence

$$C_{JM}^{(\nu)} = \sum_{\alpha, \beta, ..., \gamma} b_{\alpha m_1} b_{\beta m_2} ... b_{\gamma m_\nu} \begin{vmatrix} a_{j_1 \alpha} & a_{j_1 \beta} & \cdots & a_{j_1 \gamma} \\ a_{j_2 \alpha} & a_{j_2 \beta} & \cdots & a_{j_2 \gamma} \\ \vdots & \vdots & \cdots & \vdots \\ a_{j_\nu \alpha} & a_{j_\nu \beta} & \cdots & a_{j_\nu \gamma} \end{vmatrix}. \tag{1.3.2}$$

† Cf. (1.2.3).

Here $\alpha, \beta, ..., \gamma$ range freely between 1 and q. However, unless they are distinct, the determinant in (1.3.2) vanishes. Let $K = \{k_1, k_2, ..., k_\nu\}$ belong to S_ν^q. If $\{\alpha, \beta, ..., \gamma\}$ is a permutation of $\{k_1, k_2, ..., k_\nu\}$, then

$$\begin{vmatrix} a_{j_1\alpha} & a_{j_1\beta} & \cdots & a_{j_1\gamma} \\ a_{j_2\alpha} & a_{j_2\beta} & \cdots & a_{j_2\gamma} \\ \vdots & \vdots & \cdots & \vdots \\ a_{j_\nu\alpha} & a_{j_\nu\beta} & \cdots & a_{j_\nu\gamma} \end{vmatrix} = \epsilon_{\alpha\beta...\gamma} A_{JK}^{(\nu)},$$

where $\epsilon_{\alpha\beta...\gamma}$ has the value $+1$ or -1 according as $\{\alpha, \beta, ..., \gamma\}$ is an even or an odd permutation of $\{k_1, k_2, ..., k_\nu\}$. It follows that the contribution to the sum on the right hand side of (1.3.2) from *all* the permutations of $\{k_1, k_2, ..., k_\nu\}$ is $A_{JK}^{(\nu)} B_{KM}^{(\nu)}$. Accordingly

$$C_{JM}^{(\nu)} = \sum_K A_{JK}^{(\nu)} B_{KM}^{(\nu)} \qquad (1.3.3)$$

and we have proved

THEOREM 1. *Let A be a $p \times q$ matrix and B a $q \times t$ matrix. Then $(AB)^{(\nu)} = A^{(\nu)} B^{(\nu)}$ for every $\nu \geqslant 0$.*

Note that, because of our conventions, we can allow ν to be any non-negative integer.

We recall that R is called a *non-trivial* ring if its identity element is not zero.

EXERCISE 4. *Suppose that the ring R is non-trivial. Let A be a $p \times q$ matrix and B a $q \times p$ matrix such that AB and BA are both of them identity matrices. Show that $p = q$. (Thus A and B are unimodular matrices and each is the inverse of the other.)*

1.4   Determinantal ideals

Let A be a $p \times q$ matrix and B a $q \times t$ matrix, where p, q, t are positive integers. If now $\nu \geqslant 0$ is an integer, denote by $\mathfrak{A}_\nu(A)$ the ideal generated by the entries in $A^{(\nu)}$. Thus $\mathfrak{A}_\nu(A)$ is the ideal generated by all the $\nu \times \nu$ minors of A.

DEFINITION. *The ideals $\mathfrak{A}_\nu(A)$, where $\nu = 0, 1, 2, ...$, are called the 'determinantal ideals' of A.*

Since $A^{(0)} = \|1_R\|$, it follows that $\mathfrak{A}_0(A) = R$. Of course $\mathfrak{A}_1(A)$ is the ideal generated by the elements of A. Again

$$\mathfrak{A}_\nu(A) = 0 \quad \text{for} \quad \nu > \min(p, q) \qquad (1.4.1)$$

and $\quad R = \mathfrak{A}_0(A) \supseteq \mathfrak{A}_1(A) \supseteq \mathfrak{A}_2(A) \supseteq \mathfrak{A}_3(A) \supseteq \ldots.$ (1.4.2)

Also, because $(AB)^{(\nu)} = A^{(\nu)}B^{(\nu)}$ by Theorem 1, we have

$$\mathfrak{A}_\nu(AB) \subseteq \mathfrak{A}_\nu(A) \cap \mathfrak{A}_\nu(B) \qquad (1.4.3)$$

and this extends to a product of any finite number of matrices.

THEOREM 2. *Let A and A' be $p \times q$ matrices and suppose that $A' = UAV$, $A = U'A'V'$, where U, U' are $p \times p$ matrices and V, V' are $q \times q$ matrices. Then $\mathfrak{A}_\nu(A) = \mathfrak{A}_\nu(A')$ for all $\nu \geqslant 0$.*

Proof. Since $A' = UAV$, we have $\mathfrak{A}_\nu(A') \subseteq \mathfrak{A}_\nu(A)$ by (1.4.3) and the opposite inclusion holds similarly.

DEFINITION. *The $p \times q$ matrix A' is said to be 'equivalent' to the $p \times q$ matrix A if there exist unimodular matrices U, V such that $A' = UAV$.*

This relation is reflexive, symmetric, and transitive. Indeed if $A' = UAV$, where U, V are unimodular and therefore invertible, then $A = U^{-1}A'V^{-1}$. We can therefore apply Theorem 2 and so obtain

THEOREM 3. *If A and A' are equivalent $p \times q$ matrices, then $\mathfrak{A}_\nu(A) = \mathfrak{A}_\nu(A')$ for all $\nu \geqslant 0$.*

It is a classical result that if R is an integral domain with the property that every ideal can be generated by a single element, then the converse of Theorem 3 holds. Thus for such an integral domain two $p \times q$ matrices A and A' are equivalent if and only if $\mathfrak{A}_\nu(A) = \mathfrak{A}_\nu(A')$ for all $\nu \geqslant 0$. As we shall not be making use of this result we refer the interested reader to the literature.†

We recall that by *elementary row and column operations on A* it is customary to mean the following:

(1) multiplication of the elements of any row or column of A by one and the same unit;

(2) interchanging any two rows or columns of A;

(3) adding to any row (column) of A a multiple, by an element of R, of a different row (column).

† See (6) in the list of references at the end. The section dealing with these matters is Chapter 7, §4, no. 5.

8 *Matrices and determinants*

If we take any one of these operations, then the same effect can be produced by multiplying A by a suitable unimodular matrix. For example suppose that $1 \leqslant i,j \leqslant p$ and $i \neq j$. Denote by U the matrix produced by taking the identity matrix of order p and putting an element α, where $\alpha \in R$, in the (i,j)th position. Then $\det(U) = 1_R$, so U is unimodular, and UA is the matrix one obtains from A by adding α times the jth row to the ith row. Accordingly we have

THEOREM 4. *Let the matrix A' be obtained from A by means of elementary row and column operations. Then A and A' are equivalent and hence they have the same determinantal ideals.*

We now give a partial result concerning determinantal ideals.

LEMMA 1.† *Suppose that $A\Omega A = A$, where A is a $p \times q$ matrix and Ω a $q \times p$ matrix. Then for each $\nu \geqslant 0$ the determinantal ideal $\mathfrak{A}_\nu(A)$ is generated by an idempotent. Hence if R has no non-trivial idempotents, then either $\mathfrak{A}_\nu(A) = 0$ or $\mathfrak{A}_\nu(A) = R$.*

Proof. We may confine our attention to the case where $1 \leqslant \nu \leqslant \min(p,q)$. By Theorem 1, $A^{(\nu)}\Omega^{(\nu)}A^{(\nu)} = A^{(\nu)}$ and by definition $\mathfrak{A}_\nu(A)$ is the ideal generated by the entries in $A^{(\nu)}$. The desired result therefore follows from Exercise 3.

1.5 Some useful formulae

Throughout section (1.5) we shall be concerned with a $p \times q$ matrix $A = \|a_{jk}\|$, where p,q are positive integers. Suppose that $0 \leqslant \mu \leqslant \min(p-1,q)$ and let $M = \{m_1, m_2, \ldots, m_{\mu+1}\}$ belong to $S^p_{\mu+1}$ and $K = \{k_1, k_2, \ldots, k_\mu\}$ to S^q_μ, where the notation is as explained in section (1.3). We now define a row vector x_{MK} of length p by

$$(x_{MK})_j = \begin{cases} 0 & \text{if } j \notin M, \\ (-1)^{\alpha+1} A^{(\mu)}_{M\backslash j, K} & \text{if } j = m_\alpha. \end{cases} \tag{1.5.1}$$

† The converse of the first assertion of the lemma is also true. Cf. Chapter 4 Theorem 18.

Here by $M \backslash m_\alpha$ we mean M with the term m_α removed. Since x_{MK} is a row vector of length p we can form $x_{MK}A$ and this will be a row vector of length q. In fact for $1 \leqslant k \leqslant q$ we have

$$(x_{MK}A)_k = \begin{vmatrix} a_{m_1 k} & a_{m_1 k_1} & \cdots & a_{m_1 k_\mu} \\ a_{m_2 k} & a_{m_2 k_1} & \cdots & a_{m_2 k_\mu} \\ \vdots & \vdots & \cdots & \vdots \\ a_{m_{\mu+1} k} & a_{m_{\mu+1} k_1} & \cdots & a_{m_{\mu+1} k_\mu} \end{vmatrix} \quad (1.5.2)$$

as is readily verified by expanding the determinant by means of its first column.

There is a natural companion to this result. To describe it suppose that $0 \leqslant \nu \leqslant \min(p, q-1)$. Further suppose that $J = \{j_1, j_2, \ldots, j_\nu\}$ belongs to S^p_ν and $N = \{n_1, n_2, \ldots, n_{\nu+1}\}$ to $S^q_{\nu+1}$. We can now define a column vector y_{JN} of length q by

$$(y_{JN})_k = \begin{cases} 0 & \text{if } k \notin N, \\ (-1)^{\beta+1} A^{(\nu)}_{J, N \backslash k} & \text{if } k = n_\beta. \end{cases} \quad (1.5.3)$$

This secures that Ay_{JN} is a column vector of length p and we have

$$(Ay_{JN})_j = \begin{vmatrix} a_{jn_1} & a_{jn_2} & \cdots & a_{jn_{\nu+1}} \\ a_{j_1 n_1} & a_{j_1 n_2} & \cdots & a_{j_1 n_{\nu+1}} \\ \vdots & \vdots & \cdots & \vdots \\ a_{j_\nu n_1} & a_{j_\nu n_2} & \cdots & a_{j_\nu n_{\nu+1}} \end{vmatrix}. \quad (1.5.4)$$

On this occasion the relation can be checked by expanding the determinant on the right hand side of (1.5.4) by means of its first row.

Our next result is somewhat more complicated in character and in order to present it we shall require some additional notation. Let $t \geqslant 0$ be an integer and let $H = \{h_1, h_2, \ldots, h_t\}$. This is to be a sequence of integers between 1 and p, but on this occasion we do not postulate that $h_1, h_2, \ldots, h_t$ be distinct nor do we insist that they should form an increasing sequence. Suppose next that $1 \leqslant \mu \leqslant p$. We put

$$\omega(\mu, H) = \begin{cases} 0 & \text{if } H \text{ contains a repetition or } \mu \notin H, \\ (-1)^\alpha & \text{if } H \text{ contains no repetitions and } \mu = h_\alpha. \end{cases}$$
$$(1.5.5)$$

Again let $L = \{l_1, l_2, ..., l_t\}$ also be a sequence of integers but this time between 1 and q, where once more repetitions are allowed. For $1 \leqslant \nu \leqslant q$ we define $\omega(\nu, L)$ by means of a formula analogous to (1.5.5).

After these preliminaries put

$$\Delta_{HL} = \begin{vmatrix} a_{h_1 l_1} & a_{h_1 l_2} & \cdots & a_{h_1 l_t} \\ a_{h_2 l_1} & a_{h_2 l_2} & \cdots & a_{h_2 l_t} \\ \vdots & \vdots & \cdots & \vdots \\ a_{h_t l_1} & a_{h_t l_2} & \cdots & a_{h_t l_t} \end{vmatrix} \tag{1.5.6}$$

and for $1 \leqslant h \leqslant p$, $1 \leqslant l \leqslant q$ set

$$hH = \{h, h_1, h_2, ..., h_t\} \quad \text{and} \quad lL = \{l, l_1, l_2, ..., l_t\}. \tag{1.5.7}$$

Using these we can define a $p \times q$ matrix Θ_{HL} by

$$(\Theta_{HL})_{hl} = \Delta_{hH, lL}. \tag{1.5.8}$$

We can also define a $q \times p$ matrix Ω_{HL} as follows: if either H or L contains a repetition, then $\Omega_{HL} = 0$; on the other hand if neither H nor L contains a repetition, then the entries in Ω_{HL} are given by

$$(\Omega_{HL})_{\nu\mu} = \begin{cases} 0 & \text{if either } \mu \notin H \text{ or } \nu \notin L, \\ \omega(\mu, H)\,\omega(\nu, L)\,\Delta_{H\backslash\mu,\, L\backslash\nu} & \text{if } \mu \in H \text{ and } \nu \in L. \end{cases}$$

The significance of these various definitions is revealed by

THEOREM 5. *Let the notation be as above. Then*

$$\Theta_{HL} = \Delta_{HL}A - A\Omega_{HL}A. \tag{1.5.9}$$

Consequently if all the $(t+1) \times (t+1)$ *minors of* A *are zero, that is if* $\mathfrak{A}_{t+1}(A) = 0$, *then* $\Delta_{HL}A = A\Omega_{HL}A$.

Proof. We may suppose that neither H nor L contains a repetition. Now

$$(\Theta_{HL})_{hl} = \begin{vmatrix} a_{hl} & a_{hl_1} & \cdots & a_{hl_t} \\ a_{h_1 l} & a_{h_1 l_1} & \cdots & a_{h_1 l_t} \\ \vdots & \vdots & \cdots & \vdots \\ a_{h_t l} & a_{h_t l_1} & \cdots & a_{h_t l_t} \end{vmatrix}.$$

To obtain (1.5.9) we first expand the determinant using its first row and then we expand the cofactors of $a_{hl_1}, a_{hl_2}, ..., a_{hl_t}$ using their first columns. We give two applications of Theorem 5.

THEOREM 6. *Let the ring R have no non-trivial idempotents and let A be a $p \times q$ matrix. Then the following two statements are equivalent:*

(a) *there is a $q \times p$ matrix Ω such that $A = A\Omega A$;*

(b) *for each $\nu \geqslant 0$, the determinantal ideal $\mathfrak{A}_\nu(A)$ is either zero or the ring R itself.*

Proof. We already know, from Lemma 1, that (a) implies (b). We therefore assume (b) to be true. Choose t as large as possible so that $\mathfrak{A}_t(A) \neq 0$. Then $\mathfrak{A}_t(A) = R$ and $\mathfrak{A}_{t+1}(A) = 0$. By Theorem 5, $\Delta_{HL} A = A\Omega_{HL} A$, where the notation is the same as in the statement of that theorem. But the Δ_{HL} generate the ideal $\mathfrak{A}_t(A) = R$ and so we can find a linear combination of the Δ_{HL} whose value is 1. We use the coefficients which occur in the linear combination to form a linear combination, Ω say, of the Ω_{HL}. Then $A = A\Omega A$ as required.

THEOREM 7. *Let A be a $p \times q$ matrix and suppose that the smallest non-zero determinantal ideal of A is not annihilated by any non-zero element of R. Then the following two statements are equivalent:*

(a) *there is a $q \times p$ matrix Ω such that $A = A\Omega A$;*

(b) *for each $\nu \geqslant 0$, the determinantal ideal $\mathfrak{A}_\nu(A)$ is either zero or it is the ring R itself.*

Remark. Let t be the largest integer such that $\mathfrak{A}_t(A) \neq 0$. Then our assumption means that from $c \in R$ and $c\mathfrak{A}_t(A) = 0$ it follows that $c = 0$.

Proof. To show that (b) implies (a) we can use the argument that was employed in the proof of Theorem 6. We therefore assume (a). Let t be as in the preceding remark. By Theorem 1,

$$A^{(t)} = A^{(t)} \Omega^{(t)} A^{(t)}$$

and therefore, by Exercise 3, there exists $\alpha \in \mathfrak{A}_t(A)$ such that $(1 - \alpha)\mathfrak{A}_t(A) = 0$. It follows that $\alpha = 1$ and therefore $\mathfrak{A}_t(A) = R$.

We now see, from (1.4.2), that $\mathfrak{A}_\nu(A) = R$ for $0 \leqslant \nu \leqslant t$, and $\mathfrak{A}_\nu(A) = 0$ when $\nu > t$. This completes the proof.

1.6 The higher adjugates

In this section we study a square matrix $U = \|u_{jk}\|$ with $p(p \geqslant 1)$ rows and columns. Suppose that $1 \leqslant \mu < p$. If we put $\nu = \mu + 1$, then we have

$$\det(U) = \begin{vmatrix} u_{11}+0 & u_{12}+0 & \ldots & u_{1p}+0 \\ \vdots & \vdots & \ldots & \vdots \\ u_{\mu 1}+0 & u_{\mu 2}+0 & \ldots & u_{\mu p}+0 \\ 0+u_{\nu 1} & 0+u_{\nu 2} & \ldots & 0+u_{\nu p} \\ \vdots & \vdots & \ldots & \vdots \\ 0+u_{p1} & 0+u_{p2} & \ldots & 0+u_{pp} \end{vmatrix}$$

and therefore, with a self-explanatory notation,

$$\det(U) = \det(C_1^* + C_1^{**}, C_2^* + C_2^{**}, \ldots, C_p^* + C_p^{**}),$$

where C_i^* and C_i^{**} are the column matrices

$$\begin{Vmatrix} u_{1i} \\ \vdots \\ u_{\mu i} \\ 0 \\ \vdots \\ 0 \end{Vmatrix} \quad \text{and} \quad \begin{Vmatrix} 0 \\ \vdots \\ 0 \\ u_{\nu i} \\ \vdots \\ u_{pi} \end{Vmatrix}$$

respectively. Accordingly

$$\det(U) = \Sigma \det(C_1, C_2, \ldots, C_p),$$

where for each term in the sum C_i is chosen to be one or other of C_i^* and C_i^{**}. Thus we are concerned with 2^p matrices

$$\|C_1, C_2, \ldots, C_p\|$$

and we have to sum the determinants of these matrices. Observe that if in $\|C_1, C_2, \ldots, C_p\|$ we have $C_i = C_i^*$ for fewer than μ different values of i, then $\det(C_1, C_2, \ldots, C_p) = 0$. We reach the same conclusion if we have $C_i = C_i^{**}$ for fewer than $p - \mu$ values of i. Thus the cases that are relevant are those in which $C_i = C_i^*$ for exactly μ values of i and $C_i = C_i^{**}$ in the remaining $p - \mu$ cases.

Suppose that $K = \{k_1, k_2, ..., k_\mu\}$ belongs to S_μ^p and let $K' = \{k'_1, k'_2, ..., k'_{p-\mu}\}$ be the member of $S_{p-\mu}^p$ obtained by striking out $k_1, k_2, ..., k_\mu$ from $\{1, 2, ..., p\}$. If now

$$C_i = \begin{cases} C_i^* & \text{when } i \in K, \\ C_i^{**} & \text{when } i \in K', \end{cases}$$

then

$$\det (C_1, C_2, ..., C_p) = \text{sgn} (K, K') \, U_{\{1, ..., \mu\}K}^{(\mu)} \, U_{\{v, ..., p\}K'}^{(p-\mu)}.$$

Here $\text{sgn} (K, K')$ denotes the signature of $\{k_1, ..., k_\mu, k'_1, ..., k'_{p-\mu}\}$ considered as a permutation of $\{1, 2, ..., p\}$ and, of course, $U^{(\mu)}$ designates the μth exterior power of U. Thus

$$\det (U) = \sum_{K \in S_\mu^p} \text{sgn} (K, K') \, U_{\{1, ..., \mu\}K}^{(\mu)} U_{\{\mu+1, ..., p\}K'}^{(p-\mu)}. \quad (1.6.1)$$

This is just the general *Laplace expansion* of $\det (U)$ where we have expanded using the first μ rows.

After these preliminaries let

$$J = \{j_1, j_2, ..., j_\mu\} \quad \text{and} \quad M = \{m_1, m_2, ..., m_\mu\}$$

belong to S_μ^p and let $M' = \{n_1, n_2, ..., n_\lambda\}$ be the complement of M in $\{1, 2, ..., p\}$ so that $\lambda = p - \mu$. Consider the determinant

$$\begin{vmatrix} u_{j_1 1} & u_{j_1 2} & \cdots & u_{j_1 p} \\ \vdots & \vdots & \cdots & \vdots \\ u_{j_\mu 1} & u_{j_\mu 2} & \cdots & u_{j_\mu p} \\ u_{n_1 1} & u_{n_1 2} & \cdots & u_{n_1 p} \\ \vdots & \vdots & \cdots & \vdots \\ u_{n_\lambda 1} & u_{n_\lambda 2} & \cdots & u_{n_\lambda p} \end{vmatrix}. \quad (1.6.2)$$

This has the value zero unless $\{j_1, ..., j_\mu, n_1, ..., n_\lambda\}$ is a rearrangement of $\{1, 2, ..., p\}$, i.e. it is zero unless $J = M$. Suppose now that $J = M$. This time the value of the determinant is

$$\text{sgn} (M, M') \det (U).$$

Let us use Laplace's expansion to develop (1.6.2) by means of its first μ rows. This yields

$$\delta_{JM} \text{sgn} (M, M') \det (U) = \sum_{K \in S_\mu^p} \text{sgn} (K, K') \, U_{JK}^{(\mu)} U_{M'K'}^{(\lambda)},$$

or $\delta_{JM} \det (U) = \sum_{K \in S_\mu^p} U_{JK}^{(\mu)} \text{sgn} (M, M') \text{sgn} (K, K') \, U_{M'K'}^{(p-\mu)}. \quad (1.6.3)$

where $\delta_{JM} = 1$ if $J = M$ and is zero otherwise. Now in deriving (1.6.3) we assumed that $1 \leqslant \mu < p$. However it is easy to verify that, because of our conventions, (1.6.3) continues to hold if we enlarge the range to $0 \leqslant \mu \leqslant p$.

Let us now replace U by its transpose. Then from (1.6.3) we obtain

$$\delta_{JM} \det(U) = \sum_{K \in S_{\mu}^{p}} \operatorname{sgn}(M, M') \operatorname{sgn}(K, K') U_{K'M'}^{(p-\mu)} U_{KJ}^{(\mu)} \quad (1.6.4)$$

and this too will be valid for $0 \leqslant \mu \leqslant p$.

We next define an $S_{\mu}^{p} \times S_{\mu}^{p}$ matrix, $\operatorname{adj}^{(\mu)}U$, by

$$(\operatorname{adj}^{(\mu)}U)_{KM} = \operatorname{sgn}(K, K') \operatorname{sgn}(M, M') U_{M'K'}^{(p-\mu)}, \quad (1.6.5)$$

where $0 \leqslant \mu \leqslant p$ and $K, M \in S_{\mu}^{p}$. This matrix will be called the μth *adjugate* of U. Note that (1.6.3) and (1.6.4) together yield

THEOREM 8. *Let U be a $p \times p$ matrix, where p is a positive integer, and suppose that $0 \leqslant \mu \leqslant p$. Then*

$$U^{(\mu)}(\operatorname{adj}^{(\mu)}U) = \det(U) I = (\operatorname{adj}^{(\mu)}U) U^{(\mu)},$$

where I denotes a suitable identity matrix.

The matrix $\operatorname{adj}^{(1)}(U)$ is a $p \times p$ matrix. Indeed if $1 \leqslant j, k \leqslant p$ then (1.6.5) shows that the (k,j)th element of $\operatorname{adj}^{(1)}U$ is the cofactor of u_{jk} in the determinant of U. Thus $\operatorname{adj}^{(1)}U$ is the adjugate of U in the usual sense.

THEOREM 9. *Let U be a unimodular $p \times p$ matrix with inverse V and suppose that $0 \leqslant \mu \leqslant p$. Then $U^{(\mu)}$ is unimodular with inverse $V^{(\mu)}$ and $\det(U) V^{(\mu)} = \operatorname{adj}^{(\mu)} U$.*

Indeed the first assertion is clear and the second follows immediately from Theorem 8.

1.7 Localization of rings and matrices

In order to make full use of the results contained in the preceding sections it will be necessary to employ the techniques of localization. In this section we shall review some of the relevant parts of this theory. Other aspects will be discussed later.

Let S be a subset of R.

DEFINITION. *The subset S will be said to be 'multiplicatively closed' if* (i) *the product of two elements of S is again in S, and* (ii) $1 \in S$.

For example, if $r \in R$, then $\{r^n\}_{n \geqslant 0}$ is a multiplicatively closed subset provided we interpret r^0 as being the identity element of the ring. Note that the possibility that the zero element may be in a multiplicatively closed subset is not excluded by the definition.

Let P be an ideal of R. We say that P is a *prime ideal* if $R \backslash P$ is a multiplicatively closed subset of R or (equivalently) if R/P is an integral domain. The following theorem is of fundamental importance.

THEOREM 10. *Let S be a multiplicatively closed subset of R and let $\mathfrak{A}$ be an ideal of R such that $\mathfrak{A} \cap S$ is empty. Then there exists a prime ideal P such that $\mathfrak{A} \subseteq P$ but P does not meet S.*

Proof. Let Σ consist of all the ideals that contain $\mathfrak{A}$ but do not meet S. Evidently $\mathfrak{A} \in \Sigma$ so Σ is not empty. Partially order Σ by means of the inclusion relation. A simple verification shows that Σ is an inductive system. Consequently, by Zorn's lemma, Σ will contain at least one maximal member. It will be proved that *every maximal member of Σ is a prime ideal*. This will establish the theorem.

Let P be a maximal member of Σ. Then P does not meet S so $P \neq R$. Suppose that α_1 and α_2 are in $R \backslash P$. It is enough to show that $\alpha_1 \alpha_2$ is not in P. Now $P + R\alpha_1$ strictly contains P and therefore it is not in Σ. Hence $P + R\alpha_1$ contains an element $s_1 \in S$, say $s_1 = p_1 + r_1 \alpha_1$, where $p_1 \in P$ and $r_1 \in R$. Similarly we can find an element $s_2 \in S$ which can be expressed in the form $s_2 = p_2 + r_2 \alpha_2$ with $p_2 \in P$ and $r_2 \in R$. Accordingly

$$s_1 s_2 = (p_1 + r_1 \alpha_1)(p_2 + r_2 \alpha_2) = p + (r_1 r_2)(\alpha_1 \alpha_2),$$

where $p \in P$. However $s_1 s_2 \in S$ and therefore $s_1 s_2 \notin P$. It follows that $\alpha_1 \alpha_2 \notin P$ and with this the proof is complete.

By considering special cases we can get a great deal of information. For example, suppose that R is a non-trivial ring. Let S consist of the identity element alone and take $\mathfrak{A} = (0)$. Then Σ is

composed of all the *proper* ideals, that is all the ideals different
from R itself. Consequently our proof shows that *R possesses at
least one maximal ideal and that all maximal ideals are prime ideals.*
Again, if we take $S = \{1\}$ and let $\mathfrak{A}$ be any proper ideal, then we
discover that *every proper ideal is contained in a maximal ideal.*
Still assuming that R is non-trivial let $\alpha \in R$. Then α is a non-unit
if and only if the principal ideal $(\alpha) \neq R$. *Thus α is a non-unit if
and only if it is contained in a maximal ideal.*

We now turn our attention to the process of forming fractions.
Let S be a multiplicatively closed subset of R and consider *formal
fractions* r/s, where $r \in R$ and $s \in S$. (These are just ordered pairs,
written in a certain way, where one component comes from R
and the other from S.) With one minor difference, we can turn
these into a commutative ring in just the same way as we form
the quotient field of an integral domain. The difference is that
we treat two fractions r/s and r'/s' as being essentially the same
precisely when there exists $\sigma \in S$ such that $\sigma s'r = \sigma sr'$. The ring
obtained is denoted by R_S, and addition and multiplication in R_S
are performed according to the rules

$$\frac{r_1}{s_1} + \frac{r_2}{s_2} = \frac{s_2 r_1 + s_1 r_2}{s_1 s_2}$$

and

$$\frac{r_1}{s_1} \frac{r_2}{s_2} = \frac{r_1 r_2}{s_1 s_2}.$$

Furthermore the zero element of the new ring is $0/1$ and the
identity element is $1/1$. Now $0/1 = 1/1$ if and only if $s = 0$ for
some $s \in S$. Accordingly R_S *is a non-trivial ring if and only if* $0 \notin S$.

Next we note that we can define a ring-homomorphism
$\chi \colon R \to R_S$ by means of

$$\chi(r) = \frac{r}{1}. \tag{1.7.1}$$

We call χ the *canonical homomorphism.*

Let $\mathfrak{A}$ be an ideal of R. By $\mathfrak{A}R_S$ we mean, of course, the R_S-ideal
generated by $\chi(\mathfrak{A})$. An easy verification shows that $\mathfrak{A}R_S$ *consists
of all the elements of R_S that can be written in the form* a/s, *where*
$a \in \mathfrak{A}$ *and* $s \in S$.

E x e r c i s e 5. *Let S be a multiplicatively closed subset of R and* $\mathfrak{A}$ *an ideal of R. Show that* $\mathfrak{A}R_S = R_S$ *if and only if* $\mathfrak{A}$ *meets S.*

E x e r c i s e 6. *Let S be a multiplicatively closed subset of R, and P a prime ideal of R which does not meet S. Show that* PR_S *is a prime ideal of* R_S*, and that all prime ideals of* R_S *arise in this way.*

In fact the solution to Exercise 6 provides the following information. *There is a one–one correspondence between the prime ideals P* (*of R*) *which do not meet S and the prime ideals* Π *of* R_S. *This is such that when P and* Π *correspond* $\Pi = PR_S$ *and* $P = \chi^{-1}(\Pi)$*, where* $\chi: R \to R_S$ *is the canonical homomorphism.*

We are now ready to return to the consideration of matrices. Let $A = \|a_{jk}\|$ be a $p \times q$ matrix with entries in R, where p and q are positive integers, and let S be a multiplicatively closed subset of R. Put

$$A_S = \|a_{jk}/1\|, \tag{1.7.2}$$

where it is understood that the right hand side of (1.7.2) denotes a matrix with entries in R_S. If now $J \in S^p_\mu$ and $K \in S^q_\mu$, then

$$(A_S)^{(\mu)}_{JK} = A^{(\mu)}_{JK}/1, \tag{1.7.3}$$

and therefore

$$\mathfrak{A}_\mu(A_S) = \mathfrak{A}_\mu(A)R_S. \tag{1.7.4}$$

We must now examine a special case of fraction formation. Let P be a prime ideal of R and put $S = R \backslash P$. Then S is a multiplicatively closed set. In this particular situation it is customary to use a different notation and write R_P and A_P rather than R_S and A_S. (Since $1 \notin P$ and therefore, according to the definition we are using, P itself is not a multiplicatively closed subset of R, there is no danger of confusion.) Later we shall find occasions where it will help to change a problem involving the ring R into a similar one in which R is replaced by a ring of the form R_P. When we do this we say that we are *localizing* the problem at P. The reason that we gain some advantage by doing this is due to the fact that R_P is a ring of a special kind.

As before let P be a prime ideal of R. By Exercise 6, PR_P is a prime ideal of R_P and it contains all the other prime ideals. It follows that PR_P is a maximal ideal of R_P; indeed our remarks show that it is the only maximal ideal. This merits a

DEFINITION. *A ring which possesses exactly one maximal ideal will be called a 'quasi-local' ring.*

In this connection we mention that there is a slight confusion in the literature over terminology. Some writers say that a ring is a *local ring* if it is a Noetherian quasi-local ring, that is to say if it is a quasi-local ring in which every ideal is finitely generated. Other writers have preferred to mean by a local ring what we have called a quasi-local ring. On this occasion the description 'quasi-local ring' has been chosen as it serves as a reminder that we are dispensing with the Noetherian condition.

Note that one effect of localization is that our original ring gets replaced by a quasi-local ring. We shall now illustrate the special properties of quasi-local rings by means of two simple results. Other examples will be given later. Note that if R is a quasi-local ring with maximal ideal P and $\alpha \in R$, then α is a unit if and only if $\alpha \notin P$. Thus the maximal ideal of R consists of the elements that are not units.

THEOREM 11. *Let R be a quasi-local ring. Then R has no non-trivial idempotents.*

Proof. Let P be the maximal ideal and let e be an idempotent. Then $e(1-e) = 0$ and $e + (1-e) = 1$. Thus either $e \notin P$ or $(1-e) \notin P$, that is either e is a unit or $1-e$ is a unit. But if e is a unit then, because $e(1-e) = 0$, we have $e = 1$ whereas if $1-e$ is a unit, then $e = 0$.

THEOREM 12. *Let R be a quasi-local ring and $A = \|a_{jk}\|$ a $p \times q$ matrix with entries in R. Suppose that $\mu \geqslant 0$ and $\mathfrak{A}_\mu(A) = R$. Then A can be brought to the form*

$$\left\| \begin{array}{c|c} I_\mu & 0 \\ \hline 0 & B \end{array} \right\|$$

by means of elementary row and column operations. (I_μ denotes a $\mu \times \mu$ identity matrix.)

Proof. Let P be the maximal ideal of R. We may suppose $\mu \neq 0$ in which case $1 \leqslant \mu \leqslant \min(p,q)$ and, since $\mathfrak{A}_\mu(A) = R$, at least one a_{jk} is not in P. Thus we can find a unit among the a_{jk} and now, using elementary row and column operations, we can arrange that $a_{11} = 1$. Further elementary row and column operations will reduce A to the form

$$\left\| \begin{array}{c|ccc} 1 & 0 & \dots & 0 \\ \hline 0 & & & \\ \vdots & & A^* & \\ 0 & & & \end{array} \right\|.$$

Next we note that $\mathfrak{A}_{\mu-1}(A^*) = \mathfrak{A}_{\mu}(A) = R$ and it is clear that, if $\mu > 1$, we can reduce A^* in the same way and so arrive at a matrix of the form

$$\left\| \begin{array}{cc|ccc} 1 & 0 & 0 & \dots & 0 \\ 0 & 1 & 0 & \dots & 0 \\ \hline 0 & 0 & & & \\ \vdots & \vdots & & A^{**} & \\ 0 & 0 & & & \end{array} \right\|.$$

Proceeding in this way we obtain the desired result after μ steps. Our next result is noted for future reference.

LEMMA 2. *Let A be a $p \times q$ matrix with entries in R (an arbitrary commutative ring), let $t \geqslant 0$ be an integer, and let I_t be the identity matrix of order t. If now B is any R-matrix of the form*

$$B = \left\| \begin{array}{c|c} I_t & C \\ \hline 0 & A \end{array} \right\|$$

then $\mathfrak{A}_{\mu}(A) = \mathfrak{A}_{\mu+t}(B)$ for all $\mu \geqslant 0$.

Proof. By elementary column operations we may reduce the lemma to the case $C = 0$ in which situation it is obvious.

Solutions to the Exercises on Chapter 1

EXERCISE 1. *Let B be a $p \times p$ matrix (with entries in R) and $c_1, c_2, \dots, c_p$ elements of R such that $(c_1, c_2, \dots, c_p) B = 0$. Show that $c_i \det (B) = 0$ for $i = 1, 2, \dots, p$.*

Solution. Let C be the matrix obtained by taking the $p \times p$ identity matrix and replacing its ith row by $(c_1, c_2, \dots, c_p)$. Then

CB has a row of zeros and therefore $\det(CB) = 0$. Hence

$$c_i \det(B) = \det(C)\det(B) = \det(CB) = 0$$

as required.

EXERCISE 2. *Suppose that $AB = A$, where $A = \|a_{jk}\|$ is a $p \times q$ matrix and $B = \|b_{km}\|$ a $q \times q$ matrix with entries in R. Let $\mathfrak{A}$ be the ideal generated by the a_{jk} and $\mathfrak{B}$ the ideal generated by the b_{km}. Show that there exists $\beta \in \mathfrak{B}$ such that $(1-\beta)\mathfrak{A} = 0$.*

Solution. We have $A(I-B) = 0$, where I is the $q \times q$ identity matrix. Hence $(a_{j1}, a_{j2}, ..., a_{jq})(I-B) = 0$ and therefore, by Exercise 1, $a_{jk}\det(I-B) = 0$. Thus $\det(I-B)\mathfrak{A} = 0$. But $\det(I-B) = 1-\beta$ for a suitable β in $\mathfrak{B}$. The solution is therefore complete.

EXERCISE 3. *Suppose that $A\Omega A = A$, where A is a $p \times q$ matrix and Ω a $q \times p$ matrix. Let $\mathfrak{A}$ be the ideal generated by the entries in A. Show that there exists an idempotent α such that $\mathfrak{A} = R\alpha$. (Hence $(1-\alpha)\mathfrak{A} = 0$ and if R has no non-trivial idempotents, then either $\mathfrak{A} = 0$ or $\mathfrak{A} = R$.)*

Solution. Put $B = \Omega A$ and let $\mathfrak{B}$ be the ideal generated by the entries in B. Since $AB = A$, Exercise 2 shows that $(1-\alpha)\mathfrak{A} = 0$ for some $\alpha \in \mathfrak{B}$. But $\mathfrak{B} \subseteq \mathfrak{A}$ so $\alpha \in \mathfrak{A}$ and therefore $(1-\alpha)\alpha = 0$. Thus α is an idempotent and $R\alpha \subseteq \mathfrak{A}$. Let $x \in \mathfrak{A}$. Then $x = \alpha x + (1-\alpha)x = \alpha x$ because $(1-\alpha)\mathfrak{A} = 0$. Accordingly $x \in R\alpha$. It follows that $\mathfrak{A} = R\alpha$ and now the other assertions are clear.

EXERCISE 4. *Suppose that the ring R is non-trivial. Let A be a $p \times q$ matrix and B a $q \times p$ matrix such that AB and BA are both of them identity matrices. Show that $p = q$.*

Solution. We shall show that both $p < q$ and $q < p$ lead to contradictions. Suppose that $p < q$. Then $BA = I_q$, where I_q denotes the $q \times q$ identity matrix. Hence $B^{(q)}A^{(q)} = I_q^{(q)}$. Now $B^{(q)}$ is without columns and $A^{(q)}$ is without rows. Consequently $B^{(q)}A^{(q)}$ is the 1×1 zero matrix. However $I_q^{(q)}$ is the 1×1 identity matrix. Since R is a non-trivial ring we now have a contradiction. A similar argument leads to a contradiction if $q < p$.

EXERCISE 5. *Let S be a multiplicatively closed subset of R and $\mathfrak{A}$ an ideal of R. Show that $\mathfrak{A}R_S = R_S$ if and only if $\mathfrak{A}$ meets S.*

Solution. If $\sigma \in \mathfrak{A} \cap S$, then $\sigma/\sigma(=1/1)$ belongs to $\mathfrak{A}R_S$ and therefore $\mathfrak{A}R_S = R_S$.

Now suppose that $\mathfrak{A}R_S = R_S$. Then $1/1$ belongs to $\mathfrak{A}R_S$ and therefore $1/1 = a/s$ for some $a \in \mathfrak{A}$ and $s \in S$. It follows that there exists $s_1 \in S$ such that $s_1 s = s_1 a$. Accordingly ss_1 belongs to both $\mathfrak{A}$ and S and therefore $\mathfrak{A}$ meets S.

EXERCISE 6. *Let S be a multiplicatively closed subset of R, and P a prime ideal of R which does not meet S. Show that PR_S is a prime ideal of R_S, and that all the prime ideals of R_S arise in this way.*

Solution. By Exercise 5, $PR_S \neq R_S$. Let $r \in R$ and $s \in S$. An easy verification shows that, because P does not meet S and P is prime, we have $r/s \in PR_S$ if and only if $r \in P$. The deduction that PR_S is prime is now immediate.

Let Q be a prime ideal of R_S and let $\chi: R \to R_S$ be the canonical ring-homomorphism. Put $P' = \chi^{-1}(Q)$. Then P' is a prime ideal of R. Also, because $P'R_S \subseteq Q \subset R_S$, it follows, from Exercise 5, that P' does not meet S. Suppose now that $r/s \in Q$, where $r \in R$ and $s \in S$. Then $(s/1)(r/s) = r/1$ is also in Q and therefore $r \in P'$. Accordingly $r/s \in P'R_S$ and thus we have shown that $Q \subseteq P'R_S$. We now have $Q = P'R_S$ and with this the solution is complete. Note that $\chi^{-1}(PR_S) = P$ as this provides us with additional information.

2. *Free modules*

General remarks

In Chapter 2, R will always denote a *non-trivial* commutative ring and the main aim of the chapter will be to prepare the way for applications of the theory of matrices to the theory of modules. To this end we shall spend some time on the properties of *free* modules. From our point of view their importance lies in the fact that every module can be regarded as the cokernel of a homomorphism between two free modules. Of course the connection with the theory of matrices is particularly strong when we are dealing with a homomorphism between two free modules each having a finite base. The cokernel of such a homomorphism is known as a *finitely presented* module and this type of module will receive special attention.

Free modules over a quasi-local ring have an additional interest because of the theorem of I. Kaplansky† which says that *if R is quasi-local, then any direct summand of a free R-module is itself free*. A proof of this result is given in section (2.3). Since localization at a prime ideal always produces a quasi-local ring, there are now obvious advantages in extending the theory of localization so as to include not only R and its ideals, but also R-modules as well.

2.1 Free modules

Let E be an R-module and $\{x_i\}_{i \in I}$ a family of elements of E. This family is called a *base* for E if (i) the x_i generate E, and (ii) the x_i are linearly independent over R. Thus if $\{x_i\}_{i \in I}$ is a base for E, then for each $x \in E$ there exists a *unique* family $\{r_i\}_{i \in I}$ of elements

† See (28) in the list of references.

of R such that only a finite number of the r_i are non-zero and $x = \sum_{i \in I} r_i x_i$. A module which possesses a base is said to be a *free* module. Evidently if two modules are isomorphic and one is free, then so is the other. Also a module is free if and only if it is isomorphic to a direct sum of copies of R. To make the definition quite clear, we add that a null module is regarded as a free module with an empty base.

LEMMA 1. *Let F be a free R-module with a base $\{x_i\}_{i \in I}$ and let $\{e_i\}_{i \in I}$ be a similarly indexed family of elements of an R-module E. Then there exists a unique homomorphism $g: F \to E$ such that $g(x_i) = e_i$ for all $i \in I$.*

Proof. With a self-explanatory notation, g is given by the formula
$$g(\sum_{i \in I} r_i x_i) = \sum_{i \in I} r_i e_i.$$

LEMMA 2. *Let F be a free R-module and suppose that we have homomorphisms $h: F \to E$ and $f: G \to E$, where G and E are R-modules not subject to any special conditions. If now f is an epimorphism, then there exists a homomorphism $g: F \to G$ such that $fg = h$.*

Proof. Let $\{x_i\}_{i \in I}$ be a base for F. We can choose $y_i \in G$ so that $f(y_i) = h(x_i)$. If we now define an R-homomorphism $g: F \to G$ with $g(x_i) = y_i$ for all $i \in I$ (see Lemma 1), then g will have the required properties.

Another result of the same general character is contained in

LEMMA 3. *Let E be an R-module. Then there exists an epimorphism $\psi: F \to E$, where F is a free module. Moreover, if E can be generated by a set of m elements, where m is a non-negative integer, then F can be chosen so as to have a base of m elements.*

Proof. Let $\{e_i\}_{i \in I}$ be a family of elements which generates E and put
$$F = \bigoplus_{i \in I} R.$$

Thus F is a direct sum in which each summand is equal to R and there is one term for each member of the set I. F is a free module and its typical element is a family $\{r_i\}_{i \in I}$, where $r_i \in R$ and $r_i = 0$

for all but a finite number of values of i. We now obtain an epimorphism $\psi \colon F \to E$ by putting

$$\psi(\{r_i\}_{i \in I}) = \sum_{i \in I} r_i e_i.$$

Finally if I is a finite set with m elements, then F will have a base of m elements.

An important consequence of Lemma 3 is given by

THEOREM 1. *Let E be an R-module. Then there exists an exact sequence*
$$F_1 \xrightarrow{\ \phi\ } F_0 \xrightarrow{\ \epsilon\ } E \to 0,$$
where F_0 and F_1 are free R-modules. Thus E and $\operatorname{Coker}\phi$ *are isomorphic.*

Proof. Lemma 3 shows immediately that we can find a free module F_0 and an epimorphism $\epsilon \colon F_0 \to E$. We can now apply the same lemma to obtain an epimorphism of a free module, F_1 say, on to $\operatorname{Ker}\epsilon$.

DEFINITION. *A free module with a finite base will be called a 'finite free module'.*

If F is a free module and if it is also finitely generated then, because R is a non-trivial ring, F cannot have an infinite base. Thus a finite free module is the same as a finitely generated free module. Also a free module cannot have both a finite base and an infinite base.

THEOREM 2. *Let F be a finite free module and let $x_1, x_2, \ldots, x_p$ and $y_1, y_2, \ldots, y_q$ be bases of F. Then $p = q$.*

Proof. For $1 \leqslant i \leqslant p$ we have a relation
$$x_i = a_{i1} y_1 + a_{i2} y_2 + \ldots + a_{iq} y_q,$$
where $a_{ij} \in R$, and similarly for $1 \leqslant j \leqslant q$ we have
$$y_j = b_{j1} x_1 + b_{j2} x_2 + \ldots + b_{jp} x_p$$
with b_{ji} in R. Put $A = \|a_{ij}\|$ and $B = \|b_{ji}\|$. Then A is a $p \times q$ matrix and B a $q \times p$ matrix. Moreover AB and BA are both of them identity matrices. That $p = q$ now follows from Chapter 1 Exercise 4.

We can now make the

DEFINITION. *The 'rank' of a finite free module F is the number of elements in any base of F.*

We shall regard a free module which is not a finite free module as having infinite rank. If F is a free R-module, then $\mathrm{rank}_R(F)$ will denote its rank.

EXERCISE 1. *Let* $0 \to F' \overset{\phi}{\longrightarrow} F \overset{\psi}{\longrightarrow} F'' \to 0$ *be an exact sequence of free R-modules. Show that* $\mathrm{rank}_R(F) = \mathrm{rank}_R(F') + \mathrm{rank}_R(F'')$.

The next exercise shows that it is possible to refine the notion of rank, as it applies to free modules, so as to distinguish between free modules which have infinite rank. The solution to the exercise requires a knowledge of certain fairly basic parts of the theory of transfinite cardinals. If the reader is not familiar with these matters it is suggested that the exercise be omitted as it will not be required in the sequel.

EXERCISE 2. *Let F be a free R-module with a base* $\{e_i\}_{i \in I}$ *and F' a free R-module with a base* $\{e_j'\}_{j \in J}$. *Show that F and F' are isomorphic if and only if there is a bijection of the set I on to the set J.*

The next lemma gives the connections between the various bases of a finite free module. In it F denotes a finite free module with a base $x_1, x_2, \ldots, x_p$, $A = \|a_{ij}\|$ is a $p \times p$ matrix, and we put

$$\begin{Vmatrix} y_1 \\ y_2 \\ \vdots \\ y_p \end{Vmatrix} = A \begin{Vmatrix} x_1 \\ x_2 \\ \vdots \\ x_p \end{Vmatrix}.$$

LEMMA 4. *Let the situation be as described above. Then the following statements are equivalent:*
(i) $\det(A)$ *is a unit of* R;
(ii) A *is invertible;*
(iii) $y_1, y_2, \ldots, y_p$ *is a base for* F;
(iv) $F = Ry_1 + Ry_2 + \ldots + Ry_p$.

Proof. We already know that (i) and (ii) are equivalent.

Clearly (ii) implies (iii) and (iii) implies (iv). *Assume* (iv). Then there exists a $p \times p$ matrix B such that

$$\left\| \begin{matrix} x_1 \\ x_2 \\ \vdots \\ x_p \end{matrix} \right\| = B \left\| \begin{matrix} y_1 \\ y_2 \\ \vdots \\ y_p \end{matrix} \right\| = BA \left\| \begin{matrix} x_1 \\ x_2 \\ \vdots \\ x_p \end{matrix} \right\|.$$

Thus $BA = I$, where I is the identity $p \times p$ matrix, and therefore $\det(B)\det(A) = 1$. Accordingly $\det(A)$ is a unit of R.

2.2 Finitely presented modules

We begin this section with a theorem which is a slightly specialized version of the important result known as *Schanuel's Lemma*.

THEOREM 3. *Let* $0 \to K \to F \overset{\phi}{\longrightarrow} E \to 0$ *and*

$$0 \to K' \to F' \overset{\phi'}{\longrightarrow} E \to 0$$

be exact sequences, where F and F' are free R-modules. Then $K \oplus F'$ and $K' \oplus F$ are isomorphic modules.

Proof. We may suppose that $K \to F$ and $K' \to F'$ are inclusion mappings. Now, by Lemma 2, we can find homomorphisms $u: F \to F'$ and $v: F' \to F$ such that $\phi'u = \phi$ and $\phi v = \phi'$. Next if $x \in F$, then

$$\phi(x - vu(x)) = \phi(x) - \phi'u(x) = \phi(x) - \phi(x) = 0,$$

and therefore $x - vu(x) \in K$. Similarly if $x' \in F'$, then

$$x' - uv(x') \in K'.$$

Again if $k \in K$, then $\phi'u(k) = \phi(k) = 0$ and hence $u(k) \in K'$. Likewise when $k' \in K'$ we have $v(k') \in K$. Let us define homomorphisms

$$\xi: K \oplus F' \to K' \oplus F$$

and

$$\eta: K' \oplus F \to K \oplus F'$$

by

$$\xi(k, x') = (-u(k) + x' - uv(x'),\ k + v(x'))$$

and

$$\eta(k', x) = (-v(k') + x - vu(x),\ k' + u(x))$$

respectively. Then straightforward computations show that each of $\xi\eta$ and $\eta\xi$ is an identity mapping. Accordingly ξ and η are inverse isomorphisms. This completes the proof.

We shall now give an application of Schanuel's Lemma, but first we need some new terminology. Let E be a finitely generated R-module and let $x_1, x_2, \ldots, x_s$ be elements which generate E. Put

$$R^s = R \oplus R \oplus \ldots \oplus R \quad (s \text{ summands}).$$

Then R^s is a free module of rank s and its typical element is a sequence $(a_1, a_2, \ldots, a_s)$ composed of s elements of R. Such a sequence is called a *relation* between $x_1, x_2, \ldots, x_s$ if

$$a_1 x_1 + a_2 x_2 + \ldots + a_s x_s = 0.$$

Clearly the set of all such relations forms a submodule of R^s. It will be referred to as the *module of relations* between $x_1, x_2, \ldots, x_s$.

THEOREM 4. *Let E be a finitely generated R-module and let $x_1, x_2, \ldots, x_s$ and $y_1, y_2, \ldots, y_t$ be two finite systems of generators of E. Then the following two statements are equivalent*:

(i) *the module of relations between $x_1, x_2, \ldots, x_s$ is finitely generated*;

(ii) *the module of relations between $y_1, y_2, \ldots, y_t$ is finitely generated*.

Proof. Let F be a free module of rank s and let $u_1, u_2, \ldots, u_s$ be a base for F. Define an epimorphism $\phi: F \to E$ so that $\phi(u_i) = x_i$ for $1 \leqslant i \leqslant s$ and put $K = \operatorname{Ker} \phi$. We now have an exact sequence $0 \to K \to F \xrightarrow{\phi} E \to 0$. Note that if $a_1, a_2, \ldots, a_s$ belong to R, then $a_1 u_1 + a_2 u_2 + \ldots + a_s u_s \in K$ if and only if $(a_1, a_2, \ldots, a_s)$ is a relation between $x_1, x_2, \ldots, x_s$. It follows that K *is isomorphic to the module of relations between* $x_1, x_2, \ldots, x_s$. In a similar manner we can construct an exact sequence $0 \to K' \to F' \to E \to 0$, where F' is a finite free module and K' is isomorphic to the module of relations between $y_1, y_2, \ldots, y_t$.

Assume (i). Then K is finitely generated. Also, by Theorem 3, we have an isomorphism $K' \oplus F \approx K \oplus F'$. It follows that K' is isomorphic to a factor module of $K \oplus F'$ and, as $K \oplus F'$ is finitely generated, K' must be finitely generated as well. Thus (i) implies (ii) and the converse follows by symmetry.

DEFINITION. *An R-module E is said to be 'finitely presented' if it is possible to find a finite system $x_1, x_2, ..., x_s$ of generators such that the module of relations (between these generators) is finitely generated.*

Theorem 4 shows that if E is finitely presented, then *whatever* finite system $y_1, y_2, ..., y_t$ of generators we may choose the module of relations between $y_1, y_2, ..., y_t$ will be finitely generated.

DEFINITION. *A 'finite presentation' of an R-module E is an exact sequence $F \to G \to E \to 0$, where F and G are finite free modules.*

From the proof of Theorem 4 it will be clear that an R-module is finitely presented if and only if it has a finite presentation. To take a simple example, suppose that E is a direct summand of a finite free module F. We then have an exact sequence

$$F \to F \to E \to 0.$$

Consequently E is finitely presented.

We shall have more to say about finitely presented modules later.

2.3 Direct summands of free modules

In this section we shall prove, among other theorems, an important result which asserts that, *when R is a quasi-local ring, every direct summand of a free module is itself free.* This will require some preliminary investigations which are of interest in themselves.

LEMMA 5. *Let E be a countably generated R-module. Suppose that given any direct summand K of E and an element $x \in K$, there always exists a free direct summand of K that contains x. In these circumstances E is free.*

Proof. Let $x_1, x_2, x_3, ...$ generate E, where the sequence is infinite. We shall construct, in succession, free submodules $F_1, F_2, F_3, ...$ of E such that for all $s \geqslant 1$

(a) the sum $\sum\limits_{i=1}^{s} F_i$ is direct;

(b) $\sum\limits_{i=1}^{s} F_i$ is a direct summand of E;

(c) $x_1, x_2, ..., x_s$ are all in $\sum\limits_{i=1}^{s} F_i$.

Since $x_1 \in E$ and E is a direct summand of E, there exists a free submodule F_1, of E, such that $x_1 \in F_1$ and F_1 is a direct summand of E. This starts the construction.

Suppose now that $m \geqslant 1$ and that $F_1, F_2, ..., F_m$ have all been constructed to satisfy the above conditions. By (b),

$$E = \left(\sum_{i=1}^{m} F_i \right) \oplus K$$

for a submodule K of E. Let $x_{m+1} = y_{m+1} + z_{m+1}$, where

$$y_{m+1} \in \sum_{i=1}^{m} F_i \quad \text{and} \quad z_{m+1} \in K.$$

By hypothesis there exist modules F_{m+1} and K' for which $K = F_{m+1} \oplus K'$, and where F_{m+1} is a free module and $z_{m+1} \in F_{m+1}$. It is easily verified that $F_1, F_2, ..., F_m, F_{m+1}$ satisfy (a), (b) and (c). This shows that the sequence $F_1, F_2, F_3, ...$ can be continued indefinitely. Clearly E is the direct sum of the modules F_i ($1 \leqslant i < \infty$). Since the direct sum of a number of free modules is also free, this proves the lemma.

LEMMA 6. *Let R be a quasi-local ring with maximal ideal P, and let E be a direct summand of a free R-module F, say $F = E \oplus K$. Suppose that $x \in E$. Then there exists a free direct summand G, of E, such that $x \in G$.*

Proof. We may suppose $x \neq 0$. Let $\{\omega_i\}_{i \in I}$ be a base for F. This base is to be chosen so that when x is expressed in terms of the ω_i *the number of ω_i that have non-zero coefficients is as small as possible*. It will be supposed that the base consists of $\omega_1, \omega_2, ..., \omega_n$ and a residual set S, and that $x = a_1 \omega_1 + a_2 \omega_2 + ... + a_n \omega_n$ where the a_i are in R and none of them is zero.

We claim that $a_n \notin Ra_1 + ... + Ra_{n-1}$. For suppose that

$$a_n = r_1 a_1 + r_2 a_2 + ... + r_{n-1} a_{n-1},$$

where $r_j \in R$. Then

$$x = a_1(\omega_1 + r_1\omega_n) + a_2(\omega_2 + r_2\omega_n) + \ldots + a_{n-1}(\omega_{n-1} + r_{n-1}\omega_n).$$

But $\omega_1 + r_1\omega_n, \ldots, \omega_{n-1} + r_{n-1}\omega_n, \omega_n$ and S together form a base for F and with respect to this base the representation of x is shorter than before and so we have a contradiction. More generally the same kind of reasoning shows that if $1 \leqslant i \leqslant n$, then

$$a_i \notin Ra_1 + \ldots + Ra_{i-1} + Ra_{i+1} + \ldots + Ra_n.$$

Let $\omega_i = y_i + z_i$, where $y_i \in E$ and $z_i \in K$. Then, since $x \in E$, we have

$$x = a_1\omega_1 + a_2\omega_2 + \ldots + a_n\omega_n = a_1y_1 + a_2y_2 + \ldots + a_ny_n.$$

Now for $1 \leqslant i \leqslant n$ we can write $y_i = \sum_{j=1}^{n} c_{ij}\omega_j + t_i$, where $c_{ij} \in R$ and t_i is a linear combination of the members of S. Accordingly

$$\sum_{j=1}^{n} a_j\omega_j = \sum_{j=1}^{n} \left(\sum_{i=1}^{n} a_ic_{ij} \right) \omega_j + \sum_{i=1}^{n} a_it_i,$$

whence $a_j = \sum_{i=1}^{n} a_ic_{ij}$ or

$$\sum_{i=1}^{n} a_i(\delta_{ij} - c_{ij}) = 0,$$

where δ_{ij} is the Kronecker delta symbol. It follows that $\delta_{ij} - c_{ij} \in P$ for otherwise we would be able to express one of $a_1, a_2, \ldots, a_n$ as a linear combination of the remainder and this we know is not possible. It follows that $c_{jj} \notin P$ but $c_{ij} \in P$ whenever $i \neq j$. But this implies that the determinant of the matrix $\|c_{ij}\|$ is a unit and therefore the matrix is unimodular. Put $y_i' = \sum_{j=1}^{n} c_{ij}\omega_j$. By Lemma 4, $y_1', y_2', \ldots, y_n'$ is a base for $R\omega_1 + R\omega_2 + \ldots + R\omega_n$ and therefore $y_1', y_2', \ldots, y_n'$ and S form a base for F. It follows that $y_1, y_2, \ldots, y_n$ and S also constitute a base for F.

Let $G = Ry_1 + Ry_2 + \ldots + Ry_n$. Then G is free and $x \in G$. Also $F = G \oplus H$ for a suitable module H. Since $G \subseteq E$, we have $E = G \oplus (H \cap E)$ and therefore G is a direct summand of E. This proves the lemma.

We can now take a substantial step towards our goal.

LEMMA 7. *Let R be a quasi-local ring and E a countably generated R-module. If now E is a direct summand of a free module F, then E itself is free.*

Proof. Suppose that N is a direct summand of E and that $x \in N$. Then N is a direct summand of F. Consequently, by Lemma 6, there exists a free direct summand G, of N, such that $x \in G$. That E is free now follows from Lemma 5.

The next lemma deals with a slightly involved situation so we shall set out some of the details in this preliminary paragraph. To this end let $\{M_\lambda\}_{\lambda \in \Lambda}$ be a family of submodules of an R-module M and let M be their direct sum. The set Λ may be arbitrary *but we assume that each individual M_λ is countably generated.* More precisely we shall suppose that $x_{\lambda 1}, x_{\lambda 2}, x_{\lambda 3}, \ldots$ is a sequence of elements which generates M_λ. It will also be assumed that we have a direct sum decomposition $M = K \oplus N$, where K and N are submodules of M.

LEMMA 8. *Let the situation be as described above and suppose that Λ_0 is a subset of Λ with the property that*

$$\sum_{\lambda \in \Lambda_0} M_\lambda = \left(\left(\sum_{\lambda \in \Lambda_0} M_\lambda\right) \cap K\right) \oplus \left(\left(\sum_{\lambda \in \Lambda_0} M_\lambda\right) \cap N\right). \quad (2.3.1)$$

If now $\mu \in \Lambda \setminus \Lambda_0$, then there exists a subset Λ_1 of Λ such that
(i) $\Lambda_0 \subseteq \Lambda_1$ *and* $\Lambda_1 \setminus \Lambda_0$ *is countable;*
(ii) $\mu \in \Lambda_1$;
(iii) $\sum_{\lambda \in \Lambda_1} M_\lambda = \left(\left(\sum_{\lambda \in \Lambda_1} M_\lambda\right) \cap K\right) \oplus \left(\left(\sum_{\lambda \in \Lambda_1} M_\lambda\right) \cap N\right).$

Proof. We begin by constructing, in succession, a certain increasing sequence $S_1 \subseteq S_2 \subseteq S_3 \subseteq \ldots$ of countable subsets of $\Lambda \setminus \Lambda_0$. To this end we put $S_1 = \{\mu\}$. Now suppose that $m \geqslant 1$ and that $S_1, S_2, \ldots, S_m$ have all been constructed. For any $\lambda \in \Lambda$ and $j \geqslant 1$ let us write $x_{\lambda j}$ in the form $x_{\lambda j} = y_{\lambda j} + z_{\lambda j}$, where $y_{\lambda j} \in K$ and $z_{\lambda j} \in N$. The elements $y_{\lambda' j}$, where $\lambda' \in S_m$ and $j \geqslant 1$, form a countable set. Some of these elements may not be contained in

$$\sum_{\lambda \in (\Lambda_0 \cup S_m)} M_\lambda$$

but we can enlarge S_m to S_{m+1} by adding countably many elements of $\Lambda \backslash \Lambda_0$ so as to ensure that

$$y_{\lambda'j} \in \sum_{\lambda \in (\Lambda_0 \cup S_{m+1})} M_\lambda \qquad (2.3.2)$$

for all $\lambda' \in S_m$ and $j \geqslant 1$. This shows how the sequence $S_1, S_2, S_3, \ldots$ is to be constructed.

Put $S = \bigcup_{m=1}^{\infty} S_m$ and $\Lambda_1 = \Lambda_0 \cup S$. It is obvious that conditions (i) and (ii) of the lemma are satisfied. Suppose that $\lambda' \in \Lambda_1$ and $j \geqslant 1$. *We claim that*

$$y_{\lambda'j} \in \sum_{\lambda \in \Lambda_1} M_\lambda.$$

Indeed if $\lambda' \in \Lambda_0$ this follows from (2.3.1). If on the other hand $\lambda' \in \Lambda_1 \backslash \Lambda_0$, then $\lambda' \in S_m$ for some m and so the desired conclusion follows from (2.3.2).

We now see that if $\lambda' \in \Lambda_1$ and $j \geqslant 1$, then $x_{\lambda'j}$, $y_{\lambda'j}$ and hence also $z_{\lambda'j}$ all belong to $\sum_{\lambda \in \Lambda_1} M_\lambda$. Since the elements $x_{\lambda'j}$ ($\lambda' \in \Lambda_1, j \geqslant 1$) form a generating system for $\sum_{\lambda \in \Lambda_1} M_\lambda$, it follows that

$$\sum_{\lambda \in \Lambda_1} M_\lambda \subseteq \left(\left(\sum_{\lambda \in \Lambda_1} M_\lambda\right) \cap K\right) + \left(\left(\sum_{\lambda \in \Lambda_1} M_\lambda\right) \cap N\right).$$

But the opposite inclusion is trivial. Consequently (iii) holds as well.

At this point we shall digress a little on the subject of well-ordered sets in order to avoid assuming a familiarity with the theory of ordinals. If A is a non-empty well-ordered set, then it will have a first element θ say. Now suppose that $\alpha \in A$ and that there are elements of A which come after α. The first of these will be denoted by $\alpha + 1$. Of course there will usually be elements of A, other than θ, which are not of the form $\alpha + 1$.

Any set can be well-ordered in at least one way. Hence if Λ is any set, there will always exist a well-ordered set A whose cardinality is greater than that of the set of subsets of Λ.

THEOREM 5. *Let M be an R-module which can be expressed in the form*

$$M = \sum_{\lambda \in \Lambda} M_\lambda \quad (direct\ sum),$$

*where each M_λ is a countably generated submodule of M. Further let
K be a direct summand of M. Then K can also be expressed as a
direct sum of countably generated R-modules.*

Proof. Let A be a well-ordered set whose cardinality exceeds
that of the set of subsets of Λ. The symbol $\leqslant$ will be used to
denote the ordering relation on A and θ will designate the initial
element. We can arrange that A has no last element. This
ensures that $\alpha + 1$ is defined for all $\alpha \in A$. In what follows N will
denote a submodule of M such that $M = K \oplus N$. We shall also
well-order the set Λ.

The plan is to construct inductively an increasing series
$\{I_\alpha\}_{\alpha \in A}$ of subsets of Λ having certain properties. In particular it
will be arranged that

$$\sum_{\lambda \in I_\alpha} M_\lambda = \left(\left(\sum_{\lambda \in I_\alpha} M_\lambda\right) \cap K\right) \oplus \left(\left(\sum_{\lambda \in I_\alpha} M_\lambda\right) \cap N\right) \quad (2.3.3)$$

for all $\alpha \in A$.

We begin by defining I_θ to be the empty set. Evidently (2.3.3)
holds trivially for $\alpha = \theta$. Now suppose that $\beta \in A$, $\beta \neq \theta$ and that
we have defined an increasing series $\{I_\gamma\}_{\gamma < \beta}$ of subsets of Λ each
satisfying the equivalent of (2.3.3). We shall now define I_β
without disturbing these properties. There are two cases to
consider.

First case. Suppose that $\beta = \alpha + 1$ for some α. If $I_\alpha = \Lambda$, then we
take $I_{\alpha+1}$ to be Λ as well. If however $I_\alpha \neq \Lambda$, then we construct
$I_{\alpha+1}$ from I_α just as, in Lemma 8, we constructed Λ_1 from Λ_0. In
this construction we use the first element of Λ that is not in I_α to
play the role of μ. Note that in both cases $I_\alpha \subseteq I_{\alpha+1}$, the difference
$I_{\alpha+1} \backslash I_\alpha$ is countable, and the relation typified by (2.3.3) continues
to hold. Also $I_{\alpha+1}$ *strictly* contains I_α unless $I_\alpha = \Lambda$.

Second case. This time we suppose that β is not of the form
$\alpha + 1$ and we put

$$I_\beta = \bigcup_{\gamma < \beta} I_\gamma.$$

In this situation the continued validity of (2.3.3) follows
trivially.

From here on we may suppose that the complete series $\{I_\alpha\}_{\alpha \in A}$ has been constructed. Note that there must be an α such that $I_\alpha = \Lambda$ for otherwise the I_α would form a *strictly* increasing series and this is impossible because the cardinality of A is too large. Thus I_α becomes equal to Λ from some point onwards.

For each $\alpha \in A$ put

$$C_\alpha = \left(\sum_{\lambda \in I_\alpha} M_\lambda \right) \cap K$$

and

$$D_\alpha = \left(\sum_{\lambda \in I_\alpha} M_\lambda \right) \cap N.$$

Then

$$C_\alpha \subseteq C_{\alpha+1} \subseteq \sum_{\lambda \in I_{\alpha+1}} M_\lambda. \qquad (2.3.4)$$

But, by (2.3.3), C_α is a direct summand of $\sum_{\lambda \in I_\alpha} M_\lambda$ which in turn is a direct summand of $\sum_{\lambda \in I_{\alpha+1}} M_\lambda$. Consequently C_α is a direct summand of $\sum_{\lambda \in I_{\alpha+1}} M_\lambda$ and therefore, by (2.3.4), it is also a direct summand of $C_{\alpha+1}$. Let

$$C_{\alpha+1} = C_\alpha \oplus E_\alpha.$$

Next we have isomorphisms

$$\sum_{\lambda \in (I_{\alpha+1} \setminus I_\alpha)} M_\lambda \approx \left(\sum_{\lambda \in I_{\alpha+1}} M_\lambda \right) \Big/ \left(\sum_{\lambda \in I_\alpha} M_\lambda \right)$$

$$\approx (C_{\alpha+1}/C_\alpha) \oplus (D_{\alpha+1}/D_\alpha)$$

$$\approx E_\alpha \oplus (D_{\alpha+1}/D_\alpha).$$

But $I_{\alpha+1} \setminus I_\alpha$ is countable and each M_λ is countably generated. Accordingly

$$\sum_{\lambda \in (I_{\alpha+1} \setminus I_\alpha)} M_\lambda$$

is a countably generated module and therefore E_α is also countably generated because it is a homomorphic image of this module.

We claim that for all $\alpha \in A$

$$C_{\alpha+1} = \sum_{\gamma \leqslant \alpha} E_\gamma \quad \text{(direct sum)}.$$

This is certainly true for $\alpha = 0$. Suppose therefore that it has been proved for all elements of A that precede the element β. If $\beta = \alpha + 1$ for some α, then

$$C_{\beta+1} = C_\beta \oplus E_\beta = C_{\alpha+1} \oplus E_\beta = \left(\sum_{\gamma \leqslant \alpha} E_\gamma \right) \oplus E_\beta$$

and moreover the sum $\sum_{\gamma \leqslant \alpha} E_\gamma$ is direct. Accordingly

$$C_{\beta+1} = \sum_{\gamma \leqslant \beta} E_\gamma \quad \text{(direct sum)}$$

as required. On the other hand if β is not of the form $\alpha + 1$, then

$$C_{\beta+1} = E_\beta \oplus C_\beta = E_\beta \oplus \bigcup_{\alpha < \beta} C_\alpha = E_\beta \oplus \bigcup_{\alpha < \beta} C_{\alpha+1}$$

and therefore

$$C_{\beta+1} = E_\beta \oplus \bigcup_{\alpha < \beta} \left(\sum_{\gamma \leqslant \alpha} E_\gamma \right) = E_\beta \oplus \sum_{\gamma < \beta} E_\gamma.$$

But $\sum_{\gamma < \beta} E_\gamma$ is also a direct sum by virtue of the induction hypothesis and once again we reach the desired conclusion. This establishes our claim. Finally choose α so that $I_\alpha = I_{\alpha+1} = \Lambda$. Then

$$K = C_{\alpha+1} = \sum_{\gamma \leqslant \alpha} E_\gamma$$

and with this the theorem is proved.

Theorem 5 represents a turning point in our discussion. Before putting it to use we make the following

DEFINITION. *A module which is a direct summand of a free module is called a 'projective module'.*

Of course any module which is isomorphic to a projective module is itself projective.

Note that Lemma 7 asserts that when R is a quasi-local ring every countably generated projective R-module is free. This result is comparatively elementary and does not require Theorem 5. In particular it includes the (even more elementary) proposition that when R is a quasi-local ring every finitely generated projective module is free. Although these partial

results are absorbed by the imminent Theorem 7 they are in fact
sufficient for many applications.

EXERCISE 3. *Let P be a projective R-module. Further let
$h: P \to E$ and $f: G \to E$ be homomorphisms of R-modules and
suppose that f is surjective. Show that there exists a homomorphism
$g: P \to G$ such that $fg = h$.*

EXERCISE 4. *Let $0 \to K \to P \to E \to 0$ and $0 \to K' \to P' \to E \to 0$
be exact sequences of R-modules, where P and P' are projective.
Show that $K \oplus P'$ and $K' \oplus P$ are isomorphic modules.*

Exercise 4 is *Schanuel's Lemma* in its full generality.

THEOREM 6. *Let P be a projective R-module. Then P can be
expressed as a direct sum of a number of countably generated
projective R-modules.*

Proof. By definition P is a direct summand of a free R-module F.
Let $\{e_i\}_{i \in I}$ be a base for F. Then F is the direct sum of its submodules Re_i and each of these is singly generated and so *a fortiori*
countably generated. By Theorem 5, P contains a family
$\{P_\lambda\}_{\lambda \in \Lambda}$ of countably generated submodules such that

$$P = \sum_{\lambda \in \Lambda} P_\lambda \quad \text{(direct sum)}.$$

Since P_λ is a direct summand of P it is also a direct summand of F.
Consequently, for each $\lambda \in \Lambda$, P_λ is projective. This completes the
proof.

Our next result is the theorem of Kaplansky to which we
referred earlier.[†]

THEOREM 7. *Let R be a quasi-local ring and P a projective
R-module. Then P is a free R-module.*

Proof. By Theorem 6, we have a relation

$$P = \sum_{\lambda \in \Lambda} P_\lambda \quad \text{(direct sum)},$$

where each P_λ is a countably generated projective module.
Furthermore, by Lemma 7, P_λ is free. But any direct sum of free
modules is also a free module. Accordingly P is a free module.

† Theorems 5 and 6, which are extremely general in character, are also due
to Kaplansky and form part of his original demonstration of Theorem 7.

EXERCISE 5. *Let $f\colon K \to E$ be a monomorphism of R-modules.
Show that in order that $f(K)$ should be a direct summand of E it is
necessary and sufficient that there exists a homomorphism $g\colon E \to K$
such that gf is the identity mapping of K. Show also that, when this
is the case, $E = f(K) \oplus \operatorname{Ker} g$.*

EXERCISE 6. *Let E be an R-module and $f\colon E \to P$ an epimorphism, where P is projective. Show that $\operatorname{Ker} f$ is a direct summand of
E and that E is isomorphic to $P \oplus \operatorname{Ker} f$.*

EXERCISE 7. *Let $h\colon F \to G$ be a homomorphism, where F and G
are free R-modules. Show that the following statements are equivalent:*
  (i) *$h(F)$ is a direct summand of G;*
  (ii) *Coker h is a projective R-module;*
  (iii) *there is a homomorphism $\psi\colon G \to F$ such that $h = h\psi h$.*

2.4 Localization of modules

Let S be a multiplicatively closed subset of R. We recall that
according to the definition given in section (1.7), the identity
element of R is necessarily in S. We recall also that the ring of
fractions R_S is a non-trivial ring if and only if $0 \notin S$. For this
reason, when we are dealing with free modules it will be assumed
that our multiplicatively closed subsets do not contain the zero
element of R.

Suppose that E is an R-module and consider formal fractions
e/s, where $e \in E$ and $s \in S$. Two such fractions, say e/s and e'/s',
are to be regarded as equivalent if $\sigma s'e = \sigma se'$ for some $\sigma \in S$.
This is indeed an equivalence relation in the abstract sense. As is
customary when dealing with fractions of any kind, we shall use
e/s to denote both the fraction itself and the equivalence class to
which it belongs.

Let E_S denote the set of equivalence classes and observe that
E_S becomes an abelian group if addition is defined by means of
the formula
$$\frac{e_1}{s_1} + \frac{e_2}{s_2} = \frac{s_2 e_1 + s_1 e_2}{s_1 s_2}.$$

(The verification that this definition is compatible with the
equivalence relation between fractions is left to the reader.)

However E_S is more than just an abelian group for it has a natural structure as an R_S-module. In this structure if r/σ belongs to R_S and e/s to E_S, then the product of r/σ and e/s is given by

$$\frac{r}{\sigma}\frac{e}{s} = \frac{re}{\sigma s}.$$

Of course it is necessary to check not only that the product is well-defined, but also that the module axioms hold. These details are left as an exercise.

Let $f\colon E \to M$ be a homomorphism of R-modules. Then there is a homomorphism $f_S\colon E_S \to M_S$ of R_S-modules in which, with a self-explanatory notation,

$$f_S\left(\frac{e}{s}\right) = \frac{f(e)}{s}.$$

Observe that if $i\colon E \to E$ is the identity mapping of E, then $i_S\colon E_S \to E_S$ is the identity mapping of E_S. Again if $f\colon E \to M$ and $g\colon M \to N$ are homomorphisms of R-modules, then

$$(gf)_S = g_S f_S.$$

The reader who is familiar with the language of Category Theory will recognize that we have just observed that fraction formation using S provides a covariant functor from R-modules to R_S-modules. Our next theorem shows that, in this connection, exact sequences are preserved.

THEOREM 8. *Let* $E \xrightarrow{\ f\ } M \xrightarrow{\ g\ } N$ *be an exact sequence of* R-*modules. Then*

$$E_S \xrightarrow{\ f_S\ } M_S \xrightarrow{\ g_S\ } N_S$$

is an exact sequence of R_S-*modules.*

Proof. It is a triviality that $g_S f_S$ is a null homomorphism. Suppose that m/s, where $m \in M$ and $s \in S$, belongs to $\mathrm{Ker}\, g_S$. Then $g(m)/s = 0$ in N_S and therefore there exists $\sigma \in S$ such that $\sigma g(m) = 0$. But now $g(\sigma m) = 0$ and therefore $\sigma m = f(e)$ for some $e \in E$. Accordingly

$$f_S\left(\frac{e}{\sigma s}\right) = \frac{f(e)}{\sigma s} = \frac{\sigma m}{\sigma s} = \frac{m}{s}$$

whence $m/s \in \mathrm{Im} f_S$. This shows that $\mathrm{Ker}\, g_S \subseteq \mathrm{Im} f_S$ and completes the proof.

Suppose that K is a submodule of an R-module E and that $j \colon K \to E$ is the inclusion mapping. Then $0 \to K \xrightarrow{\;j\;} E$ is an exact sequence and therefore, by Theorem 8,

$$0 \to K_S \xrightarrow{\;js\;} E_S$$

is an exact sequence of R_S-modules. Thus we have a canonical injection $j_S \colon K_S \to E_S$ of K_S into E_S and this enables us to regard K_S as an R_S-submodule of E_S. When K_S is regarded in this way it consists of all the elements of E_S that can be expressed in the form x/s with x in K and s in S. Note that, since

$$0 \to K \xrightarrow{\;j\;} E \to E/K \to 0$$

is an exact sequence of R-modules,

$$0 \to K_S \xrightarrow{\;js\;} E_S \to (E/K)_S \to 0$$

will be an exact sequence of R_S-modules. Consequently when K_S is regarded as a submodule of E_S, we have an isomorphism

$$E_S/K_S \approx (E/K)_S \tag{2.4.1}$$

of R_S-modules.

Next suppose that $\mathfrak{A}$ is an ideal of R. Then $\mathfrak{A}$ can be regarded as a submodule of R and this enables us to consider $\mathfrak{A}_S$ as a submodule of R_S. Thus $\mathfrak{A}_S$ is an *ideal* of R_S. Since this ideal consists of all the elements of R_S that can be expressed in the form a/s with $a \in \mathfrak{A}$ and $s \in S$, it follows that

$$\mathfrak{A}_S = \mathfrak{A}R_S. \tag{2.4.2}$$

Here by $\mathfrak{A}R_S$ we mean, as on previous occasions, the ideal of R_S that is generated by $\chi(\mathfrak{A})$, where $\chi \colon R \to R_S$ is the canonical ring-homomorphism.†

LEMMA 9. *Let E be an R-module and S a multiplicatively closed subset of R. Suppose that $\{e_i\}_{i \in I}$ is a system of generators of E. Then the elements $\{e_i/1\}_{i \in I}$ of E_S generate E_S as an R_S-module.*

† See section (1.7).

Proof. Let $e \in E$ and $s \in S$. There exists a *finite* subset J of I and elements r_j $(j \in J)$ of R such that

$$e = \sum_{j \in J} r_j e_j.$$

It follows that

$$\frac{e}{s} = \sum_{j \in J} \left(\frac{r_j}{s}\right) \left(\frac{e_j}{1}\right).$$

Thus e/s belongs to the R_S-submodule of E_S generated by the elements $\{e_i/1\}_{i \in I}$. Accordingly this submodule must be E_S itself.

THEOREM 9. *Let F be a free R-module and $\{e_i\}_{i \in I}$ a base for F. Let S be a multiplicatively closed subset of R not containing the zero element. Then F_S is a free R_S-module and the elements $\{e_i/1\}_{i \in I}$ form a base for F_S. It follows that $\mathrm{rank}_R (F) = \mathrm{rank}_{R_S} (F_S)$. In particular, if F is a finite free R-module, then F_S is a finite free R_S-module.*

Proof. By Lemma 9, the family $\{e_i/1\}_{i \in I}$ generates F_S. Let J be a finite subset of I and $\{\xi_j\}_{j \in J}$ elements of R_S such that

$$\sum_{j \in J} \xi_j \frac{e_j}{1} = 0.$$

It is enough to show that all the ξ_j are zero. Since J is a finite set, we can find elements r_j $(j \in J)$ in R and a single element s in S such that $\xi_j = r_j/s$ for all j in J. If now $\sum_{j \in J} r_j e_j = e$, then $e/s = 0$. Consequently $\sigma e = 0$ for a suitable σ in S. But now

$$\sum_{j \in J} \sigma r_j e_j = 0,$$

whence $\sigma r_j = 0$ for all $j \in J$. Finally

$$\xi_j = \frac{r_j}{s} = \frac{\sigma r_j}{\sigma s} = 0$$

and with this the proof is complete.

THEOREM 10. *Let S be a multiplicatively closed subset of R not containing the zero element, and let E be a finitely presented R-module. Then E_S is a finitely presented R_S-module.*

Proof. There exists an exact sequence $G \to F \to E \to 0$, where G and F are finite free R-modules. This will induce an exact sequence

$$G_S \to F_S \to E_S \to 0$$

of R_S-modules. By Theorem 9, G_S and F_S are finite free R_S-modules. Consequently E_S is a finitely presented R_S-module.

E x e r c i s e 8. *Let K be a direct summand of an R-module E and let S be a multiplicatively closed subset of R. Show that K_S is a direct summand of the R_S-module E_S.*

T h e o r e m 11. *Let E be a projective R-module and S a multiplicatively closed subset of R not containing the zero element. Then E_S is a projective R_S-module.*

Proof. E is a direct summand of a free R-module F and therefore, by Exercise 8, E_S is a direct summand of F_S. Now Theorem 9 shows that F_S is a free R_S-module. The theorem therefore follows.

It is convenient to introduce some additional notation. To this end let E be an R-module, K a submodule of E, and $\mathfrak{A}$ an ideal of R. Put

$$K :_R E = \{r | r \in R \text{ and } rE \subseteq K\}. \tag{2.4.3}$$

It is easily verified that $K :_R E$ is an ideal of R. We also put

$$\mathrm{Ann}_R(E) = \{r | r \in R \text{ and } rE = 0\} \tag{2.4.4}$$

so that

$$\mathrm{Ann}_R(E) = 0 :_R E.$$

The ideal $\mathrm{Ann}_R(E)$ is called the *annihilator* of E. Note that

$$K :_R E = \mathrm{Ann}_R(E/K). \tag{2.4.5}$$

Finally as a complement to (2.4.3) we set

$$K :_E \mathfrak{A} = \{e | e \in E \text{ and } \mathfrak{A}e \subseteq K\} \tag{2.4.6}$$

and observe that $K :_E \mathfrak{A}$ is an R-submodule of E containing K.

T h e o r e m 12. *Let K be a submodule of a finitely generated R-module E and let S be a multiplicatively closed subset of R. Then*

$$(K :_R E) R_S = K_S :_{R_S} E_S$$

and

$$(\mathrm{Ann}_R(E)) R_S = \mathrm{Ann}_{R_S}(E_S).$$

Proof. Since E/K is finitely generated, $K:_R E = \mathrm{Ann}_R (E/K)$, and by (2.4.1)

$$K_S:_{R_S} E_S = \mathrm{Ann}_{R_S} (E_S/K_S) = \mathrm{Ann}_{R_S}((E/K)_S),$$

it is enough to establish the second assertion. Obviously if $\alpha \in \mathrm{Ann}_R (E)$, then $(\alpha/1) E_S = 0$. It follows that

$$\mathrm{Ann}_R (E) R_S \subseteq \mathrm{Ann}_{R_S} (E_S).$$

Now suppose that $r/s \in \mathrm{Ann}_{R_S}(E_S)$, where $r \in R$ and $s \in S$, and let $E = Re_1 + Re_2 + \ldots + Re_n$. Then for $1 \leqslant i \leqslant n$ we have

$$re_i/s = (r/s)(e_i/1) = 0.$$

Consequently there exists $\sigma_i \in S$ such that $\sigma_i r e_i = 0$. Put $\sigma = \sigma_1 \sigma_2 \ldots \sigma_n$. Then $\sigma \in S$ and $\sigma r e_i = 0$ for $1 \leqslant i \leqslant n$. It follows that $\sigma r \in \mathrm{Ann}_R (E)$ and therefore

$$\frac{r}{s} = \frac{\sigma r}{\sigma s} \in \mathrm{Ann}_R (E) R_S.$$

Accordingly

$$\mathrm{Ann}_{R_S} (E_S) \subseteq \mathrm{Ann}_R (E) R_S$$

and with this the proof is complete.

We turn now to localization proper. Suppose that P is a prime ideal of R and put $S = R \backslash P$ so that S is a multiplicatively closed subset of R. If now E is an R-module and $f: E \to M$ a homomorphism of R-modules, then it is customary to write E_P and f_P rather than E_S and f_S. Of course, E_P is a module over the quasi-local ring R_P.

DEFINITION. *The 'support' of the R-module E is the set of prime ideals P for which $E_P \neq 0$.*

The support of E will be denoted by $\mathrm{Supp}_R (E)$.

THEOREM 13. *Let E be a finitely generated R-module. Then $\mathrm{Supp}_R (E)$ consists precisely of the prime ildeas that contain $\mathrm{Ann}_R (E)$.*

Proof. Let P be a prime ideal. Then $E_P \neq 0$ if and only if $\mathrm{Ann}_{R_P} (E_P) \neq R_P$. But, by Theorem 12,

$$\mathrm{Ann}_{R_P}(E_P) = (\mathrm{Ann}_R (E)) R_P$$

and we know, from Chapter 1 Exercise 5, that

$$(\text{Ann}_R(E))R_P \neq R_P$$

if and only if $\text{Ann}_R(E)$ does not meet $R\backslash P$. Thus $E_P \neq 0$ when and only when $\text{Ann}_R(E) \subseteq P$.

EXERCISE 9. *Let E be a finitely presented R-module and K an arbitrary R-module. Further let S be a multiplicatively closed subset of R and $\phi\colon E_S \to K_S$ a homomorphism of R_S-modules. Show that there exists a homomorphism $f\colon E \to K$ and an element $s \in S$ such that*

$$\phi\left(\frac{e}{1}\right) = \frac{f(e)}{s}$$

for all $e \in E$.

THEOREM 14. *Let E be a finitely presented R-module. Then E is a projective R-module if and only if E_P is a free R_P-module for every maximal ideal P.*

Proof. Suppose first of all that E is a projective R-module and that P is a maximal ideal. By Theorem 11, E_P is a projective module over the quasi-local ring R_P. That E_P is a free R_P-module now follows from Theorem 7.

From here on we shall assume that, for every maximal ideal P, E_P is a free R_P-module. Construct an exact sequence

$$F \xrightarrow{\ u\ } E \to 0,$$

where F is a free R-module, and let P be a maximal ideal. There is induced an exact sequence

$$F_P \xrightarrow{\ u_P\ } E_P \to 0$$

of R_P-modules. Since E_P is R_P-free it follows, from Lemma 2, that there is an R_P-homomorphism $\phi\colon E_P \to F_P$ such that $u_P\phi = i_P$, where $i\colon E \to E$ denotes the identity mapping of E. By Exercise 9, there can be found a homomorphism $f\colon E \to F$ (depending on P) and an element $s \in R\backslash P$ such that

$$\phi\left(\frac{e}{1}\right) = \frac{f(e)}{s}$$

for all $e \in E$. Consequently $uf(e)/s = e/1$ for every e in E. Since E is finitely presented it must be finitely generated. Let $e_1, e_2, \ldots, e_n$ generate E. Then for each $j \, (1 \leqslant j \leqslant n)$ we can find $\sigma_j \in R \backslash P$ such that $\sigma_j uf(e_j) = \sigma_j se_j$. Put $\sigma = \sigma_1 \sigma_2 \ldots \sigma_n$. Then $\sigma \in R \backslash P$ and $\sigma uf(e_j) = \sigma se_j$ for $1 \leqslant j \leqslant n$. Accordingly $\sigma uf(e) = \sigma se$ for all $e \in E$.

Let $\mathfrak{A}$ be the set of elements α in R such that there is a homomorphism $E \to F$ for which the diagram

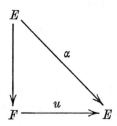

is commutative. Here the diagonal mapping is the one obtained by multiplying the elements of E by α. Clearly $\mathfrak{A}$ is an ideal of R and, by virtue of the remark at the end of the last paragraph, $\mathfrak{A}$ meets $R \backslash P$. As this holds for every maximal ideal P, it follows that $\mathfrak{A} = R$ and therefore $1 \in \mathfrak{A}$. Accordingly there is a homomorphism $\omega : E \to F$ such that $u\omega = i$. This ensures that ω is a monomorphism and it follows, from Exercise 5, that $\omega(E)$ is a direct summand of F and therefore projective. But E and $\omega(E)$ are isomorphic. Consequently E is projective as well.

EXERCISE 10. *Let R be an integral domain and E a finitely generated R-module. Further let E_P be a free R_P-module for every maximal ideal P. Show that E is a projective R-module.*

2.5 Supplementary remarks

In order to make this book as self-contained as possible, the author has deliberately avoided assuming that the reader is familiar with the theory of exterior algebras.† Nevertheless just as this theory throws light on the properties of determinants, it

---

† The sole exception occurs in Appendix C.

can also be used to illuminate some of the topics treated here. Brief indications will now be given of the way in which the connection arises.

Let $\phi: F \to G$ be a homomorphism, where F and G are finite free modules of ranks p and q respectively, and let us select a base $x_1, x_2, \ldots, x_p$ for F and a base $y_1, y_2, \ldots, y_q$ for G. Then with respect to these bases ϕ can be represented by a $p \times q$ matrix A in an obvious way. Now the nth exterior power $\overset{n}{\Lambda}F$ of the module F is a free module of rank $\binom{p}{n}$. Indeed if $J = \{j_1, j_2, \ldots, j_n\}$ belongs to S_n^p, where the notation is the same as in section (1.3), and x_J denotes the exterior product of $x_{j_1}, x_{j_2}, \ldots, x_{j_n}$, then the elements x_J ($J \in S_n^p$) form a base for $\overset{n}{\Lambda}F$. Likewise $\overset{n}{\Lambda}G$ has an induced base y_K ($K \in S_n^q$). Now the theory of exterior algebras shows that ϕ gives rise to a homomorphism

$$\overset{n}{\Lambda}\phi: \overset{n}{\Lambda}F \to \overset{n}{\Lambda}G.$$

Naturally $\overset{n}{\Lambda}\phi$ can be represented by a matrix relative to the bases x_J ($J \in S_n^p$) and y_K ($K \in S_n^q$) of $\overset{n}{\Lambda}F$ and $\overset{n}{\Lambda}G$, and it turns out that this matrix is none other than the nth exterior power $A^{(n)}$ of A. Because of this interpretation of $A^{(n)}$, the theory of exterior algebras now provides alternative methods of proving various results. For example Chapter 1 Theorem 1 follows from the fact that the operation of forming nth exterior powers is a covariant functor.

It is the view of the author that a full-blooded use of exterior algebras in the present context sometimes obscures more than it reveals, and that the most elegant presentation would result from a judicious combination of the powerful abstract theory with simple computational devices of the kind used here. The reader who wishes to become acquainted with the theory of exterior algebras will find a thorough discussion in (**16**).

Solutions to the Exercises on Chapter 2

EXERCISE 1. *Let* $0 \to F' \xrightarrow{\phi} F \xrightarrow{\psi} F'' \to 0$ *be an exact sequence of free R-modules. Show that* $\operatorname{rank}_R(F) = \operatorname{rank}_R(F') + \operatorname{rank}_R(F'')$.

Remark. It is worth noting that we do not assume that the ranks involved are finite.

Solution. Let $\{e'_i\}_{i \in I}$ be a base for F' and choose a family $\{x_j\}_{j \in J}$ of elements of F so that $\{\psi(x_j)\}_{j \in J}$ is a base for F''. It will be shown that the $\phi(e'_i)$ and the x_j jointly span F and that these elements are linearly independent.

Choose $x \in F$. There exist a finite subset J_0 of J and elements $\{r_j\}_{j \in J_0}$ of R such that

$$\psi(x) = \sum_{j \in J_0} r_j \psi(x_j).$$

Accordingly $x - \sum_{j \in J_0} r_j x_j$ belongs to $\operatorname{Ker} \psi = \phi(F')$ and therefore

$$x - \sum_{j \in J_0} r_j x_j \in \sum_{i \in I} R\phi(e'_i).$$

This shows that the families $\{\phi(e'_i)\}_{i \in I}$ and $\{x_j\}_{j \in J}$ together span F.

Now let I_1 and J_1 be finite subsets of I and J respectively and let $\{\alpha_i\}_{i \in I_1}$ and $\{\beta_j\}_{j \in J_1}$ be families of elements of R. Assume that

$$\sum_{i \in I_1} \alpha_i \phi(e'_i) + \sum_{j \in J_1} \beta_j x_j = 0.$$

Then $\sum_{j \in J_1} \beta_j \psi(x_j) = 0$. It follows that $\beta_j = 0$ for all $j \in J_1$ and that

$$\phi\left(\sum_{i \in I_1} \alpha_i e'_i \right) = \sum_{i \in I_1} \alpha_i \phi(e'_i) = 0.$$

But ϕ is a monomorphism. Accordingly $\sum_{i \in I_1} \alpha_i e'_i = 0$ and therefore $\alpha_i = 0$ for all $i \in I_1$. Thus the families $\{\phi(e'_i)\}_{i \in I}$ and $\{x_j\}_{j \in J}$ together form a base for F and the desired result is an immediate consequence of this fact.

EXERCISE 2. *Let F be a free R-module with a base $\{e_i\}_{i \in I}$ and F' a free R-module with a base $\{e'_j\}_{j \in J}$. Show that F and F' are isomorphic if and only if there is a bijection of the set I on to the set J.*

Solution. First suppose that we have a bijection $\sigma: I \to J$ of I on to J. Then, by Lemma 1, there is a homomorphism $\phi: F \to F'$ in which $\phi(e_i) = e'_{\sigma(i)}$ for all $i \in I$. This homomorphism is clearly an isomorphism.

Now suppose that we have an isomorphic mapping $\psi: F \to F'$ of F on to F'. Then $\{\psi(e_i)\}_{i \in I}$ will also be a base for F'. It follows that, for this part of the argument, we may assume that $F = F'$. Moreover, in view of Theorem 2 and the remarks which immediately precede it, we may restrict our discussion to the case where I and J are both infinite sets. The number of elements in I respectively J will be denoted by $|I|$ respectively $|J|$. Thus $|I|$ and $|J|$ are infinite cardinals.

For each $i \in I$ there is a finite subset S_i of J such that

$$e_i \in \sum_{j \in S_i} Re'_j.$$

Put
$$J' = \bigcup_{i \in I} S_i.$$

Then
$$F = \sum_{i \in I} Re_i \subseteq \sum_{j \in J'} Re'_j.$$

But if $j_0 \in J \backslash J'$, then e'_{j_0} is not in $\sum_{j \in J'} Re'_j$ because the ring R is non-trivial. It follows that $J = J'$ and therefore

$$J = \bigcup_{i \in I} S_i.$$

Since $|I|$ is infinite, the theory of infinite cardinals now shows that $|J| \leqslant |I|$ and, of course, $|I| \leqslant |J|$ for similar reasons. Accordingly $|I| = |J|$ which means precisely that there is a bijection of I on to J.

EXERCISE 3. *Let P be a projective R-module. Further let $h: P \to E$ and $f: G \to E$ be homomorphisms of R-modules and suppose that f is surjective. Show that there exists a homomorphism $g: P \to G$ such that $fg = h$.*

Solution. By definition, P is a direct summand of a free module F. Consequently there are homomorphisms $\sigma: P \to F$ and $\pi: F \to P$ such that $\pi\sigma$ is the identity mapping on P. By Lemma 2, there is a homomorphism $k\colon F \to G$ such that $fk = h\pi$. Let $g = k\sigma$ so that g is a homomorphism of P into G. Then

$$fg = fk\sigma = h\pi\sigma = h$$

as required.

EXERCISE 4. *Let* $0 \to K \to P \to E \to 0$ *and* $0 \to K' \to P' \to E \to 0$ *be exact sequences of R-modules, where P and P' are projective. Show that* $K \oplus P'$ *and* $K' \oplus P$ *are isomorphic modules.*

Solution. The argument used to prove Theorem 3 will provide a solution to this exercise provided that Exercise 3 is used in place of Lemma 2.

EXERCISE 5. *Let* $f\colon K \to E$ *be a monomorphism of R-modules. Show that in order that* $f(K)$ *should be a direct summand of E it is necessary and sufficient that there exists a homomorphism* $g\colon E \to K$ *such that* gf *is the identity mapping of K. Show also that, when this is the case,* $E = f(K) \oplus \operatorname{Ker} g$.

Solution. Assume that $g\colon E \to K$ is a homomorphism such that $gf = i$, where $i\colon K \to K$ is the identity mapping of K, and let $e \in E$. Then

$$g(e - fg(e)) = g(e) - g(e) = 0$$

and therefore, since $e = fg(e) + (e - fg(e))$, we have

$$E = f(K) + \operatorname{Ker} g.$$

Now suppose that $x \in f(K) \cap \operatorname{Ker} g$. Then $x = f(k)$ for some $k \in K$ and therefore $k = g(x)$ because $gf = i$. But $g(x) = 0$. Consequently $k = 0$ and hence $x = 0$. It follows that $E = f(K) \oplus \operatorname{Ker} g$.

Finally suppose that $E = f(K) \oplus M$ for some submodule M of E and let $\pi\colon E \to f(K)$ be the projection of E on to the first summand. Then $\pi f\colon K \to f(K)$ is an isomorphism (because f is a monomorphism) with inverse u say. Put $g = u\pi$. Then gf is the identity mapping on K.

EXERCISE 6. *Let E be an R-module and* $f\colon E \to P$ *an epimorphism, where P is projective. Show that* $\operatorname{Ker} f$ *is a direct summand of E and that E is isomorphic to* $P \oplus \operatorname{Ker} f$.

Solution. By Exercise 3, there exists a homomorphism $g\colon P \to E$ such that $fg = i$, where $i\colon P \to P$ is the identity mapping. It follows from this that g is a monomorphism and now, by Exercise 5, $E = g(P) \oplus \operatorname{Ker} f$. Since P and $g(P)$ are isomorphic, this completes the solution.

EXERCISE 7. *Let $h\colon F \to G$ be a homomorphism where F and G are free R-modules. Show that the following statements are equivalent:*

  (i) *$h(F)$ is a direct summand of G;*

  (ii) Coker h *is a projective R-module;*

  (iii) *there is a homomorphism $\psi\colon G \to F$ such that $h = h\psi h$.*

Solution. It is clear that (i) implies (ii) and Exercise 6 shows that (ii) implies (i).

Assume (i). Then $G = h(F) \oplus N$ for some submodule N of G. Next $h(F)$ is a projective R-module and therefore, by Exercise 6, $F = \operatorname{Ker} h \oplus M$ for a suitable submodule M of F. Note that if $j\colon M \to F$ is the inclusion mapping, then hj is an isomorphism of M on to $h(F)$ and this will have an inverse $u\colon h(F) \to M$ say.

Let $\pi\colon G \to h(F)$ be the projection of G on to $h(F)$ that is associated with the decomposition $G = h(F) \oplus N$ and put $\psi = ju\pi$. Then for $x \in F$ we have

$$h\psi h(x) = hju\pi h(x) = hju(h(x)) = h(x).$$

This shows that (i) implies (iii).

Finally *assume* (iii) and let $\eta\colon h(F) \to G$ be the inclusion mapping. If now $y \in h(F)$, say $y = h(x)$ where $x \in F$, then

$$h\psi\eta(y) = h\psi h(x) = h(x) = y.$$

Thus $(h\psi)\,\eta\colon h(F) \to G$ and it induces the identity mapping on $h(F)$. Since $\operatorname{Im}(h\psi) \subseteq h(F)$, we may conclude, using Exercise 5, that $h(F)$ is a direct summand of G. This shows that (iii) implies (i) and completes the solution.

EXERCISE 8. *Let K be a direct summand of an R-module E and let S be a multiplicatively closed subset of R. Show that K_S is a direct summand of the R_S-module E_S.*

Solution. Let $j\colon K \to E$ be the inclusion mapping. By Exercise 5, there exists a homomorphism $g\colon E \to K$ such that gj is the identity mapping of K. But then $g_S j_S$ is the identity mapping of K_S. A further application of Exercise 5 shows that K_S is a direct summand of the R_S-module E_S.

EXERCISE 9. *Let E be a finitely presented R-module and K an arbitrary R-module. Further let S be a multiplicatively closed subset of R and $\phi\colon E_S \to K_S$ a homomorphism of R_S-modules. Show that there exists a homomorphism $f\colon E \to K$ and an element $s \in S$ such that*

$$\phi\left(\frac{e}{1}\right) = \frac{f(e)}{s}$$

for all $e \in E$.

Solution. Let $e_1, e_2, \ldots, e_q$ generate E. Then the elements $e_j/1$ ($1 \leqslant j \leqslant q$) will generate the R_S-module E_S. We now choose $k_1, k_2, \ldots, k_q$ in K and s' in S so that

$$\phi\left(\frac{e_j}{1}\right) = \frac{k_j}{s'}$$

for $1 \leqslant j \leqslant q$.

The module of relations between $e_1, e_2, \ldots, e_q$ is finitely generated. Suppose that this module is generated by the relations $(a_{i1}, a_{i2}, \ldots, a_{iq})$, where i ranges from 1 to p. Then

$$\sum_{j=1}^{q} \left(\frac{a_{ij}}{1}\right)\left(\frac{e_j}{1}\right) = \frac{\sum_{j=1}^{q} a_{ij}e_j}{1} = 0$$

and therefore, applying ϕ to this equation, we obtain

$$\frac{\sum_{j=1}^{q} a_{ij}k_j}{s'} = 0.$$

It follows that there exists $\sigma \in S$ such that

$$\sigma\left(\sum_{j=1}^{q} a_{ij}k_j\right) = 0$$

for $1 \leqslant i \leqslant p$.

Suppose that $a_1 e_1 + a_2 e_2 + \ldots + a_q e_q = 0$ where $a_j \in R$. Then there exist elements $r_1, r_2, \ldots, r_p$ in R such that

$$\sum_{i=1}^{p} r_i a_{ij} = a_j$$

and therefore

$$\sum_{j=1}^{q} a_j(\sigma k_j) = 0.$$

This shows that there is a homomorphism $f\colon E \to K$ with the property that $f(e_j) = \sigma k_j$ for $1 \leqslant j \leqslant q$. Accordingly

$$\phi\left(\frac{e_j}{1}\right) = \frac{k_j}{s'} = \frac{f(e_j)}{\sigma s'}.$$

If therefore we put $s = \sigma s'$, then

$$\phi\left(\frac{e}{1}\right) = \frac{f(e)}{s}$$

for all $e \in E$.

EXERCISE 10. *Let R be an integral domain and E a finitely generated R-module. Further let E_P be a free R_P-module for every maximal ideal P. Show that E is a projective R-module.*

Solution. Construct an exact sequence $F \xrightarrow{u} E \to 0$, where F is a free R-module, and let P be a maximal ideal. Then there is induced an exact sequence

$$F_P \xrightarrow{u_P} E_P \to 0$$

of R_P-modules. Since E_P is free by hypothesis, Lemma 2 shows that there is an R_P-homomorphism $\phi\colon E_P \to F_P$ such that $u_P \phi = i_P$, where $i\colon E \to E$ denotes the identity mapping of E.

Let $e_1, e_2, \ldots, e_q$ generate E and choose $x_1, x_2, \ldots, x_q$ in F and s in $R \backslash P$ so that

$$\phi\left(\frac{e_j}{1}\right) = \frac{x_j}{s}$$

for $1 \leqslant j \leqslant q$. Now assume that $a_1, a_2, \ldots, a_q$ belong to R and that $a_1 e_1 + a_2 e_2 + \ldots + a_q e_q = 0$. Then

$$\frac{a_1 x_1 + a_2 x_2 + \ldots + a_q x_q}{s} = 0$$

and therefore there exists $\sigma \in R \backslash P$ such that

$$\sigma(a_1 x_1 + a_2 x_2 + \ldots + a_q x_q) = 0.$$

But F is a free R-module and R is an integral domain. Consequently σ does not annihilate any non-zero element of F and therefore $a_1 x_1 + a_2 x_2 + \ldots + a_q x_q = 0$.

These remarks show that there is a homomorphism $f \colon E \to F$ (depending on P) with the property that $f(e_j) = x_j$ for $1 \leqslant j \leqslant q$. Accordingly

$$\phi \left(\frac{e_j}{1} \right) = \frac{f(e_j)}{s} \quad (1 \leqslant j \leqslant q)$$

and hence $\phi(e/1) = f(e)/s$ for all $e \in E$. The rest of the argument proceeds exactly as in the proof of Theorem 14.

3. The invariants of Fitting and MacRae

General remarks

It may assist the reader if we give a brief survey of the main topics treated in this chapter. It will be shown how, with the aid of determinantal ideals, it is possible to associate with each finitely generated module a family of ideals. These ideals, which are invariants of the module, are known as its *Fitting invariants* after H. Fitting who investigated their properties in (**20**). The Fitting invariants of an R-module E form an increasing sequence of ideals. The smallest of them, here called the *initial Fitting invariant*, has an intimate connection with the annihilator of E (Theorem 5).

Suppose that the module E has the property that there exists an exact sequence

$$0 \to F_n \to F_{n-1} \to \ldots \to F_1 \to F_0 \to E \to 0,$$

where $F_0, F_1, \ldots, F_n$ are finite free modules. If we put

$$\mathrm{Char}_R(E) = \sum_{\nu=0}^{n} (-1)^\nu \mathrm{rank}_R(F_\nu),$$

then, as is demonstrated in Theorem 20, $\mathrm{Char}_R(E)$ depends only on E.

Among the modules for which $\mathrm{Char}_R(E)$ is defined special interest attaches to those for which $\mathrm{Char}_R(E) = 0$. However for technical reasons we shall restrict our attention (at this stage) to those for which $\mathrm{Ann}_R(E)$ contains a non-zerodivisor. The connection between these two conditions will take us some time to unravel, but we set out the main facts here.

Let $\mathfrak{A}$ be an ideal of R and x an indeterminate. Then $\mathfrak{A}$ generates an ideal $\mathfrak{A}R[x]$ in the polynomial ring $R[x]$. We shall say that $\mathfrak{A}$ contains a *latent non-zerodivisor* if $\mathfrak{A}R[x]$ contains a non-zerodivisor. Of course if $\mathfrak{A}$ contains a non-zerodivisor, then it contains a latent non-zerodivisor but, as will be shown in section (5.3), the converse need not hold.†

Now suppose that $\mathrm{Char}_R(E)$ is defined. Then $\mathrm{Char}_R(E) = 0$ if and only if $\mathrm{Ann}_R(E)$ contains a latent non-zerodivisor (see Chapter 5 Exercise 6). Let us assume that $\mathrm{Ann}_R(E)$ actually contains a non-zerodivisor. Then following R. E. MacRae (32) it is possible to introduce another very interesting invariant which is multiplicative for short exact sequences (Theorem 26) and which is closely related to the initial Fitting invariant. In Chapter 7 this theory will be extended to all modules for which $\mathrm{Char}_R(E) = 0$.

Our immediate aim is to take a first look at these different invariants. A more searching investigation is delayed until we have uncovered more of the properties of exact sequences

$$0 \to F_n \to F_{n-1} \to \ldots \to F_1 \to F_0$$

in which all the terms are finite free modules. Such sequences will form a recurring topic from here onwards.

Throughout Chapter 3, R will denote a non-trivial commutative ring with identity element and when we speak of a $p \times q$ matrix $A = \|a_{jk}\|$ with entries in R, it is to be understood that p and q denote positive integers.

3.1 The Fitting invariants

Let E be a finitely generated R-module and let $e_1, e_2, \ldots, e_q$ be elements which generate E. We shall use $\mathbf{e}$ to denote the sequence $(e_1, e_2, \ldots, e_q)$ and we shall write

$$\mathbf{e} = (e_1, e_2, \ldots, e_q). \tag{3.1.1}$$

Now let $\nu \geqslant 0$ be an integer and put

$$\mathfrak{A}_\nu(\mathbf{e}|E) = \sum_A \mathfrak{A}_\nu(A), \tag{3.1.2}$$

where (i) A stands for an arbitrary (finite) matrix with q columns each of whose rows is a relation between $e_1, e_2, \ldots, e_q$, and (ii)

† However the converse does hold if R is *Noetherian*, i.e. if every ideal of R is finitely generated.

$\mathfrak{A}_\nu(A)$ denotes the νth determinantal ideal† of A. Accordingly $\mathfrak{A}_\nu(\mathbf{e}|E)$ is the ideal generated by the totality of $\nu \times \nu$ minors of all such matrices. Note that, by (1.4.1) and (1.4.2), we have

$$R = \mathfrak{A}_0(\mathbf{e}|E) \supseteq \mathfrak{A}_1(\mathbf{e}|E) \supseteq \mathfrak{A}_2(\mathbf{e}|E) \supseteq \dots \qquad (3.1.3)$$

and $\qquad \mathfrak{A}_\nu(\mathbf{e}|E) = 0 \quad$ when $\quad \nu > q. \qquad (3.1.4)$

We pause for a moment to make some general observations. There will, of course, be infinitely many matrices A which contribute to the sum in (3.1.2). Let us see how we can reduce this number. The relations between $e_1, e_2, \dots, e_q$ form an R-module. Let $\mathfrak{S}$ be a set of relations (between $e_1, e_2, \dots, e_q$) which generates this module. *Then in evaluating* $\sum\limits_A \mathfrak{A}_\nu(A)$ *it is sufficient to consider matrices A whose rows are members of the set $\mathfrak{S}$.*

To take an example suppose that E is *finitely presented*. Then we can find a single matrix, say

$$A = \begin{Vmatrix} a_{11} & a_{12} & \dots & a_{1q} \\ a_{21} & a_{22} & \dots & a_{2q} \\ \vdots & \vdots & \dots & \vdots \\ a_{p1} & a_{p2} & \dots & a_{pq} \end{Vmatrix},$$

where each row is a relation between $e_1, e_2, \dots, e_q$ and where every relation between $e_1, e_2, \dots, e_q$ is a linear combination of the rows of A. In this situation the remarks at the end of the last paragraph are applicable and they show that

$$\mathfrak{A}_\nu(\mathbf{e}|E) = \mathfrak{A}_\nu(A) \qquad (3.1.5)$$

for all $\nu \geqslant 0$.

We return to the consideration of an arbitrary finitely generated R-module E and suppose that $E = Re_1 + Re_2 + \dots + Re_q$. Assume that S is a multiplicatively closed subset of R and that $0 \notin S$. Then

$$E_S = R_S(e_1/1) + R_S(e_2/1) + \dots + R_S(e_q/1).$$

We put $\qquad \mathbf{e}/1 = (e_1/1, e_2/1, \dots, e_q/1).$

LEMMA 1. *Let the situation be as described above. Then*

$$\mathfrak{A}_\nu(\mathbf{e}/1|E_S) = \mathfrak{A}_\nu(\mathbf{e}|E)\, R_S \qquad (3.1.6)$$

for all $\nu \geqslant 0$.

† See section (1.4).

Proof. We can construct an exact sequence

$$G \xrightarrow{\phi} F \xrightarrow{\psi} E \to 0, \qquad (3.1.7)$$

where (i) F is a free R-module with a base $u_1, u_2, ..., u_q$, (ii) for $1 \leqslant k \leqslant q$ we have $\psi(u_k) = e_k$, and (iii) G is a free R-module with a base $\{v_j\}_{j \in J}$. Now

$$\phi(v_j) = a_{j1}u_1 + a_{j2}u_2 + ... + a_{jq}u_q$$

for suitable elements a_{jk} in R. It follows that $\{(a_{j1}, a_{j2}, ..., a_{jq})\}_{j \in J}$ is a family of relations between $e_1, e_2, ..., e_q$ which generates the whole module of such relations.

Denote by Ω the set of all matrices A having the property that each row of A is one of the relations $(a_{j1}, a_{j2}, ..., a_{jq})$. Then, as we saw above,

$$\mathfrak{A}_\nu(\mathbf{e}|E) = \sum_{A \in \Omega} \mathfrak{A}_\nu(A) \qquad (3.1.8)$$

for all $\nu \geqslant 0$ and hence

$$\mathfrak{A}_\nu(\mathbf{e}|E) R_S = (\sum_{A \in \Omega} \mathfrak{A}_\nu(A)) R_S = \sum_{A \in \Omega} (\mathfrak{A}_\nu(A) R_S). \qquad (3.1.9)$$

Let us return to (3.1.7). This induces an exact sequence

$$G_S \xrightarrow{\phi_S} F_S \xrightarrow{\psi_S} E_S \longrightarrow 0.$$

Here F_S and G_S are free R_S-modules with bases $\{u_k/1\}_{1 \leqslant k \leqslant q}$ and $\{v_j/1\}_{j \in J}$ respectively. Also $\psi_S(u_k/1) = e_k/1$ and

$$\phi_S\left(\frac{v_j}{1}\right) = \left(\frac{a_{j1}}{1}\right)\left(\frac{u_1}{1}\right) + \left(\frac{a_{j2}}{1}\right)\left(\frac{u_2}{1}\right) + ... + \left(\frac{a_{jq}}{1}\right)\left(\frac{u_q}{1}\right).$$

Furthermore, if the matrix A_S with entries in R_S is defined as in (1.7.2), then $\{A_S\}_{A \in \Omega}$ is just the set of all matrices whose rows belong to the family $\{(a_{j1}/1, a_{j2}/1, ..., a_{jq}/1)\}_{j \in J}$. Accordingly the argument which yielded (3.1.8) can be used again this time to show that

$$\mathfrak{A}_\nu(\mathbf{e}/1|E_S) = \sum_{A \in \Omega} \mathfrak{A}_\nu(A_S).$$

But, by (1.7.4), $\mathfrak{A}_\nu(A_S) = \mathfrak{A}_\nu(A) R_S$ and so

$$\mathfrak{A}_\nu(\mathbf{e}/1|E_S) = \sum_{A \in \Omega} (\mathfrak{A}_\nu(A) R_S) = \mathfrak{A}_\nu(\mathbf{e}|E) R_S$$

by virtue of (3.1.9). The proof is now complete.

We continue to assume that E is a finitely generated R-module and that $e_1, e_2, ..., e_q$ generate E. Now suppose that $\epsilon_1, \epsilon_2, ..., \epsilon_t$ are additional elements of E. We set

$$\mathbf{e} = (e_1, e_2, ..., e_q), \quad \boldsymbol{\epsilon} = (\epsilon_1, \epsilon_2, ..., \epsilon_t),$$

and
$$\mathbf{e}\boldsymbol{\epsilon} = (e_1, ..., e_q, \epsilon_1, ..., \epsilon_t).$$

Of course $e_1, ..., e_q, \epsilon_1, ..., \epsilon_t$ is also a system of generators of E.

LEMMA 2. *For all* $\nu \geqslant 0$, *we have* $\mathfrak{A}_\nu(\mathbf{e}|E) = \mathfrak{A}_{t+\nu}(\mathbf{e}\boldsymbol{\epsilon}|E)$. *Moreover* $\mathfrak{A}_\mu(\mathbf{e}\boldsymbol{\epsilon}|E) = R$ *for* $0 \leqslant \mu \leqslant t$.

Proof. Suppose that $1 \leqslant k \leqslant t$. Since $\epsilon_k \in Re_1 + Re_2 + ... + Re_q$, we have an equation of the form

$$c_{k1}e_1 + c_{k2}e_2 + ... + c_{kq}e_q + \epsilon_k = 0,$$

where $c_{km} \in R$. Now let $A = \|a_{ij}\|$ be a $p \times q$ matrix whose rows are relations between $e_1, e_2, ..., e_q$ and put

$$B = \left\| \begin{matrix} c_{11} & \cdots & c_{1q} & 1 & \cdots & 0 \\ \vdots & & \vdots & \vdots & & \vdots \\ c_{t1} & \cdots & c_{tq} & 0 & \cdots & 1 \\ a_{11} & \cdots & a_{1q} & 0 & \cdots & 0 \\ \vdots & & \vdots & \vdots & & \vdots \\ a_{p1} & \cdots & a_{pq} & 0 & \cdots & 0 \end{matrix} \right\|.$$

Then the rows of B are relations between $e_1, ..., e_q, \epsilon_1, ..., \epsilon_t$. Since $\mathfrak{A}_\mu(B) = R$ for $1 \leqslant \mu \leqslant t$, this shows that $\mathfrak{A}_\mu(\mathbf{e}\boldsymbol{\epsilon}|E) = R$ for $1 \leqslant \mu \leqslant t$. Next if $\nu \geqslant 0$, then, by Chapter 1 Lemma 2,

$$\mathfrak{A}_\nu(A) = \mathfrak{A}_{\nu+t}(B) \subseteq \mathfrak{A}_{\nu+t}(\mathbf{e}\boldsymbol{\epsilon}|E),$$

whence $\mathfrak{A}_\nu(\mathbf{e}|E) \subseteq \mathfrak{A}_{\nu+t}(\mathbf{e}\boldsymbol{\epsilon}|E)$.

In order to derive the reverse inclusion we consider a matrix D with $q+t$ columns each of whose rows is a relation between $e_1, ..., e_q, \epsilon_1, ..., \epsilon_t$. Put

$$Q = \left\| \begin{matrix} c_{11} & \cdots & c_{1q} & 1 & \cdots & 0 \\ \vdots & & \vdots & \vdots & & \vdots \\ c_{t1} & \cdots & c_{tq} & 0 & \cdots & 1 \\ \hline & & D & & & \end{matrix} \right\|$$

and observe that, by means of elementary row operations alone, it is possible to bring Q to the form

$$Q' = \left\| \begin{array}{ccc|ccc} c_{11} & \cdots & c_{1q} & 1 & \cdots & 0 \\ \vdots & & \vdots & \vdots & & \vdots \\ c_{t1} & \cdots & c_{tq} & 0 & \cdots & 1 \\ \hline & A' & & & 0 & \end{array} \right\|.$$

Since the rows of Q (and therefore the rows of Q' as well) are relations between $e_1, \ldots, e_q, \epsilon_1, \ldots, \epsilon_t$, it follows that the rows of A' are relations between $e_1, e_2, \ldots, e_q$ and therefore $\mathfrak{A}_\nu(A') \subseteq \mathfrak{A}_\nu(\mathbf{e}|E)$. But, by Chapter 1 Lemma 2,

$$\mathfrak{A}_\nu(A') = \mathfrak{A}_{\nu+t}(Q') = \mathfrak{A}_{\nu+t}(Q) \supseteq \mathfrak{A}_{\nu+t}(D)$$

which shows that $\mathfrak{A}_{\nu+t}(D) \subseteq \mathfrak{A}_\nu(\mathbf{e}|E)$. Finally by allowing D to vary we deduce that $\mathfrak{A}_{\nu+t}(\mathbf{e}\boldsymbol{\epsilon}|E) \subseteq \mathfrak{A}_\nu(\mathbf{e}|E)$ and with this the proof is complete.

In order to apply Lemma 2 we assume that $\epsilon_1, \epsilon_2, \ldots, \epsilon_t$ as well as $e_1, e_2, \ldots, e_q$, is a system of generators for E. If now

$$0 \leqslant \mu \leqslant \min(q, t),$$

then Lemma 2 shows that

$$\mathfrak{A}_{q-\mu}(\mathbf{e}|E) = \mathfrak{A}_{t+q-\mu}(\mathbf{e}\boldsymbol{\epsilon}|E) = \mathfrak{A}_{t-\mu}(\boldsymbol{\epsilon}|E). \qquad (3.1.10)$$

Furthermore, it follows, from the same result, that if $q < \nu \leqslant t$, then

$$\mathfrak{A}_{t-\nu}(\boldsymbol{\epsilon}|E) = \mathfrak{A}_{q+t-\nu}(\mathbf{e}\boldsymbol{\epsilon}|E) = R. \qquad (3.1.11)$$

In view of these relations we put

$$\mathfrak{F}_\mu(E) = \begin{cases} \mathfrak{A}_{q-\mu}(\mathbf{e}|E) & \text{when} \quad 0 \leqslant \mu \leqslant q, \\ R & \text{when} \quad \mu > q, \end{cases} \qquad (3.1.12)$$

and note that our discussion immediately yields

THEOREM 1. *Let E be a finitely generated R-module and let ideals $\mathfrak{F}_\mu(E)$, where $\mu = 0, 1, 2, \ldots$, be defined by means of the formula* (3.1.12). *Then these ideals do not depend on the particular set of generators $\mathbf{e} = (e_1, e_2, \ldots, e_q)$ that is used to compute them.*

Theorem 1 shows that the ideals $\mathfrak{F}_\mu(E)$ have a rather special relationship with the module E. For this reason we make the

DEFINITION. *Let E be a finitely generated R-module. Then the ideals $\mathfrak{F}_\mu(E)$ of Theorem 1 are called the 'Fitting invariants' of E.*

It is clear that isomorphic finitely generated modules have the same Fitting invariants.

We shall now restate some of our earlier conclusions in terms of Fitting invariants.

THEOREM 2. *Let E be a finitely generated R-module. Then its Fitting invariants form an increasing sequence*

$$\mathfrak{F}_0(E) \subseteq \mathfrak{F}_1(E) \subseteq \mathfrak{F}_2(E) \subseteq \dots$$

of ideals. Furthermore if E can be generated by q elements, then $\mathfrak{F}_q(E) = R$.

This theorem is an immediate consequence of (3.1.3) and (3.1.12).

THEOREM 3. *Let E be a finitely generated R-module and S a multiplicatively closed subset of R not containing the zero element. Then for every integer $\mu \geqslant 0$ we have*

$$\mathfrak{F}_\mu(E)\, R_S = \mathfrak{F}_\mu(E_S),$$

where by $\mathfrak{F}_\mu(E_S)$ is meant the μth Fitting invariant of the R_S-module E_S.

Theorem 3 is obtained by combining Lemma 1 with (3.1.12).

Let us suppose, for a moment, that E is a finitely presented module and that $e_1, e_2, \dots, e_q$ generate E. Then, as we have noted previously, there exists a matrix A with q columns and having the properties that each row of A is a relation between $e_1, e_2, \dots, e_q$ and, furthermore, every relation between $e_1, e_2, \dots, e_q$ is a linear combination of the rows of A. Now in these circumstances (3.1.5) is valid. Consequently

$$\mathfrak{F}_\mu(E) = \begin{cases} \mathfrak{A}_{q-\mu}(A) & \text{for } 0 \leqslant \mu \leqslant q, \\ R & \text{for } \mu > q. \end{cases} \tag{3.1.13}$$

In particular we have proved

THEOREM 4. *Let E be a finitely presented R-module. Then each of its Fitting invariants is a finitely generated ideal.*

EXERCISE 1. *Determine the Fitting invariants of a free R-module of rank q.*

Let E be a finitely generated R-module. By Theorem 2, $\mathfrak{F}_0(E)$ is the smallest of its Fitting invariants.

DEFINITION. *We put $\mathfrak{F}(E) = \mathfrak{F}_0(E)$ and call $\mathfrak{F}(E)$ the 'initial Fitting invariant' of E.*

There is an intimate connection between $\mathfrak{F}(E)$ and the annihilator, $\mathrm{Ann}_R(E)$, of E. Before we proceed to establish this connection, we recall that if $\mathfrak{A}$ and $\mathfrak{A}'$ are ideals of R, then their product $\mathfrak{A}\mathfrak{A}'$ consists of all elements that can be written in the form $a_1 a_1' + a_2 a_2' + \ldots + a_m a_m'$, where $a_i \in \mathfrak{A}$ and $a_i' \in \mathfrak{A}'$. Multiplication of ideals is commutative and associative. Note that if $\mathfrak{A}''$ is also an ideal, then $(\mathfrak{A}\mathfrak{A}')\mathfrak{A}'' = \mathfrak{A}(\mathfrak{A}'\mathfrak{A}'')$ is composed of all elements which can be expressed as sums

$$\alpha_1 \alpha_1' \alpha_1'' + \alpha_2 \alpha_2' \alpha_2'' + \ldots + \alpha_n \alpha_n' \alpha_n'',$$

where $\alpha_i \in \mathfrak{A}$, $\alpha_i' \in \mathfrak{A}'$ and $\alpha_i'' \in \mathfrak{A}''$. Indeed this latter observation extends to any finite product of ideals. Finally $R\mathfrak{A} = \mathfrak{A}$ for every ideal $\mathfrak{A}$.

THEOREM 5. *Let E be an R-module which can be generated by q elements. Then*

$$[\mathrm{Ann}_R(E)]^q \subseteq \mathfrak{F}(E) \subseteq \mathrm{Ann}_R(E),$$

where $\mathfrak{F}(E)$ is the initial Fitting invariant of E. In particular if E can be generated by a single element we have $\mathfrak{F}(E) = \mathrm{Ann}_R(E)$.

Proof. Let $E = Re_1 + Re_2 + \ldots + Re_q$ and suppose that

$$c_i \in \mathrm{Ann}_R(E)$$

for $i = 1, 2, \ldots, q$. Then each row of the matrix

$$C = \begin{Vmatrix} c_1 & 0 & \ldots & 0 \\ 0 & c_2 & \ldots & 0 \\ \vdots & \vdots & \ldots & \vdots \\ 0 & 0 & \ldots & c_q \end{Vmatrix}$$

is a relation between $e_1, e_2, \ldots, e_q$ and therefore

$$c_1 c_2 \ldots c_q = \det(C) \in \mathfrak{A}_q(C) \subseteq \mathfrak{F}(E).$$

This shows that $[\mathrm{Ann}_R(E)]^q \subseteq \mathfrak{F}(E)$.

Now assume that a $q \times q$ matrix A has the property that each row is a relation between $e_1, e_2, ..., e_q$. Then $\det(A) e_i = 0$ for $i = 1, 2, ..., q$, as may be seen by multiplying the equation

$$A \left\| \begin{array}{c} e_1 \\ e_2 \\ \vdots \\ e_q \end{array} \right\| = 0$$

by the adjugate of A. Thus $\det(A) \in \text{Ann}_R(E)$. Since $\mathfrak{F}(E)$ is generated by elements of the form $\det(A)$, it follows that $\mathfrak{F}(E) \subseteq \text{Ann}_R(E)$.

EXERCISE 2. *Let* $0 \to E' \to E \to E'' \to 0$ *be an exact sequence of finitely generated R-modules and let* $\mu \geqslant 0$ *and* $\nu \geqslant 0$ *be integers. Show that* $\mathfrak{F}_\mu(E'') \mathfrak{F}_\nu(E') \subseteq \mathfrak{F}_{\mu+\nu}(E)$ *and hence that*

$$\mathfrak{F}(E'') \mathfrak{F}(E') \subseteq \mathfrak{F}(E).$$

EXERCISE 3. *Let E' and E'' be finitely generated R-modules and let $k \geqslant 0$ be an integer. Prove that*

$$\mathfrak{F}_k(E' \oplus E'') = \sum_{\mu+\nu=k} \mathfrak{F}_\mu(E') \mathfrak{F}_\nu(E'').$$

EXERCISE 4. *Let* $\mathfrak{A}_1, \mathfrak{A}_2, ..., \mathfrak{A}_s$ *be ideals of R and put*

$$E = (R/\mathfrak{A}_1) \oplus (R/\mathfrak{A}_2) \oplus ... \oplus (R/\mathfrak{A}_s).$$

Show that $\mathfrak{F}(E) = \mathfrak{A}_1 \mathfrak{A}_2 ... \mathfrak{A}_s$.

At this point we shall temporarily leave the subject of Fitting invariants in order to introduce other ideas that have to do with matrices and modules. These will provide us with certain basic information which will enable us not only to develop some additional properties of Fitting invariants, but which will also help prepare the way for the study of modules which possess finite free resolutions of finite length.†

3.2 The ranks of a matrix

Throughout section (3.2) $A = \|a_{jk}\|$ will denote a $p \times q$ matrix with entries in R, where p and q are positive integers, and $E \neq 0$

† See section (3.3) for the relevant definition.

will denote an R-module. As on previous occasions $\mathfrak{A}_\nu(A)$ will denote the νth determinantal ideal of A.

We shall now define *the rank† of A relative to E*. This will be a non-negative integer and it will be denoted by $\operatorname{rank}_R(A, E)$; in fact

$$\operatorname{rank}_R(A, E) = \textit{the largest } \nu \textit{ such that } \mathfrak{A}_\nu(A)\, E \neq 0. \qquad (3.2.1)$$

In the case where $E = R$ we write $\operatorname{rank}_R(A)$ instead of the more cumbersome $\operatorname{rank}_R(A, R)$. Accordingly

$$\operatorname{rank}_R(A) = \textit{the largest } \nu \textit{ such that } \mathfrak{A}_\nu(A) \neq 0. \qquad (3.2.2)$$

We recall that, by (1.4.2),

$$R = \mathfrak{A}_0(A) \supseteq \mathfrak{A}_1(A) \supseteq \mathfrak{A}_2(A) \supseteq \ldots$$

and that, by (1.4.1),

$$\mathfrak{A}_\nu(A) = 0 \quad \text{for} \quad \nu > \min(p, q).$$

It follows that $\operatorname{rank}_R(A, E)$ is well-defined and

$$0 \leqslant \operatorname{rank}_R(A, E) \leqslant \operatorname{rank}_R(A) \leqslant \min(p, q). \qquad (3.2.3)$$

We shall also need the notion of the *reduced rank of A relative to E*. For this concept the notation $\operatorname{red.rank}_R(A, E)$ will be used and this time the definition is provided by

$$\operatorname{red.rank}_R(A, E) = \textit{the largest } \nu \textit{ such that } 0:_E \mathfrak{A}_\nu(A) = 0. \qquad (3.2.4)$$

Since $E \neq 0$, it follows that if $0:_E \mathfrak{A}_\nu(A) = 0$ then we must have $\mathfrak{A}_\nu(A)\, E \neq 0$ as well. Consequently (3.2.3) can be lengthened to

$$0 \leqslant \operatorname{red.rank}_R(A, E) \leqslant \operatorname{rank}_R(A, E) \leqslant \operatorname{rank}_R(A) \leqslant \min(p, q). \qquad (3.2.5)$$

As the case $E = R$ has special interest, it is convenient to modify the notation in this situation and put

$$\operatorname{red.rank}_R(A) = \operatorname{red.rank}_R(A, R). \qquad (3.2.6)$$

Note that if A^{T} is the transpose of A, then

$$\operatorname{rank}_R(A^{\mathrm{T}}, E) = \operatorname{rank}_R(A, E), \qquad (3.2.7)$$

and

$$\operatorname{red.rank}_R(A^{\mathrm{T}}, E) = \operatorname{red.rank}_R(A, E), \qquad (3.2.8)$$

† There is no connection with the notion of the rank of a free module.

since $\mathfrak{A}_\nu(A^T) = \mathfrak{A}_\nu(A)$ for all $\nu \geqslant 0$. Also if U is a unimodular $p \times p$ matrix and V a unimodular $q \times q$ matrix, then

$$\operatorname{rank}_R(A, E) = \operatorname{rank}_R(UAV, E) \qquad (3.2.9)$$

and $\qquad \operatorname{red.rank}_R(A, E) = \operatorname{red.rank}_R(UAV, E) \qquad (3.2.10)$

because, by Chapter 1 Theorem 3, $\mathfrak{A}_\nu(UAV) = \mathfrak{A}_\nu(A)$.
Our next result is known as *McCoy's Theorem.*†

THEOREM 6. *Let* $A = \|a_{jk}\|$ *be a* $p \times q$ *R-matrix and* $E \neq 0$ *an R-module. Then in order that there should exist elements* $e_1, e_2, ..., e_q$, *of* E, *not all zero and such that*

$$a_{11}e_1 + a_{12}e_2 + ... + a_{1q}e_q = 0$$
$$a_{21}e_1 + a_{22}e_2 + ... + a_{2q}e_q = 0$$
$$\vdots \qquad \vdots \qquad ... \qquad \vdots$$
$$a_{p1}e_1 + a_{p2}e_2 + ... + a_{pq}e_q = 0$$

it is necessary and sufficient to have $\operatorname{red.rank}_R(A, E) < q$.

Remark. Since $\operatorname{red.rank}_R(A, E) < q$ if and only if

$$0 :_E \mathfrak{A}_q(A) \neq 0,$$

Theorem 6 can be restated as follows: *the equations have a non-trivial solution in* E *precisely when there is a non-zero element of* E *that is annihilated by all the* $q \times q$ *minors of* A.

Proof. First assume that $\operatorname{red.rank}_R(A, E) < q$. Put

$$\nu = \operatorname{red.rank}_R(A, E)$$

so that $0 \leqslant \nu \leqslant \min(p, q-1)$. Then $0 :_E \mathfrak{A}_{\nu+1}(A) \neq 0$ and therefore we can choose $e \in E$ so that $e \neq 0$ but $\mathfrak{A}_{\nu+1}(A)e = 0$. Of course, by the definition of reduced rank, $\mathfrak{A}_\nu(A)e \neq 0$.
Suppose that $J = \{j_1, j_2, ..., j_\nu\}$ belongs to S_ν^p and

$$N = \{n_1, n_2, ..., n_{\nu+1}\}$$

to $S_{\nu+1}^q$, where the notation is that introduced in section (1.3), and define y_{JN} as in (1.5.3). Then y_{JN} is a column vector of length q

† See N. H. McCoy [(**33**) Chapter 8, p. 159] and H. Flanders (**21**).

with entries in R so we may define $e_1, e_2, ..., e_q$ by

$$y_{JN} e = \left\| \begin{array}{c} e_1 \\ e_2 \\ \vdots \\ e_q \end{array} \right\| .$$

Accordingly $a_{j1} e_1 + a_{j2} e_2 + ... + a_{jq} e_q$ is just $(Ay_{JN})_j e$ and this is zero because, by (1.5.4), $(Ay_{JN})_j \in \mathfrak{A}_{\nu+1}(A)$. Thus

$$a_{j1} e_1 + a_{j2} e_2 + ... + a_{jq} e_q = 0$$

for $j = 1, 2, ..., p$. Moreover, by (1.5.3), we can arrange that (apart from sign) any assigned $\nu \times \nu$ minor of A occurs as one of the entries in y_{JN}. But $\mathfrak{A}_\nu(A) e \ne 0$. Consequently we can arrange that $y_{JN} e \ne 0$.

Next assume that there are elements $e_1, e_2, ..., e_q$ in E which satisfy the equations and where e_k (say) is non-zero. We wish to show that $0 :_E \mathfrak{A}_q(A) \ne 0$. If $q > p$, then $\mathfrak{A}_q(A) = 0$ and there is no problem. We may therefore assume that $q \leqslant p$. Let D be the $q \times q$ minor obtained from the first q rows of A. Consideration of the first q equations shows that $De_k = 0$. Similarly if Δ is any $q \times q$ minor of A we must have $\Delta e_k = 0$. Thus e_k belongs to $0 :_E \mathfrak{A}_q(A)$ and with this the proof is complete.

COROLLARY. *Let $A = \|a_{jk}\|$ be a $p \times q$ matrix and let $E \ne 0$ be an R-module. If now $p < q$, then there exist elements $e_1, e_2, ..., e_q$ in E which are not all zero and which satisfy $a_{j1} e_1 + a_{j2} e_2 + ... + a_{jq} e_q = 0$ for $j = 1, 2, ..., p$.*

As before let $A = \|a_{jk}\|$ be a $p \times q$ matrix and $E \ne 0$ an R-module. The case where the rank of A relative to E equals the reduced rank of A relative to E is particularly important. Since it will concern us a great deal we introduce the

DEFINITION. *The matrix A is said to be 'stable relative to E' if* red.$\mathrm{rank}_R(A, E) = \mathrm{rank}_R(A, E)$. *We shall simply say that A is 'stable' if it is stable relative to R.*

We shall leave for the moment the explanation of why the adjective *stable* has been chosen to describe this situation. Let us observe, however, that *if A is stable relative to E, then so is its transpose and so too is any matrix equivalent to A.* Indeed the first of these assertions follows from (3.2.7) and (3.2.8) and the second from (3.2.9) and (3.2.10). Let us also note that we have, in fact, encountered stable matrices before namely in Chapter 1 Theorem 7 though we did not use the term at that stage. If we restate the result in question using the new terminology it becomes

THEOREM 7. *Let A be a stable $p \times q$ matrix with entries in R. Then the following statements are equivalent:*

(a) *there is a $q \times p$ matrix Ω such that $A\Omega A = A$;*

(b) *for each $\nu \geqslant 0$ the determinantal ideal $\mathfrak{A}_\nu(A)$ is either the zero ideal or it is the whole ring.*

Another theorem which is readily available and which illustrates the relevance of the notion of stability is

THEOREM 8. *Let $A = \|a_{jk}\|$ be a $p \times q$ matrix and $E \neq 0$ an R-module. Then the equations*

$$
\begin{aligned}
a_{11}e_1 + a_{12}e_2 + \ldots + a_{1q}e_q &= 0 \\
a_{21}e_1 + a_{22}e_2 + \ldots + a_{2q}e_q &= 0 \\
\vdots \qquad \vdots \qquad \ldots \qquad \vdots \\
a_{p1}e_1 + a_{p2}e_2 + \ldots \quad a_{pq}e_q &= 0
\end{aligned}
$$

have no non-trivial solution in E if and only if $0:_E \mathfrak{A}_q(A) = 0$. Moreover when this is the case we have (i) $q \leqslant p$, (ii) A *is stable relative to E, and* (iii) $\operatorname{rank}_R(A, E) = q$.

Proof. The first assertion is merely a restatement of Theorem 6. Now suppose that $0:_E \mathfrak{A}_q(A) = 0$. Then

$$
q \leqslant \text{red.rank}_R(A, E) \leqslant \operatorname{rank}_R(A, E) \leqslant \min(p, q) \leqslant q,
$$

and (i), (ii), (iii) all follow from these inequalities.

We next examine how the ranks of a matrix behave in relation to the operation of forming fractions. For this purpose the information contained in the next exercise will be useful.

EXERCISE 5.† *Let K be a submodule of an R-module E, let $\mathfrak{A}$ be a finitely generated ideal, and let S be a multiplicatively closed subset of R. Show that*

$$
(K:_E \mathfrak{A})_S = K_S:_{E_S} \mathfrak{A} R_S
$$

† Compare with Chapter 2 Theorem 12.

Now suppose that A is a $p \times q$ matrix, that $E \neq 0$ is an R-module, and that S is a multiplicatively closed subset of R with the property that $E_S \neq 0$. Put $\mu = \mathrm{red.rank}_R(A, E)$ and $\nu = \mathrm{rank}_{R_S}(A_S, E_S)$, where A_S is defined as in (1.7.2). Then $\mathfrak{A}_\nu(A_S) E_S \neq 0$. But, by (1.7.4), $\mathfrak{A}_\nu(A_S) = \mathfrak{A}_\nu(A) R_S$. Consequently we must have $\mathfrak{A}_\nu(A) E \neq 0$ and this shows that

$$\mathrm{rank}_{R_S}(A_S, E_S) \leqslant \mathrm{rank}_R(A, E). \qquad (3.2.11)$$

On the other hand
$$0:_E \mathfrak{A}_\mu(A) = 0$$

and $\mathfrak{A}_\mu(A)$ is a finitely generated ideal. Accordingly, by Exercise 5,
$$0:_{E_S} \mathfrak{A}_\mu(A) R_S = 0,$$

which, since $\mathfrak{A}_\mu(A) R_S = \mathfrak{A}_\mu(A_S)$, proves that

$$\mathrm{red.rank}_R(A, E) \leqslant \mathrm{red.rank}_{R_S}(A_S, E_S). \qquad (3.2.12)$$

Thus when we form fractions the reduced rank goes up whereas the ordinary rank comes down. As a consequence we have

THEOREM 9. *Let $E \neq 0$ be an R-module and A a matrix which is stable relative to E. If now S is a multiplicatively closed subset of R such that $E_S \neq 0$, then A_S is stable relative to the R_S-module E_S and $\mathrm{rank}_R(A, E) = \mathrm{rank}_{R_S}(A_S, E_S)$.*

In fact it is because stable matrices have the properties described in this theorem that they have been so named.

3.3 Finite free resolutions

Let E be an R-module. If there exists an exact sequence

$$0 \to F_n \to F_{n-1} \to \ldots \to F_1 \to F_0 \to E \to 0 \qquad (3.3.1)$$

in which each F_i is a finite free R-module, then we say that E has a *finite free resolution of finite length*, and we say that (3.3.1) is a *finite free resolution of E of length n*. In order to develop a theory of such resolutions it is convenient to introduce a number of auxiliary definitions.

An exact sequence

$$0 \to \Pi_n \to \Pi_{n-1} \to \ldots \to \Pi_1 \to \Pi_0 \to E \to 0, \qquad (3.3.2)$$

where $\Pi_0, \Pi_1, \ldots, \Pi_n$ are projective R-modules, is called a *projective resolution (of E) of length n.* Modules which admit projective resolutions of finite length are often said to be of *finite homological dimension.*

Let Π be a projective R-module.

DEFINITION. *The projective R-module Π is said to be ' supplementable' if there exists a finite free module F such that $\Pi \oplus F$ is also a finite free module.*†

Naturally if in the exact sequence (3.3.2) each of $\Pi_0, \Pi_1, \ldots, \Pi_n$ is a supplementable projective module, then (3.3.2) is called a *supplementable projective resolution* of E.

LEMMA 3. *Let E be an R-module and $n \geqslant 1$ an integer. Then E has a finite free resolution of length n if and only if it has a supplementable projective resolution of length n.*

Proof. Let

$$0 \to \Pi_n \to \Pi_{n-1} \to \ldots \to \Pi_1 \to \Pi_0 \to E \to 0$$

be a supplementable projective resolution of E and suppose that $0 \leqslant k < n$. Choose finite free modules F_k and F_{k+1} so that $\Pi_k \oplus F_k$ and $\Pi_{k+1} \oplus F_{k+1}$ are themselves finite free modules. If

$$F = F_k \oplus F_{k+1},$$

then we can construct (in an obvious manner) an exact sequence of the form

$$0 \to \Pi_n \to \Pi_{n-1} \to \ldots \to \Pi_{k+1} \oplus F \to \Pi_k \oplus F \to \ldots \to \Pi_0 \to E \to 0.$$

Note that the terms with suffixes k and $k+1$ are now finite free modules and the remaining terms are exactly as before.

Using this device we first secure that Π_0 and Π_1 are finite free modules. After this we arrange that Π_0, Π_1, Π_2 are finite free modules. A further application will result in Π_0, Π_1, Π_2 and Π_3 being finite free modules. And so on. Eventually we obtain a finite free resolution of E of length n. This proves half the lemma and the other half is trivial.

† Supplementable projective modules are also known as *stably free* modules. An example of a non-free supplementable projective module will be found in Appendix A.

Note that a supplementable projective module has a finite free resolution of length one.

LEMMA 4. *Let Π_1 and Π_2 be R-modules. If now any two of Π_1, Π_2 and $\Pi_1 \oplus \Pi_2$ are supplementable projective modules, then so is the third.*

The proof is obvious so we shall give no details.

Let E_1 and E_2 be R-modules.

DEFINITION. *We shall say that E_1 and E_2 are 'strongly equivalent' if there exist finite free modules F_1 and F_2 such that $E_1 \oplus F_1$ and $E_2 \oplus F_2$ are isomorphic.*

This is clearly an equivalence relation between modules in the abstract sense. Note that if Π_1 and Π_2 are supplementable projective modules and $E_1 \oplus \Pi_1$ is isomorphic to $E_2 \oplus \Pi_2$, then E_1 and E_2 will necessarily be strongly equivalent.

THEOREM 10. *Let E_1 and E_2 be strongly equivalent R-modules and let*
$$0 \to E_1' \to \Pi_1 \to E_1 \to 0$$
and
$$0 \to E_2' \to \Pi_2 \to E_2 \to 0$$
be exact sequences, where Π_1 and Π_2 are supplementable projective modules. Then E_1' and E_2' are strongly equivalent.

Proof. We have an isomorphism $E_1 \oplus \Pi_1^* \approx E_2 \oplus \Pi_2^*$, where Π_1^* and Π_2^* are supplementable projective modules. Further we can construct exact sequences
$$0 \to E_1' \to \Pi_1 \oplus \Pi_1^* \to E_1 \oplus \Pi_1^* \to 0$$
and
$$0 \to E_2' \to \Pi_2 \oplus \Pi_2^* \to E_2 \oplus \Pi_2^* \to 0.$$

Accordingly, by Chapter 2 Exercise 4, $E_2' \oplus \Pi_1 \oplus \Pi_1^*$ and $E_1' \oplus \Pi_2 \oplus \Pi_2^*$ are isomorphic. The theorem now follows because, by Lemma 4, the projective modules $\Pi_1 \oplus \Pi_1^*$ and $\Pi_2 \oplus \Pi_2^*$ are supplementable.

THEOREM 11. *Let E be an R-module. Further let*
$$0 \to \Pi_n \to \Pi_{n-1} \to \dots \to \Pi_0 \to E \to 0 \qquad (3.3.3)$$
be a supplementable projective resolution of E and
$$\Pi_m' \to \Pi_{m-1}' \to \dots \to \Pi_0' \to E \to 0 \qquad (3.3.4)$$

an exact sequence, where each Π'_j is a supplementable projective module. If now $m < n$, then (3.3.4) can be continued so as to provide a supplementable projective resolution of E of length n.

Proof. If $m > 0$ put

$$E_m = \operatorname{Ker}(\Pi_m \to \Pi_{m-1}), \quad E'_m = \operatorname{Ker}(\Pi'_m \to \Pi'_{m-1})$$

and in case $m = 0$ put $E_0 = \operatorname{Ker}(\Pi_0 \to E)$, $E'_0 = \operatorname{Ker}(\Pi'_0 \to E)$. By means of repeated applications of Theorem 10 we see that E_m and E'_m are strongly equivalent, say $E_m \oplus P \approx E'_m \oplus P'$, where P and P' are supplementable projective modules.

First suppose that $m = n - 1$. Then E_m is isomorphic to Π_n. It follows that $E_m \oplus P$ and hence $E'_m \oplus P'$ is a supplementable projective module. Consequently, by Lemma 4, the same holds for E'_m and therefore

$$0 \to E'_m \to \Pi_m \to \dots \to \Pi'_1 \to \Pi'_0 \to E \to 0$$

is a resolution with the required properties.

Now assume that $m < n - 1$. Then E_m is a homomorphic image of Π_{m+1} and now we see that the isomorphic modules $E_m \oplus P$ and $E'_m \oplus P'$ are homomorphic images of the supplementable projective module $\Pi_{m+1} \oplus P$. It follows that E'_m is a homomorphic image of $\Pi_{m+1} \oplus P$ and therefore we have an exact sequence

$$\Pi_{m+1} \oplus P \to \Pi'_m \to \dots \to \Pi'_0 \to E \to 0.$$

If $m + 1 = n - 1$, the proof is complete by virtue of the first part of the argument; if $m + 1 < n - 1$, then we may continue in the same way. Eventually a suitable resolution is obtained.

COROLLARY. *Let $0 \to E' \to \Pi \to E \to 0$ be an exact sequence, where Π is a supplementable projective module. If now E has a supplementable projective resolution of length n $(n \geqslant 1)$ then E' has a supplementable projective resolution of length $n - 1$.*

The reader will probably have noticed that we have been adapting the techniques of the comparatively familiar theory of projective resolutions in order to create tools for dealing with finite free resolutions. We shall now examine some of the connections between the two theories.

EXERCISE 6. *Let P be a projective R-module. Show that there exists a free module F such that $P \oplus F$ is isomorphic to F.*

Let E_1 and E_2 be R-modules.

DEFINITION. *The modules E_1 and E_2 are said to be 'equivalent' if there exist free R-modules F_1 and F_2 such that $E_1 \oplus F_1$ and $E_2 \oplus F_2$ are isomorphic.*

It is clear that, once again, we have to do with an equivalence relation in the abstract sense. Note that if $E_1 \oplus P_1$ and $E_2 \oplus P_2$ are isomorphic, where P_1 and P_2 are projective, then (by virtue of Exercise 6) E_1 and E_2 are equivalent.

THEOREM 12. *Let E_1 and E_2 be equivalent R-modules and let*

$$0 \to E_1' \to P_1 \to E_1 \to 0$$

and $$0 \to E_2' \to P_2 \to E_2 \to 0$$

be exact sequences, where P_1 and P_2 are projective. Then E_1' and E_2' are equivalent.

The proof is a trivial modification of that of Theorem 10.

THEOREM 13. *Suppose that*

$$0 \to P_n \to P_{n-1} \to \dots \to P_1 \to P_0 \to E \to 0$$

is a projective resolution of the R-module E and that

$$P_m' \to P_{m-1}' \to \dots \to P_1' \to P_0' \to E \to 0 \qquad (3.3.5)$$

is an exact sequence with each P_j' projective. If now $m < n$, then (3.3.5) can be continued so as to provide a projective resolution of E of length n.

This time the argument used to prove Theorem 11 can be adapted. On this occasion the details are simpler.

COROLLARY. *Let $0 \to E' \to P \to E \to 0$ be an exact sequence of R-modules, where P is projective. If now E has a projective resolution of length n and $n \geqslant 1$, then E' has a projective resolution of length $n-1$.*

THEOREM 14. *If P is a projective R-module, then the following statements are equivalent:*

(a) *P is supplementable;*

(b) *P has a finite free resolution of finite length.*

Proof. Obviously (*a*) implies (*b*). Assume (*b*) and let

$$0 \to \Pi_n \to \Pi_{n-1} \to \ldots \to \Pi_1 \to \Pi_0 \to P \to 0$$

be a supplementable projective resolution of P of length n. We wish to show that P is supplementable. For this we use induction on n.

If $n = 0$ there is no problem. We shall therefore assume that $n \geqslant 1$ and that the assertion under consideration has been proved for smaller values of the inductive variable. Put

$$P' = \mathrm{Ker}\,(\Pi_0 \to P)$$

so that we have an exact sequence $0 \to P' \to \Pi_0 \to P \to 0$. By Chapter 2 Exercise 6, we have an isomorphism $\Pi_0 \approx P \oplus P'$. Accordingly $P \oplus P'$ is a supplementable projective module and P' is projective. Next the inductive hypothesis applied to P' shows that it is supplementable and now an appeal to Lemma 4 completes the proof.

3.4 Restricted projective dimension

If E is an R-module, then $\mathrm{Pd}_R\,(E)$ will denote its *projective dimension*. We recall the definition. First suppose that $E \neq 0$. Then if E possesses a projective resolution of finite length, $\mathrm{Pd}_R\,(E)$ is the length of the shortest such resolution; otherwise $\mathrm{Pd}_R\,(E) = \infty$. To complete the definition we put $\mathrm{Pd}_R\,(0) = -1$. Note that E is a projective module if and only if $\mathrm{Pd}_R\,(E) \leqslant 0$.

Let $0 \to E' \to \Pi \to E \to 0$ be an exact sequence with Π projective. By Theorem 13 Cor., if $\mathrm{Pd}_R\,(E) \geqslant 1$ (that is if E is not projective), then $\mathrm{Pd}_R\,(E') = \mathrm{Pd}_R\,(E) - 1$; whereas if $\mathrm{Pd}_R\,(E) \leqslant 0$, then E' is isomorphic to a direct summand of Π and hence $\mathrm{Pd}_R\,(E') \leqslant 0$ as well.

The *restricted projective dimension* will be denoted by $\mathrm{Pd}_R^*\,(E)$. This is defined in the same way as the ordinary projective dimension save that we use only supplementable projective resolutions. In view of Lemma 3, $\mathrm{Pd}_R^*\,(E) < \infty$ is a necessary and sufficient condition for E to have a finite free resolution of finite length; and if $\mathrm{Pd}_R^*\,(E) = n$ with $1 \leqslant n < \infty$, then E possesses a finite free resolution of length n. Note that $\mathrm{Pd}_R^*(E) \leqslant 0$ when and only when E is a supplementable projective module. Also if

$$0 \to E' \to \Pi \to E \to 0$$

is an exact sequence and Π is a supplementable projective module, then $\mathrm{Pd}_R^*(E') = \mathrm{Pd}_R^*(E) - 1$ provided that $\mathrm{Pd}_R^*(E) \geqslant 1$; on the other hand if $\mathrm{Pd}_R^*(E) \leqslant 0$, then $\mathrm{Pd}_R^*(E') \leqslant 0$ as well. These assertions follow from Theorem 11 Cor. and Lemma 4 respectively.

For an arbitrary R-module E we have $\mathrm{Pd}_R(E) \leqslant \mathrm{Pd}_R^*(E)$. Observe that if F is a free module of infinite rank, then $\mathrm{Pd}_R(F) = 0$ but $\mathrm{Pd}_R^*(F) = \infty$. However, we do have

THEOREM 15. *Suppose that* $\mathrm{Pd}_R^*(E) < \infty$. *Then*

$$\mathrm{Pd}_R^*(E) = \mathrm{Pd}_R(E).$$

Proof. Put $\mathrm{Pd}_R^*(E) = n$. We use induction on n and begin by observing that the assertion is certainly true if $n \leqslant 0$. It will therefore be supposed that $n \geqslant 1$ and that the theorem has been proved for modules that have smaller restricted projective dimension.

By the definition of $\mathrm{Pd}_R^*(E)$, there exists a supplementable projective resolution

$$0 \to \Pi_n \to \Pi_{n-1} \to \dots \to \Pi_1 \to \Pi_0 \to E \to 0$$

of length n. E is not a projective module. (For otherwise, by Theorem 14, E would be supplementable and we should have $n = \mathrm{Pd}_R^*(E) \leqslant 0$ contrary to hypothesis.) Put $E' = \mathrm{Ker}\,(\Pi_0 \to E)$. Then $\mathrm{Pd}_R^*(E') = \mathrm{Pd}_R^*(E) - 1$ and $\mathrm{Pd}_R(E') = \mathrm{Pd}_R(E) - 1$. The induction hypothesis now shows that $\mathrm{Pd}_R^*(E') = \mathrm{Pd}_R(E')$ and the theorem follows at once.

Suppose that the exact sequence

$$0 \to E' \xrightarrow{\;f\;} E \xrightarrow{\;g\;} E'' \to 0$$

is composed of *finitely generated* modules. We can construct epimorphisms $\psi' : \Pi' \to E'$ and $\psi'' : \Pi'' \to E''$, where Π' and Π'' are supplementable projective modules. (In fact we can take them to be finite free modules.) Put $\Pi = \Pi' \oplus \Pi''$ so that Π is also a supplementable projective module and choose a homomorphism $\omega\colon \Pi'' \to E$ with the property that $g\omega = \psi''$. (This is possible, by Chapter 2 Exercise 3, because Π'' is projective and g is an epimorphism.) Let us next define a homomorphism $\psi\colon \Pi \to E$ by $\psi(p', p'') = f\psi'(p') + \omega(p'')$. Then the diagram

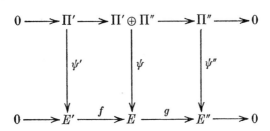

is commutative and ψ is an epimorphism. (The mappings in the upper row are the canonical homomorphisms.) It is easy to see that this diagram can be enlarged to give a new commutative

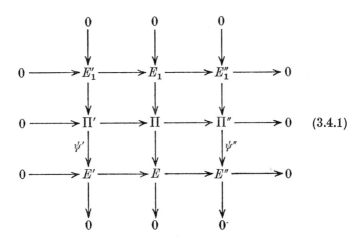

(3.4.1)

diagram in which both the rows and the columns are exact.

THEOREM 16. *Let* $0 \to E' \to E \to E'' \to 0$ *be an exact sequence of R-modules, where* $\mathrm{Pd}_R^*(E') < \infty$ *and* $\mathrm{Pd}_R^*(E) < \infty$. *Then* $\mathrm{Pd}_R^*(E'') < \infty$ *as well.*

Proof. We use induction on $n = \mathrm{Pd}_R^*(E)$. If $n \leqslant 0$, then E is a supplementable projective module and the assertion is trivial. Now suppose that $n \geqslant 1$ and make the obvious induction hypothesis. The assumptions that $\mathrm{Pd}_R^*(E')$ and $\mathrm{Pd}_R^*(E)$ are finite ensure that E' and E are finitely generated. Since E is finitely generated, E'' must be finitely generated as well.

We can now construct a commutative diagram (3.4.1) in which the rows and columns are exact and Π', Π, Π'' are supplementable projective modules. This ensures that

$$\mathrm{Pd}_R^*(E_1') < \infty \quad \text{and} \quad \mathrm{Pd}_R^*(E_1) = \mathrm{Pd}_R^*(E) - 1 = n - 1.$$

Induction now shows that $\mathrm{Pd}_R^*(E_1'') < \infty$ and the theorem follows.

The next theorem is a well known result on ordinary projective dimension. A proof has been included for the sake of completeness.

THEOREM 17. *Let* $0 \to E' \to E \to E'' \to 0$ *be an exact sequence of R-modules. In these circumstances*

   (i) *if* $\mathrm{Pd}_R(E') < \mathrm{Pd}_R(E)$, *then* $\mathrm{Pd}_R(E'') = \mathrm{Pd}_R(E)$;

  (ii) *if* $\mathrm{Pd}_R(E') > \mathrm{Pd}_R(E)$, *then* $\mathrm{Pd}_R(E'') = \mathrm{Pd}_R(E') + 1$;

 (iii) *if* $\mathrm{Pd}_R(E') = \mathrm{Pd}_R(E)$, *then* $\mathrm{Pd}_R(E'') \leqslant \mathrm{Pd}_R(E) + 1$.

Proof. Construct epimorphisms $\psi': \Pi' \to E'$ and $\psi'': \Pi'' \to E''$, where Π' and Π'' are projective modules, and put $\Pi = \Pi' \oplus \Pi''$ so that Π is projective as well. We can then construct a commutative diagram (3.4.1) in which all the rows and columns are exact. The only difference between the former and the present situation is that Π', Π'' and Π need not be supplementable.

If $\min[\mathrm{Pd}_R(E'), \mathrm{Pd}_R(E)] = \infty$ there is no problem. We may therefore suppose that

$$\min[\mathrm{Pd}_R(E'), \mathrm{Pd}_R(E)] = n$$

(say) is finite and use induction on n.

First suppose that $n \leqslant 0$. If E is projective, then the assertions in the theorem are all of them clear. We have therefore to consider the case where E' is projective but E is not. In this situation E'' cannot be projective. (For otherwise, by Chapter 2 Exercise 6, E would be isomorphic to $E' \oplus E''$ and so projective.) Moreover, since E' is projective, we can arrange that in the diagram (3.4.1) $E_1' = 0$. This makes E_1 isomorphic to E_1''. Thus

$$\mathrm{Pd}_R(E) = \mathrm{Pd}_R(E_1) + 1 = \mathrm{Pd}_R(E_1'') + 1 = \mathrm{Pd}_R(E'')$$

and we have disposed of the case $n \leqslant 0$.

From here on we shall suppose that $1 \leqslant n < \infty$. Then

$$\mathrm{Pd}_R(E_1) = \mathrm{Pd}_R(E) - 1, \quad \mathrm{Pd}_R(E_1') = \mathrm{Pd}_R(E') - 1$$

and, provided that E'' is not projective, $\mathrm{Pd}_R(E_1'') = \mathrm{Pd}_R(E'') - 1$. Induction therefore yields the desired conclusions.

Finally suppose that E'' is projective. In this situation we can arrange that $E_1'' = 0$ and then E_1' and E_1 will be isomorphic. Thus we have $\mathrm{Pd}_R(E) = \mathrm{Pd}_R(E') = n \geqslant 1$ and we are in case (iii). But $\mathrm{Pd}_R(E'') \leqslant 0$ because E'' is projective so once again we arrive at the right conclusion. This completes the inductive step and establishes the theorem.

THEOREM 18. *Let $0 \to E' \to E \to E'' \to 0$ be an exact sequence and suppose that $\mathrm{Pd}_R^*(E') < \infty$ and $\mathrm{Pd}_R^*(E) < \infty$. In these circumstances*

(i) *if $\mathrm{Pd}_R^*(E') < \mathrm{Pd}_R^*(E)$, then $\mathrm{Pd}_R^*(E'') = \mathrm{Pd}_R^*(E)$;*
(ii) *if $\mathrm{Pd}_R^*(E') > \mathrm{Pd}_R^*(E)$, then $\mathrm{Pd}_R^*(E'') = \mathrm{Pd}_R^*(E') + 1$;*
(iii) *if $\mathrm{Pd}_R^*(E') = \mathrm{Pd}_R^*(E)$, then $\mathrm{Pd}_R^*(E'') \leqslant \mathrm{Pd}_R^*(E) + 1$.*

Proof. By Theorem 16, we also have $\mathrm{Pd}_R^*(E'') < \infty$. Theorem 18 therefore follows by combining Theorems 15 and 17.

EXERCISE 7. *Let $0 \to E' \to E \to E'' \to 0$ be an exact sequence of R-modules, where $\mathrm{Pd}_R^*(E') < \infty$ and $\mathrm{Pd}_R^*(E'') < \infty$. Show that $\mathrm{Pd}_R^*(E) < \infty$.*

3.5 The Euler characteristic

THEOREM 19. *Suppose that*

$$0 \to F_n \to F_{n-1} \to \ldots \to F_1 \to F_0 \to 0$$

is an exact sequence of finite free modules. Then

$$\sum_{\mu=0}^{n} (-1)^\mu \mathrm{rank}_R(F_\mu) = 0.$$

Proof. If we localize at any prime ideal of R, not only does an exact sequence stay exact, but each finite free module becomes a finite free module of the same rank. We may therefore assume that R is a quasi-local ring.

Put $P_j = \text{Im}\,(F_{j+1} \to F_j)$. Then $P_0 = F_0$ and we have exact sequences

$$
\begin{array}{ccccc}
0 & \to P_1 & \to F_1 & \to P_0 & \to 0 \\
0 & \to P_2 & \to F_2 & \to P_1 & \to 0 \\
\vdots & \vdots & \vdots & \vdots & \vdots \\
0 & \to P_{n-1} & \to F_{n-1} & \to P_{n-2} \to & 0.
\end{array}
$$

By Chapter 2 Exercise 6, F_1 is isomorphic to $P_1 \oplus P_0$. This shows that P_1 is projective. But R is a quasi-local ring. Consequently P_1 is a free module in view of Theorem 7 of Chapter 2. But P_1 is also finitely generated because it is a homomorphic image of F_2. This proves that P_1 is a finite free module. Now consider the exact sequence $0 \to P_2 \to F_2 \to P_1 \to 0$. This time similar considerations show that P_2 is a finite free module. Thus step by step we deduce that $P_0, P_1, ..., P_{n-1}$ are all of them finite free modules. Note that P_{n-1} is isomorphic to F_n.

The exact sequence $0 \to P_\mu \to F_\mu \to P_{\mu-1} \to 0$ and Chapter 2 Exercise 1 together show that

$$\text{rank}_R\,(F_\mu) = \text{rank}_R\,(P_\mu) + \text{rank}_R\,(P_{\mu-1}).$$

Since $\text{rank}_R\,(P_0) = \text{rank}_R\,(F_0)$ and $\text{rank}_R\,(P_{n-1}) = \text{rank}_R\,(F_n)$, the theorem follows.

THEOREM 20. *Let E be an R-module and let*

$$0 \to F_n \to F_{n-1} \to ... \to F_1 \to F_0 \to E \to 0$$

and $\qquad 0 \to F'_m \to F'_{m-1} \to ... \to F'_1 \to F'_0 \to E \to 0$

be finite free resolutions of E. Then

$$\sum_{\nu=0}^{n} (-1)^\nu \text{rank}_R\,(F_\nu) = \sum_{\mu=0}^{m} (-1)^\mu \text{rank}_R\,(F'_\mu).$$

Proof. Put $k = \text{Pd}_R^*\,(E)$. The argument will use induction on k. Note that by including null modules if necessary we may suppose that $n > 1$ and $m > 1$.

Consider first the case where $\mathrm{Pd}_R^*(E) \leqslant 0$. E is then a finitely generated projective module. If we localize at an arbitrary prime ideal, E will become a finite free module and we shall have two exact sequences of finite free modules. Since under localization the ranks of F_ν and F_μ' remain unchanged, this case of the theorem follows from Theorem 19.

We shall now suppose that $k \geqslant 1$ and that the theorem has been proved for modules with a restricted projective dimension that is smaller than k. Put $K = \mathrm{Ker}\,(F_0 \to E)$ and $K' = \mathrm{Ker}\,(F_0' \to E)$. The exact sequences $0 \to K \to F_0 \to E \to 0$ and $0 \to K' \to F_0' \to E \to 0$ together with Theorem 3 of Chapter 2 show that $K \oplus F_0'$ and $K' \oplus F_0$ are isomorphic. Next

$$\mathrm{Pd}_R^*(K \oplus F_0') = \mathrm{Pd}_R^*(K' \oplus F_0) < \mathrm{Pd}_R^*(E) = k$$

and we can construct exact sequences

$$0 \to F_n \to F_{n-1} \to \dots \to F_2 \to F_1 \oplus F_0' \to K \oplus F_0' \to 0$$

and

$$0 \to F_m' \to F_{m-1}' \to \dots \to F_2' \to F_1' \oplus F_0 \to K' \oplus F_0 \to 0.$$

Since $K \oplus F_0'$ and $K' \oplus F_0$ are isomorphic we are in a situation where the induction hypothesis can be applied. Its application shows that

$$-\mathrm{rank}_R\,(F_1 \oplus F_0') + \sum_{\nu=2}^{n} (-1)^\nu \,\mathrm{rank}_R\,(F_\nu)$$

is equal to

$$-\mathrm{rank}_R\,(F_1' \oplus F_0) + \sum_{\mu=2}^{m} (-1)^\mu \,\mathrm{rank}_R\,(F_\mu').$$

However, by Chapter 2 Exercise 1,

$$\mathrm{rank}_R\,(F_1 \oplus F_0') = \mathrm{rank}_R\,(F_1) + \mathrm{rank}_R\,(F_0')$$

and $\quad \mathrm{rank}_R\,(F_1' \oplus F_0) = \mathrm{rank}_R\,(F_1') + \mathrm{rank}_R\,(F_0).$

The theorem follows on substituting these values.

Suppose that

$$0 \to F_n \to F_{n-1} \to \dots \to F_1 \to F_0 \to E \to 0$$

is a finite free resolution of E of finite length. Put

$$\mathrm{Char}_R\,(E) = \sum_{\nu=0}^{n} (-1)^\nu \,\mathrm{rank}_R\,(F_\nu). \qquad (3.5.1)$$

By Theorem 20, $\mathrm{Char}_R(E)$ depends only on E and not on the choice of the finite free resolution. It will be referred to as the *Euler characteristic* of E. Later we shall show that $\mathrm{Char}_R(E)$ is always non-negative.[†] Of particular interest are those modules for which it has the value zero. We are not yet in a position to characterize the modules which have this property,[‡] but the following simple result will serve for the present. We recall that an element α, of R, is called *nilpotent* if $\alpha^m = 0$ for some positive integer m.

LEMMA 5. *Suppose that* $\mathrm{Pd}_R^*(E) < \infty$. *If now* $\mathrm{Ann}_R(E)$ *contains an element which is not nilpotent, then* $\mathrm{Char}_R(E) = 0$. *In particular* $\mathrm{Char}_R(E) = 0$ *if* $\mathrm{Ann}_R(E)$ *contains a non-zerodivisor.*

Proof. Let $0 \to F_n \to F_{n-1} \to \ldots \to F_1 \to F_0 \to E \to 0$ be a finite free resolution of E and let α be an element of $\mathrm{Ann}_R(E)$ which is not nilpotent. Then $\{\alpha^n\}_{n \geqslant 0}$ is a multiplicatively closed subset of R not meeting the zero ideal, hence, by Chapter 1 Theorem 10, there exists a prime ideal P such that $\alpha \notin P$. If we localize at P, then $\alpha/1$ is a unit of R_P and it annihilates E_P. Consequently $E_P = 0$.

By localization at P we arrive at an exact sequence

$$0 \to (F_n)_P \to (F_{n-1})_P \to \ldots \to (F_1)_P \to (F_0)_P \to 0$$

of finite free R_P-modules and in addition

$$\mathrm{rank}_R(F_\mu) = \mathrm{rank}_{R_P}((F_\mu)_P).$$

It therefore follows, from Theorem 19, that

$$\mathrm{Char}_R(E) = \sum_{\mu=0}^{n} (-1)^\mu \mathrm{rank}_{R_P}((F_\mu)_P) = 0.$$

EXERCISE 8. *Let* $0 \to E' \to E \to E'' \to 0$ *be an exact sequence, where each of* E', E, E'' *possesses a finite free resolution of finite length. Show that* $\mathrm{Char}_R(E) = \mathrm{Char}_R(E') + \mathrm{Char}_R(E'')$.

---

† See section (4.3).

‡ In fact $\mathrm{Char}_R(E) = 0$ if and only if $\mathrm{Ann}_R(E) \neq 0$. See Chapter 4 Theorem 12.

THEOREM 21. *Let* $0 \to F_1 \xrightarrow{\phi} F_0 \to E \to 0$ *be an exact sequence, where* F_1 *and* F_0 *are finite free modules. Then*

$$\operatorname{rank}_R(F_1) \leqslant \operatorname{rank}_R(F_0).$$

Furthermore the following two statements are equivalent:

(i) $\operatorname{Char}_R(E) = 0$;

(ii) $\operatorname{Ann}_R(E)$ *contains a non-zerodivisior.*

When (i) *and* (ii) *hold, the initial Fitting invariant* $\mathfrak{F}(E)$ *is a principal ideal generated by a non-zerodivisor.*

Proof. By adding a suitable finite free module to both F_0 and F_1 we may suppose that $F_0 \neq 0$ and $F_1 \neq 0$. Put $p = \operatorname{rank}_R(F_1)$ and $q = \operatorname{rank}_R(F_0)$. Then F_1 has a base $x_1, x_2, \ldots, x_p$ and F_0 a base $y_1, y_2, \ldots, y_q$. Let

$$\phi(x_j) = \sum_{k=1}^{q} a_{jk} y_k \quad (1 \leqslant j \leqslant p),$$

where $a_{jk} \in R$, and set $A = \|a_{jk}\|$.

Now suppose that $r_1, r_2, \ldots, r_p$ belong to R. Then

$$\phi(r_1 x_1 + r_2 x_2 + \ldots + r_p x_p) = \sum_{k=1}^{q} \left(\sum_{j=1}^{p} r_j a_{jk} \right) y_k.$$

However ϕ is a monomorphism. It follows that if

$$r_1 a_{1k} + r_2 a_{2k} + \ldots + r_p a_{pk} = 0$$

for $k = 1, 2, \ldots, q$, then $r_1 = r_2 = \ldots = r_p = 0$. But in this situation we can apply Theorem 8 to the transpose of A. This tells us *that* $p \leqslant q$, *that* A *is a stable matrix, and that* $\operatorname{rank}_R(A) = p$.

The images $z_1, z_2, \ldots, z_q$ (say) of $y_1, y_2, \ldots, y_q$ in E generate E. Further each row of A is a relation between $z_1, z_2, \ldots, z_q$ and every relation between these elements is a linear combination of the rows of A. It therefore follows, by (3.1.13), that $\mathfrak{F}(E) = \mathfrak{A}_q(A)$.

We already know, from Lemma 5, that (ii) implies (i). From here on we shall assume that (i) holds. With this assumption $p = q$ and A is a $p \times p$ matrix. Since A is stable and

$$\operatorname{rank}_R(A) = p = q,$$

we have $0 :_R \mathfrak{A}_q(A) = 0$. But $\mathfrak{F}(E) = \mathfrak{A}_q(A)$ and this is simply the principal ideal generated by $\det(A)$. Accordingly

$$0 :_R (\det(A)) = 0.$$

This shows that $\det(A)$ is not a zerodivisor. Finally, by Theorem 5, $\det(A) \in \mathfrak{F}(E) \subseteq \mathrm{Ann}_R(E)$ and now all the assertions contained in the theorem have been proved.

It is convenient to make the following

DEFINITION. *An R-module K will be called an 'elementary module' if* (i) *it possesses a finite free resolution of length one, and* (ii) *its Euler characteristic is zero.*

Let K be an elementary R-module. The proof of Theorem 21 shows that we can find elements $z_1, z_2, ..., z_p$ which generate K, together with a $p \times p$ matrix A whose rows are relations between $z_1, z_2, ..., z_p$ and with the property that every relation between $z_1, z_2, ..., z_p$ is a linear combination of the rows of A. $\mathfrak{F}(K)$ is then the ideal generated by $\det(A)$, and $\det(A)$ is a non-zerodivisor contained in $\mathrm{Ann}_R(K)$.

THEOREM 22. *Let* $0 \to M \xrightarrow{\phi} E \xrightarrow{\psi} K \to 0$ *be an exact sequence of finitely generated R-modules, and suppose that K is an elementary module. Then* $\mathfrak{F}(E) = \mathfrak{F}(M)\mathfrak{F}(K)$.

Proof. We may suppose that M is a submodule of E. Choose elements $z_1, z_2, ..., z_p$ of K and a matrix A as in the paragraph immediately preceding the statement of the theorem. If we now select $e_1, e_2, ..., e_p$ in E so that $\psi(e_i) = z_i$ and also elements $\omega_1, \omega_2, ..., \omega_q$ in M so that $M = R\omega_1 + R\omega_2 + ... + R\omega_q$, then $e_1, ..., e_p, \omega_1, ..., \omega_q$ will be a system of generators for E.

Let $\alpha_1 z_1 + \alpha_2 z_2 + ... + \alpha_p z_p = 0$, where $\alpha_i \in R$. Then

$$\alpha_1 e_1 + \alpha_2 e_2 + ... + \alpha_p e_p \in M$$

and therefore we can find $\beta_1, \beta_2, ..., \beta_q$ in R so that

$$(\alpha_1, ..., \alpha_p, \beta_1, ..., \beta_q)$$

is a relation between $e_1, ..., e_p, \omega_1, ..., \omega_q$. This remark shows that there exists a $p \times q$ matrix B such that each row of $\|A|B\|$ is a relation between $e_1, ..., e_p, \omega_1, ..., \omega_q$. Note that if

$$\alpha_1' e_1 + ... + \alpha_p' e_p + \beta_1' \omega_1 + ... + \beta_q' \omega_q = 0,$$

then $(\alpha_1', \alpha_2', ..., \alpha_p')$ is a relation between $z_1, z_2, ..., z_p$ and therefore a linear combination of the rows of A.

Now suppose that D is a matrix with $p+q$ columns each of whose rows is a relation between $e_1, \ldots, e_p, \omega_1, \ldots, \omega_q$. Put

$$T = \left\| \frac{A \mid B}{D} \right\|.$$

Then by elementary row operations T can be brought to the form

$$T' = \left\| \frac{A \mid B}{0 \mid B'} \right\|$$

and we have $\quad \mathfrak{A}_{p+q}(D) \subseteq \mathfrak{A}_{p+q}(T) = \mathfrak{A}_{p+q}(T').$

Next each $(p+q) \times (p+q)$ minor of T' is either zero or it can be expressed in the form $\gamma \det(A)$, where γ is a $q \times q$ minor of B' and therefore an element of $\mathfrak{A}_q(B')$. However each row of B' is a relation between $\omega_1, \omega_2, \ldots, \omega_q$ and so $\mathfrak{A}_q(B') \subseteq \mathfrak{F}(M)$. Moreover $\mathfrak{F}(K)$ is generated by $\det(A)$. Thus the $(p+q) \times (p+q)$ minors of T' belong to $\mathfrak{F}(M)\mathfrak{F}(K)$ and therefore

$$\mathfrak{A}_{p+q}(D) \subseteq \mathfrak{A}_{p+q}(T') \subseteq \mathfrak{F}(M)\mathfrak{F}(K).$$

But by definition $\quad \mathfrak{F}(E) = \sum_D \mathfrak{A}_{p+q}(D),$

where the sum is taken over all possible choices of D. Accordingly $\mathfrak{F}(E) \subseteq \mathfrak{F}(M)\mathfrak{F}(K)$ and now the proof is complete because the opposite inclusion follows by Exercise 2.

EXERCISE 9. *Let R be an integral domain with the property that each ideal can be generated by a single element. Show that every sub-module of a finite free module is a finite free module, and hence that every finitely generated module has a finite free resolution of length one. Deduce that an R-module is elementary if and only if it is finitely generated and has a non-zero annihilator.*

Let E be an R-module. An exact sequence

$$0 \to K_n \to K_{n-1} \to \ldots \to K_1 \to K_0 \to E \to 0, \qquad (3.5.2)$$

where each K_j is an elementary module, will be called an *elementary resolution* of E of length n.

THEOREM 23. *If E is an R-module, then the following two statements are equivalent:*
(a) *E admits an elementary resolution of finite length;*
(b) *$\mathrm{Pd}_R^*(E) < \infty$ and $\mathrm{Ann}_R(E)$ contains a non-zerodivisor.*

Proof. Assume (*a*). Let (3.5.2) be an elementary resolution of E. Put $J_s = \text{Im}\,(K_s \to K_{s-1})$ for $s = 1, 2, ..., n$. Since J_n is isomorphic to K_n, we have $\text{Pd}_R^*(J_n) < \infty$. Next, by repeated applications of Theorem 18 to the exact sequences

$$0 \to J_s \to K_{s-1} \to J_{s-1} \to 0,$$

we deduce that $\text{Pd}_R^*(J_s) < \infty$ for $1 \leqslant s \leqslant n$ and the same theorem applied to
$$0 \to J_1 \to K_0 \to E \to 0$$

now shows that $\text{Pd}_R^*(E) < \infty$. Finally, since K_0 is an elementary module, Theorem 21 shows that $\alpha K_0 = 0$ for some $\alpha \in R$ which is not a zerodivisor. But then $\alpha \in \text{Ann}_R(E)$ and we have shown that (*a*) implies (*b*).

Assume (*b*). We shall deduce (*a*) by employing induction on $\text{Pd}_R^*(E)$.

First suppose that $\text{Pd}_R^*(E) \leqslant 1$. Then E has a finite free resolution of length one and Theorem 21 shows that $\text{Char}_R(E) = 0$. Thus E is itself an elementary module and (*a*) is certainly true in this case. From here on we shall suppose that $\text{Pd}_R^*(E) > 1$ and that the assertion that (*b*) implies (*a*) has been proved in the case of modules with smaller restricted projective dimension.

Since $\text{Pd}_R^*(E) < \infty$, E is finitely generated and therefore we can construct an exact sequence $0 \to E' \to F \to E \to 0$, where F is a finite free module and $E' \to F$ is an inclusion mapping. Choose $\alpha \in \text{Ann}_R(E)$ so that α is not a zerodivisor. Then $\alpha F \subseteq E'$ and we have an exact sequence

$$0 \to E'/\alpha F \to F/\alpha F \to E \to 0. \qquad (3.5.3)$$

Next multiplication by α produces an exact sequence

$$0 \to F \to F \to F/\alpha F \to 0$$

and from this we conclude that $F/\alpha F$ is an elementary module. Now $\text{Pd}_R^*(E') = \text{Pd}_R^*(E) - 1$ and, since αF and F are isomorphic,

$$\text{Pd}_R^*(\alpha F) = \text{Pd}_R^*(F) = 0.$$

Hence, by Theorem 16, $\text{Pd}_R^*(E'/\alpha F) < \infty$. In fact

$$\text{Pd}_R^*(E'/\alpha F) < \text{Pd}_R^*(E).$$

(For otherwise it would follow, from (3.5.3) and Theorem 18, that
$$\mathrm{Pd}_R^*(E) = \mathrm{Pd}_R^*(E'/\alpha F) + 1 > \mathrm{Pd}_R^*(E)$$
giving a contradiction.) Evidently α annihilates $E'/\alpha F$ and therefore, by virtue of the induction hypothesis, $E'/\alpha F$ has an elementary resolution of finite length. But then (3.5.3) implies that E has an elementary resolution of finite length and with this the proof is complete.

Before we can proceed with the study of modules which possess elementary resolutions of finite length we must say a little on the subject of *invertible ideals*. This will enable us to introduce a new invariant for such modules which will subsequently play an important rôle.

Let Σ consist of all the elements of R that are *not* zerodivisors. Then Σ is a multiplicatively closed subset of R. Put $Q = R_\Sigma$. Q is called the *full ring of fractions* of R. (Of course, if R is an integral domain, then Q is just its quotient field.) The canonical ring-homomorphism† $R \to R_\Sigma = Q$ is on this occasion an injective mapping. Consequently R may be regarded as a subring of Q.

By a *fractional ideal* of R is meant an R-submodule $\mathfrak{A}$, of Q, with the additional property that there exists an element $\delta \in R$ which is not a zerodivisor and which is such that $\delta \mathfrak{A} \subseteq R$. For example, an ordinary ideal of R is a fractional ideal because in this case we may take δ to be the identity element. When fractional ideals are under consideration it is usual to refer to ordinary ideals as *integral ideals*. Note that if $\xi_1, \xi_2, \ldots, \xi_m$ belong to Q, then $R\xi_1 + R\xi_2 + \ldots + R\xi_m$ is a fractional ideal.

Let $\mathfrak{A}$ and $\mathfrak{B}$ be fractional ideals. We define their product $\mathfrak{A}\mathfrak{B}$ exactly as in the case of integral ideals. Thus the fractional ideal $\mathfrak{A}\mathfrak{B}$ consists of all elements that can be written as finite sums $\alpha_1\beta_1 + \alpha_2\beta_2 + \ldots + \alpha_n\beta_n$, where $\alpha_i \in \mathfrak{A}$ and $\beta_i \in \mathfrak{B}$. Multiplication of fractional ideals is commutative and associative, and for any fractional ideal $\mathfrak{A}$ we have $R\mathfrak{A} = \mathfrak{A}R = \mathfrak{A}$. Thus the fractional ideals of R form a commutative semi-group with R as its neutral element. Observe that if $\xi_1, \xi_2, \ldots, \xi_m$ and $\eta_1, \eta_2, \ldots, \eta_n$ belong to Q, then
$$\left(\sum_{i=1}^{m} R\xi_i \right) \left(\sum_{j=1}^{n} R\eta_j \right) = \sum_{i,j} R\xi_i\eta_j.$$

† See section (1.7).

A fractional ideal $\mathfrak{A}$ is said to be *invertible* if there exists a fractional ideal $\mathfrak{B}$ such that

$$\mathfrak{A}\mathfrak{B} = \mathfrak{B}\mathfrak{A} = R.$$

If $\mathfrak{B}$ exists, it is unique. It is called the *inverse* of $\mathfrak{A}$ and is denoted by $\mathfrak{A}^{-1}$. Thus $\mathfrak{A}\mathfrak{A}^{-1} = R$ and $(\mathfrak{A}^{-1})^{-1} = \mathfrak{A}$. For example, if $\alpha \in R$ and α is not a zerodivisor, then $R\alpha$ is invertible and its inverse is $R(1/\alpha)$. Again, if K is an elementary module, then, by Theorem 21, $\mathfrak{F}(K)$ is an invertible ideal.

EXERCISE 10. *Let $\mathfrak{A}$ be an integral ideal of R. Show that $\mathfrak{A}$ is invertible if and only if it is both projective (as an R-module) and contains a non-zerodivisor.*

THEOREM 24. *Suppose that*

$$0 \to K_n \to K_{n-1} \to \ldots \to K_1 \to K_0 \to 0$$

is an exact sequence of elementary R-modules. Then

$$\prod_{i=0}^{n} \mathfrak{F}(K_i)^{(-1)^i} = R.$$

Proof. We first dispose of the small values of n. To this end we observe that the theorem is true when $n = 0, 1$ and Theorem 22 shows that it is also true for $n = 2$. From here on it will be assumed that $n > 2$.

Put $J_s = \mathrm{Im}(K_s \to K_{s-1})$. We claim that $J_n, J_{n-1}, \ldots, J_1$ are all of them elementary modules. First $\mathrm{Pd}_R^*(J_s) < \infty$ as may be seen by repeating the argument at the beginning of the proof of Theorem 23. Next $J_1 = K_0$ is elementary by hypothesis and therefore the exact sequence

$$0 \to J_2 \to K_1 \to J_1 \to 0$$

shows that $\mathrm{Pd}_R^*(J_2) \leqslant 1$. (Otherwise, by Theorem 18, it would follow that $\mathrm{Pd}_R^*(J_1) \geqslant 3$ and this is not the case.) Since K_1 is an elementary module, it is annihilated by a non-zerodivisor and this non-zerodivisor will annihilate J_2. It follows that J_2 is an elementary module. Similar arguments applied to the exact sequence

$$0 \to J_3 \to K_2 \to J_2 \to 0$$

lead to the conclusion that J_3 is an elementary module. And so on. This establishes our claim.

Let us apply Theorem 22 to the exact sequence

$$0 \to J_{s+1} \to K_s \to J_s \to 0.$$

This shows that

$$\mathfrak{F}(K_s)^{(-1)^s} = \mathfrak{F}(J_s)^{(-1)^s} \mathfrak{F}(J_{s+1})^{-(-1)^{s+1}}$$

and now the theorem follows by multiplication.

3.6 MacRae's invariant

Suppose that the R-module E has an elementary resolution

$$0 \to K_n \to K_{n-1} \to \ldots \to K_1 \to K_0 \to E \to 0 \qquad (3.6.1)$$

of finite length and put

$$\mathfrak{G}(E) = \mathfrak{F}(K_0)\,\mathfrak{F}(K_1)^{-1}\,\mathfrak{F}(K_2)\,\mathfrak{F}(K_3)^{-1}\ldots . \qquad (3.6.2)$$

By Theorem 21, $\mathfrak{F}(K_j)$ is an integral ideal generated by a non-zerodivisor. Consequently

$$\mathfrak{G}(E) = R\frac{\alpha}{\beta}, \qquad (3.6.3)$$

where $\alpha, \beta \in R$ and are non-zerodivisors. Thus $\mathfrak{G}(E)$ appears here as a rather simple kind of invertible fractional ideal. The notation is reasonable because, as will shortly be established (see Theorem 25), $\mathfrak{G}(E)$ depends only on E and not on the choice of the resolution (3.6.1). Anticipating this result we shall refer to $\mathfrak{G}(E)$ as the *MacRae invariant*† of E. Note that if $\mathrm{Pd}_R^*(E) \leqslant 1$, ensuring that E itself is an elementary module, then the fact that all elementary resolutions of E lead to the same value for $\mathfrak{G}(E)$ follows from Theorem 24. Indeed this observation shows that *when K is an elementary module* $\mathfrak{G}(K) = \mathfrak{F}(K)$.

Following MacRae, we shall prove the next two theorems simultaneously.

THEOREM 25. *Suppose that*

$$0 \to K_n \to K_{n-1} \to \ldots \to K_1 \to K_0 \to E \to 0$$

and $\qquad 0 \to L_m \to L_{m-1} \to \ldots \to L_1 \to L_0 \to E \to 0$

are two elementary resolutions of the same module E. Then

$$\mathfrak{F}(K_0)\,\mathfrak{F}(K_1)^{-1}\mathfrak{F}(K_2)\,\mathfrak{F}(K_3)^{-1}\ldots = \mathfrak{F}(L_0)\,\mathfrak{F}(L_1)^{-1}\mathfrak{F}(L_2)\,\mathfrak{F}(L_3)^{-1}\ldots .$$

† See R. E. MacRae (32) for the original treatment of this ideal and its properties.

THEOREM 26. *Let $0 \to E' \to E \to E'' \to 0$ be an exact sequence of R-modules, where each of E', E, E'' possesses an elementary resolution of finite length. Then $\mathfrak{G}(E) = \mathfrak{G}(E')\mathfrak{G}(E'')$.*

Proof of Theorems 25 and 26. Let $k \geqslant 1$ be an integer. In what follows S_k and T_k will stand for the assertions '*Theorem 25 holds if $\mathrm{Pd}_R^*(E) \leqslant k$*' and '*Theorem 26 holds if none of $\mathrm{Pd}_R^*(E')$, $\mathrm{Pd}_R^*(E)$, $\mathrm{Pd}_R^*(E'')$ exceeds k*' respectively. It has already been observed that S_1 holds. Consider T_1. This has to do with the case where E', E, E'' are elementary modules and therefore

$$\mathfrak{G}(E') = \mathfrak{F}(E'), \quad \mathfrak{G}(E) = \mathfrak{F}(E) \quad \text{and} \quad \mathfrak{G}(E'') = \mathfrak{F}(E'').$$

The truth of T_1 is therefore a consequence of Theorem 22.

Now suppose that $m > 1$ and that both S_{m-1} and T_{m-1} have been established. It will suffice to show that S_m and T_m hold as well.

Let us begin with S_m. For the situation envisaged here we have $\mathrm{Pd}_R^*(E) \leqslant m$. Put $U = \mathrm{Ker}\,(K_0 \to E)$ and $V = \mathrm{Ker}\,(L_0 \to E)$. Then U and V possess elementary resolutions of finite length and therefore, by Theorem 23, $\mathrm{Pd}_R^*(U) < \infty$ and $\mathrm{Pd}_R^*(V) < \infty$. Let X be the submodule of $K_0 \oplus L_0$ consisting of all pairs (ξ, η), where $\xi \in K_0$, $\eta \in L_0$ and they have the same image under the mappings $K_0 \to E$ and $L_0 \to E$ respectively. Denote by $j\colon X \to K_0 \oplus L_0$ the inclusion mapping and by $\pi\colon K_0 \oplus L_0 \to K_0$ the projection on to the first summand. Then we have an exact sequence

$$0 \to V \to X \xrightarrow{\pi j} K_0 \to 0. \tag{3.6.4}$$

Note that, by Exercise 7, $\mathrm{Pd}_R^*(X) < \infty$.

We know that $\mathrm{Pd}_R^*(L_0) \leqslant 1$. Consequently Theorem 18 applied to the exact sequence

$$0 \to V \to L_0 \to E \to 0$$

shows that $\mathrm{Pd}_R^*(V) < m$ and now the same theorem applied to (3.6.4) yields $\mathrm{Pd}_R^*(X) < m$. Next it is possible to find a non-zerodivisor which annihilates both K_0 and L_0 and which therefore annihilates X. Consequently, by Theorem 23, X possesses an elementary resolution of finite length. Indeed because of this and our various inequalities, we are able to apply T_{m-1} to (3.6.4). This shows that

$$\mathfrak{G}(X) = \mathfrak{G}(K_0)\,\mathfrak{G}(V) = \mathfrak{F}(K_0)\,\mathfrak{G}(V).$$

(Observe that S_{m-1} ensures that $\mathfrak{G}(X)$ and $\mathfrak{G}(V)$ are unambiguous.) Accordingly

$$\mathfrak{G}(X) = \mathfrak{F}(K_0)\,\mathfrak{F}(L_1)\,\mathfrak{F}(L_2)^{-1}\,\mathfrak{F}(L_3)\,\mathfrak{F}(L_4)^{-1}...$$

and in an exactly similar manner we can show that

$$\mathfrak{G}(X) = \mathfrak{F}(L_0)\,\mathfrak{F}(K_1)\,\mathfrak{F}(K_2)^{-1}\,\mathfrak{F}(K_3)\,\mathfrak{F}(K_4)^{-1}....$$

Comparison of the two expressions for $\mathfrak{G}(X)$ immediately yields the relation described in Theorem 25. This means that we have derived S_m.

We now turn our attention to T_m and therefore suppose that

$$\max\,[\mathrm{Pd}_R^*(E'),\,\mathrm{Pd}_R^*(E),\,\mathrm{Pd}_R^*(E'')] \leqslant m.$$

Let us construct epimorphisms $\psi': F' \to E'$, $\psi'': F'' \to E''$, where F' and F'' are finite free modules, and put $F = F' \oplus F''$ so that F is a finite free module as well. After this we can construct, in a familiar manner,† a commutative diagram

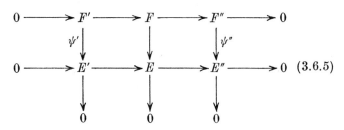

$$\hspace{12cm} (3.6.5)$$

whose rows and columns are exact. But each of E', E, E'' has an annihilator that contains a non-zerodivisor since we are supposing that we are in the situation contemplated in Theorem 26. We can therefore find $\alpha \in R$ so that α annihilates all of E', E, E'' and is not a zerodivisor. It is now a simple matter to derive from (3.6.5) a further commutative diagram

† See the discussion associated with (3.4.1).

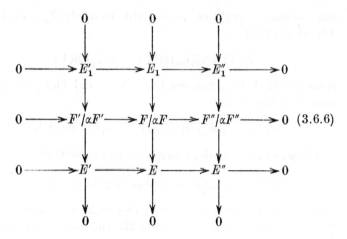

$$(3.6.6)$$

This also has exact rows and columns. Evidently $F'/\alpha F'$, $F/\alpha F$ and $F''/\alpha F''$ are elementary modules. Next, since

$$E_1 = \mathrm{Ker}\,(F/\alpha F \to E),$$

the argument previously used in connection with (3.5.3) shows that E_1 possesses an elementary resolution of finite length. We can use this elementary resolution of E_1 to compute $\mathfrak{G}(E)$ and so obtain

$$\mathfrak{G}(E) = \mathfrak{F}(F/\alpha F)\,\mathfrak{G}(E_1)^{-1}.$$

(Note that, by Theorem 18 applied to the centre column of (3.6.6), $\mathrm{Pd}_R^*(E_1) < m$. Consequently, by virtue of S_{m-1}, there is no ambiguity surrounding $\mathfrak{G}(E_1)^{-1}$.)

By similar arguments we can show that E_1' and E_1'' have elementary resolutions of finite length and that both

$$\mathrm{Pd}_R^*(E_1') < m \quad \text{and} \quad \mathrm{Pd}_R^*(E_1'') < m.$$

Accordingly
$$\mathfrak{G}(E') = \mathfrak{F}(F'/\alpha F')\,\mathfrak{G}(E_1')^{-1}$$

and
$$\mathfrak{G}(E'') = \mathfrak{F}(F''/\alpha F'')\,\mathfrak{G}(E_1'')^{-1}.$$

But, by Theorem 22,

$$\mathfrak{F}(F/\alpha F) = \mathfrak{F}(F'/\alpha F')\,\mathfrak{F}(F''/\alpha F'')$$

and, because T_{m-1} holds by hypothesis,

$$\mathfrak{G}(E_1) = \mathfrak{G}(E_1')\,\mathfrak{G}(E_1'').$$

Thus
$$\mathfrak{G}(E')\,\mathfrak{G}(E'') = \mathfrak{F}(F/\alpha F)\,\mathfrak{G}(E_1)^{-1} = \mathfrak{G}(E)$$

and now T_m has been established. As already observed, this completes the proof of Theorems 25 and 26.

Finally, for future reference, we record a fact which was noted earlier.

THEOREM 27. *When E is an elementary module* $\mathfrak{G}(E) = \mathfrak{F}(E)$.

Theorem 27 shows, in particular, that when E is an elementary module $\mathfrak{G}(E)$ is an integral ideal and not just a fractional ideal. Indeed MacRae proves in (**32**) that when R is a Noetherian ring $\mathfrak{G}(E)$ is always an integral ideal. We shall return to this point later.†

Let S be a multiplicatively closed subset of R not containing the zero element and let $\mathfrak{A}$ be a fractional ideal of R. We can then choose an element α, of R, so that $\alpha\mathfrak{A} \subseteq R$ and α is not a zero-divisor. As $\alpha/1$ is not a zerodivisor in R_S, we may put

$$\mathfrak{A}R_S = [(\alpha\mathfrak{A})R_S]\left[R_S\frac{\alpha}{1}\right]^{-1}. \qquad (3.6.7)$$

Evidently $\mathfrak{A}R_S$ is a fractional ideal of R_S and an easy verification shows that it does not depend on the choice of α. Again if $\mathfrak{B}$ is also a fractional ideal of R, then

$$(\mathfrak{A}\mathfrak{B})R_S = (\mathfrak{A}R_S)(\mathfrak{B}R_S). \qquad (3.6.8)$$

It follows that if $\mathfrak{A}$ is an invertible fractional ideal of R, then $\mathfrak{A}R_S$ is an invertible fractional ideal of R_S and

$$\mathfrak{A}^{-1}R_S = (\mathfrak{A}R_S)^{-1}. \qquad (3.6.9)$$

Let K be an elementary R-module. Then there exists an exact sequence $0 \to G \to F \to K \to 0$, where F, G are finite free modules of the same rank. This exact sequence induces an exact sequence

$$0 \to G_S \to F_S \to K_S \to 0$$

of R_S-modules, where F_S, G_S are finite free R_S-modules of the same rank. Accordingly, *if K is an elementary R-module, then* K_S *is an elementary* R_S*-module.*

THEOREM 28. *Let E be an R-module which admits an elementary resolution of finite length and let S be a multiplicatively*

† See section (6.2).

closed subset of R not containing the zero element. Then the R_S-module E_S admits an elementary resolution of finite length, and $\mathfrak{G}(E)\,R_S = \mathfrak{G}(E_S)$.

Proof. Let

$$0 \to K_n \to \ldots \to K_1 \to K_0 \to E \to 0$$

be an elementary resolution of E. Then

$$0 \to (K_n)_S \to \ldots \to (K_1)_S \to (K_0)_S \to E_S \to 0$$

is an elementary resolution of the R_S-module E_S. Now

$$\mathfrak{G}(E) = \mathfrak{F}(K_0)\,\mathfrak{F}(K_1)^{-1}\,\mathfrak{F}(K_2)\,\mathfrak{F}(K_3)^{-1}\ldots$$

and therefore, by (3.6.8) and (3.6.9),

$$\mathfrak{G}(E)\,R_S = \prod_{j=0}^{n} \{\mathfrak{F}(K_j)\,R_S\}^{(-1)^j}.$$

However, by Theorem 3, $\mathfrak{F}(K_j)\,R_S = \mathfrak{F}((K_j)_S)$. It follows that

$$\mathfrak{G}(E)\,R_S = \prod_{j=0}^{n} \{\mathfrak{F}((K_j)_S)\}^{(-1)^j} = \mathfrak{G}(E_S)$$

as required.

Solutions to the Exercises on Chapter 3

EXERCISE 1. *Determine the Fitting invariants of a free R-module of rank q.*

Solution. Let F be a free R-module with a base $e_1, e_2, \ldots, e_q$ and let A be the zero row matrix of length q. Since the only relation between $e_1, e_2, \ldots, e_q$ is the trivial one, we have, by (3.1.13),

$$\mathfrak{F}_\mu(F) = \begin{cases} \mathfrak{A}_{q-\mu}(A) & \text{for} \quad 0 \leqslant \mu \leqslant q, \\ R & \text{for} \quad \mu > q. \end{cases}$$

Accordingly

$$\mathfrak{F}_\mu(F) = \begin{cases} 0 & \text{for} \quad 0 \leqslant \mu < q, \\ R & \text{for} \quad \mu \geqslant q. \end{cases}$$

EXERCISE 2. *Let $0 \to E' \to E \to E'' \to 0$ be an exact sequence of finitely generated R-modules and let $\mu \geqslant 0$ and $\nu \geqslant 0$ be integers. Show that $\mathfrak{F}_\mu(E'')\,\mathfrak{F}_\nu(E') \subseteq \mathfrak{F}_{\mu+\nu}(E)$ and hence that*

$$\mathfrak{F}(E'')\,\mathfrak{F}(E') \subseteq \mathfrak{F}(E).$$

Solution. We may suppose that E' is a submodule of E and that $E' \to E$ is the inclusion mapping. Let $\omega_1, \omega_2, ..., \omega_q$ generate E'. Further let $e_1, e_2, ..., e_p$ belong to E and denote by $e''_1, e''_2, ..., e''_p$ their images in E''. We choose $e_1, e_2, ..., e_p$ so that

$$E'' = Re''_1 + Re''_2 + ... + Re''_p.$$

This ensures that $e_1, ..., e_p, \omega_1, ..., \omega_q$ is a system of generators for E. Note that if $\alpha_1 e''_1 + \alpha_2 e''_2 + ... + \alpha_p e''_p = 0$, where $\alpha_i \in R$, then $\alpha_1 e_1 + \alpha_2 e_2 + ... + \alpha_p e_p \in E'$ and therefore

$$\alpha_1 e_1 + ... + \alpha_p e_p + \beta_1 \omega_1 + ... + \beta_q \omega_q = 0$$

for suitable elements $\beta_1, \beta_2, ..., \beta_q$ in R.

We first consider the case where $0 \leqslant \mu \leqslant p$ and $0 \leqslant \nu \leqslant q$. Let A be a matrix with p columns whose rows are relations between $e''_1, e''_2, ..., e''_p$ and C a matrix with q columns whose rows are relations between $\omega_1, \omega_2, ..., \omega_q$. The remark at the end of the preceding paragraph shows that we can find a matrix B, with q columns and the same number of rows as A, such that the rows of the matrix

$$\| \, A \mid B \, \|$$

are relations between $e_1, ..., e_p, \omega_1, ..., \omega_q$.

Let D be a $(p - \mu) \times (p - \mu)$ minor of A and Δ a $(q - \nu) \times (q - \nu)$ minor of C. Then $D\Delta$ is a $(p + q - \mu - \nu) \times (p + q - \mu - \nu)$ minor of the matrix

$$\left\| \frac{A \mid B}{0 \mid C} \right\|.$$

This shows that $D\Delta \in \mathfrak{F}_{\mu+\nu}(E)$. However $\mathfrak{F}_\mu(E'')$ is generated by elements such as D and $\mathfrak{F}_\nu(E')$ by elements such as Δ. Accordingly $\mathfrak{F}_\mu(E'') \mathfrak{F}_\nu(E') \subseteq \mathfrak{F}_{\mu+\nu}(E)$ in this case.

If $\mu > p$ and $\nu > q$, then $\mathfrak{F}_{\mu+\nu}(E)$, $\mathfrak{F}_\mu(E'')$ and $\mathfrak{F}_\nu(E')$ all equal R and there is no problem. Suppose that $\mu > p$ and $0 \leqslant \nu \leqslant q$. Then $\mathfrak{F}_\mu(E'') = R = \mathfrak{F}_p(E'')$ and therefore

$$\mathfrak{F}_\mu(E'') \mathfrak{F}_\nu(E') = \mathfrak{F}_p(E'') \mathfrak{F}_\nu(E') \subseteq \mathfrak{F}_{p+\nu}(E) \subseteq \mathfrak{F}_{\mu+\nu}(E)$$

by virtue of the case already considered and a similar device disposes of the case $0 \leqslant p \leqslant \mu$ and $\nu > q$.

The concluding assertion in the statement of the exercise is obtained by taking $\mu = \nu = 0$.

Exercise 3. *Let E' and E'' be finitely generated R-modules and let $k \geqslant 0$ be an integer. Prove that*

$$\mathfrak{F}_k(E' \oplus E'') = \sum_{\mu+\nu=k} \mathfrak{F}_\mu(E') \mathfrak{F}_\nu(E'').$$

Solution. Let $e'_1, e'_2, \ldots, e'_p$ generate E' and $e''_1, e''_2, \ldots, e''_q$ generate E''. Then $e'_1, \ldots, e'_p, e''_1, \ldots, e''_q$ generate $E = E' \oplus E''$. By Exercise 2,

$$\sum_{\mu+\nu=k} \mathfrak{F}_\mu(E') \mathfrak{F}_\nu(E'') \subseteq \mathfrak{F}_k(E).$$

Now if $k \geqslant p+q$, then $\mathfrak{F}_k(E) = R$ and we can choose $\mu \geqslant p$, $\nu \geqslant q$ so that $\mu + \nu = k$. Since we then have $\mathfrak{F}_\mu(E') = R = \mathfrak{F}_\nu(E'')$ we at once obtain the opposite inclusion. From here on we shall therefore assume that $0 \leqslant k < p+q$.

In what follows Q will be used to denote a matrix of the form

$$Q = \left\| \begin{array}{c|c} A' & 0 \\ \hline 0 & A'' \end{array} \right\|,$$

where the rows of A' are relations between $e'_1, e'_2, \ldots, e'_p$ and the rows of A'' relations between $e''_1, e''_2, \ldots, e''_q$. Note that if

$$\alpha'_1 e'_1 + \ldots + \alpha'_p e'_p + \alpha''_1 e''_1 + \ldots + \alpha''_q e''_q = 0,$$

then $(\alpha'_1, \alpha'_2, \ldots, \alpha'_p)$ is a relation between $e'_1, e'_2, \ldots, e'_p$ and $(\alpha''_1, \alpha''_2, \ldots, \alpha''_q)$ is a relation between $e''_1, e''_2, \ldots, e''_q$. Hence the rows of matrices such as Q generate the module of relations between $e'_1, \ldots, e'_p, e''_1, \ldots, e''_q$ and therefore

$$\mathfrak{F}_k(E) = \sum_Q \mathfrak{A}_{p+q-k}(Q).$$

Let us take a particular matrix Q and let D be one of its $(p+q-k) \times (p+q-k)$ minors. If

$$Q = \left\| \begin{array}{c|c} A' & 0 \\ \hline 0 & A'' \end{array} \right\|,$$

then

$$D = \left| \begin{array}{c|c} B' & 0 \\ \hline 0 & B'' \end{array} \right|,$$

where B' is an $r \times (p-\mu)$ submatrix of A' and B'' an $s \times (q-\nu)$ submatrix of A''. (Here $\mu+\nu = k$ and $r+s = p+q-k$.) However,

unless $r = p - \mu$, and therefore $s = q - \nu$, we must have $D = 0$.
Suppose then that $r = p - \mu$ and $s = q - \nu$. We now have

$$D = \det(B') \det(B'') \in \mathfrak{F}_\mu(E') \, \mathfrak{F}_\nu(E'').$$

Thus, in any event,

$$D \in \sum_{\mu + \nu = k} \mathfrak{F}_\mu(E') \, \mathfrak{F}_\nu(E'').$$

But $\mathfrak{F}_k(E)$ is generated by elements such as D. Accordingly

$$\mathfrak{F}_k(E) \subseteq \sum_{\mu + \nu = k} \mathfrak{F}_\mu(E') \, \mathfrak{F}_\nu(E'')$$

and with this the solution is complete.

E x e r c i s e 4. *Let $\mathfrak{A}_1, \mathfrak{A}_2, \ldots, \mathfrak{A}_s$ be ideals of R and put*

$$E = (R/\mathfrak{A}_1) \oplus (R/\mathfrak{A}_2) \oplus \ldots \oplus (R/\mathfrak{A}_s).$$

Show that $\mathfrak{F}(E) = \mathfrak{A}_1 \mathfrak{A}_2 \ldots \mathfrak{A}_s$.

Solution. In view of Exercise 3 it is enough to show that if $\mathfrak{A}$ is
an ideal of R, then $\mathfrak{F}(R/\mathfrak{A}) = \mathfrak{A}$. But $R/\mathfrak{A}$ can be generated by a
single element. Consequently, by Theorem 5,

$$\mathfrak{F}(R/\mathfrak{A}) = \operatorname{Ann}_R (R/\mathfrak{A}) = \mathfrak{A}.$$

E x e r c i s e 5. *Let K be a submodule of an R-module E, let $\mathfrak{A}$ be a
finitely generated ideal, and let S be a multiplicatively closed subset
of R. Show that*

$$(K :_E \mathfrak{A})_S = K_S :_{E_S} \mathfrak{A} R_S.$$

Solution. Put $M = K :_E \mathfrak{A}$. Then $\mathfrak{A}M \subseteq K$, $M_S \subseteq E_S$ and
$(\mathfrak{A}R_S) M_S \subseteq K_S$. Consequently

$$M_S \subseteq K_S :_{E_S} (\mathfrak{A}R_S).$$

By hypothesis $\mathfrak{A}$ is finitely generated, say

$$\mathfrak{A} = Ra_1 + Ra_2 + \ldots + Ra_m.$$

Let e/s, where $e \in E$ and $s \in S$, belong to $K_S :_{E_S} (\mathfrak{A}R_S)$. Then

$$\frac{a_i e}{s} = \frac{a_i}{1} \frac{e}{s} \in K_S.$$

It follows that there exists $\sigma \in S$ such that $\sigma a_i e \in K$ for
$i = 1, 2, \ldots, m$. We now have $\mathfrak{A}\sigma e \subseteq K$ and therefore $\sigma e \in M$.
Accordingly

$$\frac{e}{s} = \frac{\sigma e}{\sigma s} \in M_S$$

and we have shown that

$$K_{S:E_S}(\mathfrak{A}R_S) \subseteq M_S.$$

This completes the solution.

EXERCISE 6. *Let P be a projective R-module. Show that there exists a free module F such that $P \oplus F$ is isomorphic to F.*

Solution. There exists a module Q such that $P \oplus Q$ is a free module. Put $F = P \oplus Q \oplus P \oplus Q \oplus P \oplus Q \oplus \ldots,$

where there is a countable infinity of summands. Since F is isomorphic to $Q \oplus P \oplus Q \oplus P \oplus Q \oplus P \oplus \ldots,$

$P \oplus F$ is isomorphic to

$$P \oplus Q \oplus P \oplus Q \oplus P \oplus Q \oplus \ldots = F.$$

EXERCISE 7. *Let $0 \to E' \to E \to E'' \to 0$ be an exact sequence of R-modules, where $\mathrm{Pd}_R^*(E') < \infty$ and $\mathrm{Pd}_R^*(E'') < \infty$. Show that $\mathrm{Pd}_R^*(E) < \infty$.*

Solution. Put $k = \max[\mathrm{Pd}_R^*(E'), \mathrm{Pd}_R^*(E'')] < \infty$. If $k \leqslant 0$, then E' and E'' are supplementable projective modules and E is isomorphic to $E' \oplus E''$. Consequently, by Lemma 4, E itself is a supplementable projective module and therefore $\mathrm{Pd}_R^*(E) \leqslant 0$.

Now suppose that $k \geqslant 1$. The hypotheses ensure that E', E'' (and hence also E) are finitely generated. We can therefore construct a commutative diagram†

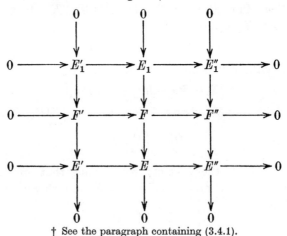

† See the paragraph containing (3.4.1).

with exact rows and columns, and where F', F, F'' are finite free modules. It is now enough to show that $\mathrm{Pd}_R^*(E_1) < \infty$. But

$$\max[\mathrm{Pd}_R^*(E_1'),\ \mathrm{Pd}_R^*(E_1'')] < \max[\mathrm{Pd}_R^*(E'),\ \mathrm{Pd}_R^*(E'')]$$

and so induction with respect to k gives the desired conclusion.

EXERCISE 8. *Let $0 \to E' \to E \to E'' \to 0$ be an exact sequence, where each of E', E, E'' possesses a finite free resolution of finite length. Show that $\mathrm{Char}_R(E) = \mathrm{Char}_R(E') + \mathrm{Char}_R(E'')$.*

Solution. We use induction on

$$k = \min[\mathrm{Pd}_R^*(E'),\ \mathrm{Pd}_R^*(E),\ \mathrm{Pd}_R^*(E'')].$$

First suppose that $k \leqslant 0$. Then we can find finite free modules F', F, F'', G', G, G'' such that $E' \oplus F' = G'$, $E \oplus F = G$, and $E'' \oplus F'' = G''$. From the exact sequence $0 \to F \to G \to E \to 0$ we obtain $\mathrm{Char}_R(E) = \mathrm{rank}_R(G) - \mathrm{rank}_R(F)$ and there are similar expressions for $\mathrm{Char}_R(E')$ and $\mathrm{Char}_R(E'')$. Next we can construct an exact sequence

$$0 \to E' \oplus F' \oplus F \to E \oplus F' \oplus F \oplus F'' \to E'' \oplus F'' \to 0,$$

i.e. an exact sequence

$$0 \to G' \oplus F \to G \oplus F' \oplus F'' \to G'' \to 0.$$

This shows that

$$\mathrm{rank}_R(G) + \mathrm{rank}_R(F') + \mathrm{rank}_R(F'')$$
$$= \mathrm{rank}_R(G') + \mathrm{rank}_R(G'') + \mathrm{rank}_R(F),$$

whence $\mathrm{Char}_R(E) = \mathrm{Char}_R(E') + \mathrm{Char}_R(E'')$. We have thus disposed of the case $k \leqslant 0$. The method used in the solution of Exercise 7 can now be easily adapted to enable the inductive step to be carried out.

EXERCISE 9. *Let R be an integral domain with the property that each ideal can be generated by a single element. Show that every submodule of a finite free module is a finite free module, and hence that every finitely generated module has a finite free resolution of length one. Deduce that an R-module is elementary if and only if it is finitely generated and has a non-zero annihilator.*

Solution. Let F be a free R-module with a finite base $e_1, e_2, \ldots, e_p$ and let K be a submodule of F. We shall show that K is a finite free module.

For $0 \leqslant j \leqslant p$ put

$$K_j = K \cap (Re_1 + Re_2 + \ldots + Re_j).$$

We then have

$$0 = K_0 \subseteq K_1 \subseteq K_2 \subseteq \ldots \subseteq K_p = K.$$

Suppose that $j \geqslant 1$ and let $x \in K_j$. Then $x = r_1 e_1 + r_2 e_2 + \ldots + r_j e_j$ for unique elements $r_1, r_2, \ldots, r_j$ in R. Define a homomorphism $f_j \colon K_j \to R$ by $f_j(x) = r_j$. We now have $\mathrm{Ker} f_j = K_{j-1}$, and $\mathrm{Im} f_j = \mathfrak{A}_j$ (say) is an ideal of R. Since R is an integral domain and $\mathfrak{A}_j$ is generated by a single element, either $\mathfrak{A}_j = 0$ or $\mathfrak{A}_j$ is isomorphic to R. In either event $\mathfrak{A}_j$ is a finite free module.

From the exact sequence

$$0 \to K_{j-1} \to K_j \to \mathfrak{A}_j \to 0$$

it follows, by Chapter 2 Exercise 6, that K_{j-1} is a direct summand of K_j, say

$$K_j = K_{j-1} \oplus F_j \quad (1 \leqslant j \leqslant p).$$

Here F_j is isomorphic to $\mathfrak{A}_j$ and so it is a finite free module. But now

$$K = K_p = F_p \oplus F_{p-1} \oplus \ldots \oplus F_1$$

and we have shown that K is also a finite free module.

Let E be a finitely generated R-module. We can construct an exact sequence $0 \to K \to F \to E \to 0$ where F is a finite free module and then, by the first part of the proof, K will be a finite free module as well. Accordingly E has a finite free resolution of length one. Now R is an *integral domain*. Consequently $\mathrm{Ann}_R(E)$ contains a non-zerodivisor if and only if $\mathrm{Ann}_R(E) \neq 0$. Thus if E is finitely generated and $\mathrm{Ann}_R(E) \neq 0$, then E is an elementary module. Of course, the converse of this latter statement is true by virtue of the more general considerations contained in the main text.

EXERCISE 10. *Let $\mathfrak{A}$ be an integral ideal of R. Show that $\mathfrak{A}$ is invertible if and only if it is both projective (as an R-module) and contains a non-zerodivisor.*

Solution. First suppose that $\mathfrak{A}$ is invertible. Then there exists a fractional ideal $\mathfrak{B}$ such that $\mathfrak{A}\mathfrak{B} = R$. Thus $1 \in \mathfrak{A}\mathfrak{B}$, say $1 = a_1 b_1 + a_2 b_2 + \ldots + a_m b_m$, where $a_i \in \mathfrak{A}$ and $b_i \in \mathfrak{B}$. Evidently $R a_1 + R a_2 + \ldots + R a_m \subseteq \mathfrak{A}$. Now let $a \in \mathfrak{A}$. Then

$$a = a_1(ab_1) + a_2(ab_2) + \ldots + a_m(ab_m) \in R a_1 + R a_2 + \ldots + R a_m$$

because $ab_j \in \mathfrak{A}\mathfrak{B} \subseteq R$. This shows that $\mathfrak{A} = R a_1 + R a_2 + \ldots + R a_m$. Define homomorphisms $\psi \colon R^m \to \mathfrak{A}$ and $\phi \colon \mathfrak{A} \to R^m$ by

$$\psi(r_1, r_2, \ldots, r_m) = r_1 a_1 + r_2 a_2 + \ldots + r_m a_m$$

and $$\phi(a) = (ab_1, ab_2, \ldots, ab_m).$$

Then $$\psi\phi(a) = ab_1 a_1 + ab_2 a_2 + \ldots + ab_m a_m = a$$

so $\psi\phi$ is the identity mapping of $\mathfrak{A}$. Accordingly ϕ is a mono-morphism and, by Chapter 2 Exercise 5, $\mathfrak{A}$ is isomorphic to a direct summand of R^m and therefore it is projective. Finally we can choose $\beta \in R$ so that β is not a zerodivisor and $\beta b_j \in R$ for $j = 1, 2, \ldots, m$. This ensures that

$$\beta = a_1(\beta b_1) + a_2(\beta b_2) + \ldots + a_m(\beta b_m) \in \mathfrak{A}.$$

We must now establish the converse. Assume therefore that $\mathfrak{A}$ is projective and that it contains a non-zerodivisor α. Construct an epimorphism $\omega \colon F \to \mathfrak{A}$, where F is a free R-module with a base $\{e_i\}_{i \in I}$ and put $\omega(e_i) = a_i$. Since $\mathfrak{A}$ is projective there exists an R-homomorphism $\theta \colon \mathfrak{A} \to F$ such that $\omega\theta$ is the identity mapping of $\mathfrak{A}$.

Let $a \in \mathfrak{A}$. Then $\theta(a) = \sum_{i \in I} r_i e_i$ for a unique family $\{r_i\}_{i \in I}$ of elements of R. Define an R-homomorphism $f_i \colon \mathfrak{A} \to R$ by $f_i(a) = r_i$. Then, for each $a \in \mathfrak{A}$, there is only a finite subset of I on which $f_i(a)$ is not zero; moreover

$$a = \omega\theta(a) = \sum_{i \in I} f_i(a) a_i.$$

Put $q_i = f_i(\alpha)/\alpha$. (This is an element of the full ring of fractions of R.) Then there exists a finite subset I_0, of I, such that $q_i = 0$ if $i \in I \setminus I_0$. Also, since

$$\alpha = \sum_{i \in I} f_i(\alpha) a_i,$$

we have
$$1 = \sum_{i \in I_0} a_i q_i.$$

Accordingly, if
$$\mathfrak{B} = \sum_{i \in I_0} R q_i,$$

then $\mathfrak{B}$ is a fractional ideal and $\mathfrak{A}\mathfrak{B} \supseteq R$. Lastly, for $a \in \mathfrak{A}$,

$$aq_i = \frac{af_i(\alpha)}{\alpha} = \frac{f_i(a\alpha)}{\alpha} = f_i(a) \in R$$

and this shows that $\mathfrak{A}\mathfrak{B} \subseteq R$. Thus in fact $\mathfrak{A}\mathfrak{B} = R$ and with this the solution is complete.

4. Stability and finite free resolutions

General remarks

We are now ready to begin the deeper study of finite free resolutions. The results that we shall obtain will enable us to provide solutions to some of the problems that have been left over from Chapter 3, but in the process we shall encounter certain new difficulties of an arithmetical nature. In order to deal with these it will be necessary to develop, from first principles, a general theory of *grade*. This is a topic to which Chapter 5 will be devoted.

Throughout the present chapter, R will denote a non-trivial commutative ring with an identity element. Since R is commutative, there is no essential difference between left R-modules and right R-modules. If therefore E is an R-module, $r \in R$ and $e \in E$, it is immaterial whether we write the product of r and e as re or as er. We shall take advantage of this fact.

4.1 Further results on matrices

In section (4.1) we shall be concerned with a $p \times q$ matrix A and a $q \times t$ matrix B (each with entries in R), where p, q, t denote positive integers. (Later we shall need to introduce conventions to cover situations in which the set of rows or columns, of one or other of these matrices, is empty. However it is convenient to avoid these minor complications for the time being.) Because A has q columns and B has q rows, it is possible to form the product AB. We shall be particularly concerned with cases where $AB = 0$.

Note that if S is a multiplicatively closed subset of R not containing zero and $AB = 0$, then $A_S B_S = 0$ as well. Here, of course A_S and B_S are the usual matrices† with entries in R_S that are derived from A and B.

EXERCISE 1. *Let E be an R-module and $\mathfrak{A}$ the ideal generated by the finite set of elements $\alpha_1, \alpha_2, \ldots, \alpha_p$. If now for each i $(1 \leqslant i \leqslant p)$ some power of α_i annihilates E, show that there exists a positive integer m such that $\mathfrak{A}^m E = 0$.*

THEOREM 1. *Let A be a $p \times q$ R-matrix and B a $q \times t$ R-matrix such that $AB = 0$. Further let $\mu \geqslant 0$, $\nu \geqslant 0$ be integers such that $\mu + \nu > q$. Then some power of the determinantal ideal $\mathfrak{A}_\mu(A)$ annihilates $\mathfrak{A}_\nu(B)$.*

Proof. We may suppose that $0 \leqslant \mu \leqslant \min(p, q)$ because otherwise $\mathfrak{A}_\mu(A) = 0$. Let D be a $\mu \times \mu$ minor of A. Since $\mathfrak{A}_\mu(A)$ is generated by a finite number of elements such as D, it is enough, in view of Exercise 1, to show that some power of D annihilates $\mathfrak{A}_\nu(B)$.

Assume the contrary. Then $\{D^n\}_{n \geqslant 0}$ is a multiplicatively closed subset of R not meeting $\operatorname{Ann}_R\{\mathfrak{A}_\nu(B)\}$. Accordingly, by Chapter 1 Theorem 10, there exists a prime ideal P such that $D \notin P$ and $\operatorname{Ann}_R\{\mathfrak{A}_\nu(B)\} \subseteq P$.

Let us localize at P and then, in order to simplify the notation, drop P as a suffix. By availing ourselves of (1.7.4) and Chapter 2 Theorem 12, we arrive at the following situation: R is a quasi-local ring, $AB = 0$, $\mathfrak{A}_\mu(A) = R$, and $\mathfrak{A}_\nu(B) \neq 0$ (because the annihilator of $\mathfrak{A}_\nu(B)$ is a proper ideal). To complete the proof it will suffice to derive a contradiction.

By Chapter 1 Theorem 12, there exist unimodular matrices U and V such that

$$UAV = \left\| \begin{array}{c|c} I_\mu & 0 \\ \hline 0 & A^* \end{array} \right\|$$

for a suitable matrix A^*, where I_μ denotes the identity matrix of order μ. By replacing A and B by UAV and $V^{-1}B$ respectively we may, without loss of generality, suppose that

$$A = \left\| \begin{array}{c|c} I_\mu & 0 \\ \hline 0 & A^* \end{array} \right\|.$$

† See (1.7.2).

But $AB = 0$. Consequently the first μ rows of B are composed of zeros. However B has q rows and $\mu + \nu > q$. It follows that $\mathfrak{A}_\nu(B) = 0$ and now we have the desired contradiction.

THEOREM 2. *Suppose that $AB = 0$, where A is a $p \times q$ matrix and B a $q \times t$ matrix each with entries in R. Further let $E \neq 0$ be an R-module. Then*

$$\mathrm{red.\,rank}_R (A, E) + \mathrm{rank}_R (B, E) \leqslant q.$$

Hence if A is stable relative to E we have

$$\mathrm{rank}_R (A, E) + \mathrm{rank}_R (B, E) \leqslant q.$$

Proof. Put $\mu = \mathrm{red.\,rank}_R (A, E)$, $\nu = \mathrm{rank}_R (B, E)$ and assume that $\mu + \nu > q$. By Theorem 1, there exists a positive integer m such that $\mathfrak{A}_\mu^m(A)\,\mathfrak{A}_\nu(B) = 0$, and therefore

$$\mathfrak{A}_\mu^m(A)\{\mathfrak{A}_\nu(B)\,E\} = 0.$$

However, this implies that $\mathfrak{A}_\nu(B)\,E = 0$ because $\mathfrak{A}_\mu(A)$ does not annihilate any non-zero element of E. Now we cannot have $\mathfrak{A}_\nu(B)\,E = 0$ because $\nu = \mathrm{rank}_R (B, E)$. Thus we have arrived at a contradiction and the theorem is proved.

If E is an R-module and $p \geqslant 1$ is an integer we put

$$E^p = E \oplus E \oplus \ldots \oplus E \quad (p \text{ summands}). \tag{4.1.1}$$

(This is an extension of the notation R^p that was introduced in section (2.2).) Suppose now that A is a $p \times q$ R-matrix. Then A gives rise to a homomorphism $E^p \to E^q$ in which $(e_1, e_2, \ldots, e_p)$ is mapped into

$$\left(\sum_{j=1}^{p} a_{j1} e_j, \ \sum_{j=1}^{p} a_{j2} e_j, \ \ldots, \ \sum_{j=1}^{p} a_{jq} e_j \right);$$

in other terms $E^p \to E^q$ is the mapping in which

$$(e_1, e_2, \ldots, e_p) \mapsto (e_1, e_2, \ldots, e_p)\,A, \tag{4.1.2}$$

where we have used the fact that if $r \in R$ and $e \in E$ then their product can be written either as re or as er.

Next let B be a $q \times t$ R-matrix. This will give rise to a homomorphism $E^q \to E^t$. Observe that if $AB = 0$, then the result of combining $E^p \to E^q$ with $E^q \to E^t$ will be the null homomorphism of E^p into E^t.

DEFINITION. *We shall say that '$AB = 0$ exactly on E' if* (i) $AB = 0$, *and* (ii) *for every $\eta \in E^q$ such that $\eta B = 0$ there exists $\xi \in E^p$ such that $\xi A = \eta$.*

EXERCISE 2. *Suppose that $AB = 0$ exactly on the R-module E and that S is a multiplicatively closed subset of R not containing zero. Show that $A_S B_S = 0$ exactly on the R_S-module E_S.*

LEMMA 1. *Suppose that $AB = 0$ exactly on E ($E \neq 0$), where A is a $p \times q$ matrix and B a $q \times t$ matrix. Put $\mu = \operatorname{rank}_R (A, E)$ and let $\nu \geqslant 0$ be an integer such that $\mu + \nu < q$. If now*

$$e \in E \quad and \quad \mathfrak{A}_{\nu+1}(B)\, e = 0,$$

then some power of $\mathfrak{A}_\mu(A)$ annihilates $\mathfrak{A}_\nu(B)\, e$.

Proof. Let D be a $\mu \times \mu$ minor of A. Then, by Exercise 1, it is enough to show that some power of D annihilates $\mathfrak{A}_\nu(B)\, e$. We shall assume the contrary and derive a contradiction.

In the situation envisaged the multiplicatively closed set $\{D^n\}_{n \geqslant 0}$ does not meet $\operatorname{Ann}_R \{\mathfrak{A}_\nu(B)\, e\}$. Consequently, by Chapter 1 Theorem 10, there exists a prime ideal P such that $D \notin P$ and

$$\operatorname{Ann}_R \{\mathfrak{A}_\nu(B)\, e\} \subseteq P.$$

Now $\mathfrak{A}_\nu(B)\, e$ is a finitely generated R-module. Accordingly, by Chapter 2 Theorem 13, P belongs to its support and therefore $(\mathfrak{A}_\nu(B)\, e)_P \neq 0$. This implies, among other things, that $E_P \neq 0$. Note that $\mathfrak{A}_\mu(A_P) = R_P$ and therefore $\mathfrak{A}_\mu(A_P)\, E_P \neq 0$. Thus

$$\operatorname{rank}_{R_P}(A_P, E_P) \geqslant \mu = \operatorname{rank}_R (A, E).$$

However, by (3.2.11), we have the opposite inequality. Thus $\operatorname{rank}_{R_P}(A_P, E_P) = \mu$. Let us observe, before proceeding, that $e/1$ is a non-zero element of E_P.

After these preliminaries we are ready to localize at P and subsequently drop P as a suffix. We then find ourselves in the following situation: R is a quasi-local ring, $AB = 0$ exactly on the non-zero R-module E, $\mu = \operatorname{rank}_R (A, E)$ and $\mathfrak{A}_\mu(A) = R$; further $\mu + \nu < q$ and E contains an element e such that $\mathfrak{A}_\nu(B)\, e \neq 0$ but $\mathfrak{A}_{\nu+1}(B)\, e = 0$. To complete the proof we shall derive a contradiction.

Because R is quasi-local and $\mathfrak{A}_\mu(A) = R$, it follows, from Chapter 1 Theorem 12, that there exist unimodular matrices U, V such that UAV has the form

$$UAV = \left\| \begin{array}{c|c} I_\mu & 0 \\ \hline 0 & A^* \end{array} \right\|,$$

where, as usual, I_μ denotes the identity matrix of order μ. However it is easily verified that $(UAV)(V^{-1}B) = 0$ exactly on E. We may therefore replace A and B by UAV and $V^{-1}B$ respectively and thus arrange that

$$A = \left\| \begin{array}{c|c} I_\mu & 0 \\ \hline 0 & A^* \end{array} \right\|.$$

But $\mathrm{rank}_R(A, E) = \mu$. Consequently each entry in the matrix A^* annihilates E.

Since $\mu + \nu < q$ and $\mathfrak{A}_\nu(B)e \neq 0$, we have both $\nu < q$ and $\nu \leqslant t$. Let $M \in S_{\nu+1}^q$, $K \in S_\nu^t$ and define x_{MK} as in (1.5.1) save that on this occasion we use B and ν where previously we had A and μ. Then $(ex_{MK})B = 0$ because, by (1.5.2), the components of $x_{MK}B$ belong to $\mathfrak{A}_{\nu+1}(B)$ and we know that $\mathfrak{A}_{\nu+1}(B)e = 0$. It follows, because $AB = 0$ exactly on E, that there exists $(\xi_1, \xi_2, \ldots, \xi_p)$ in E^p such that
$$(\xi_1, \xi_2, \ldots, \xi_p)A = ex_{MK},$$

and therefore
$$(\xi_1, \ldots, \xi_\mu, 0, \ldots, 0) = ex_{MK}. \tag{4.1.3}$$

To complete the proof let $J \in S_\nu^q$ and (as before) $K \in S_\nu^t$. Since $\nu < q - \mu$ it is possible to find j so that $\mu + 1 \leqslant j \leqslant q$ and $j \notin J$. Let M in $S_{\nu+1}^q$ be obtained by adjoining j to J. Then, by (4.1.3),

$$0 = e(x_{MK})_j = \pm eB_{JK}^{(\nu)}.$$

But $B_{JK}^{(\nu)}$ is an arbitrary $\nu \times \nu$ minor of B. Thus it has been shown that $\mathfrak{A}_\nu(B)e = 0$ and this is the contradiction we have been seeking.

THEOREM 3. *Let A be a $p \times q$ matrix and B a $q \times t$ matrix such that $AB = 0$ exactly on the non-zero R-module E. If now A is stable relative to E, then B is also stable relative to E and*

$$\mathrm{rank}_R(A, E) + \mathrm{rank}_R(B, E) = q.$$

Proof. Put $\mu = \operatorname{rank}_R(A, E)$ and $\nu = \operatorname{red.rank}_R(B, E)$. Then, by Theorem 2,
$$\mu + \nu \leqslant \mu + \operatorname{rank}_R(B, E) \leqslant q.$$

Thus it will suffice to show that $\mu + \nu = q$. We shall assume that $\mu + \nu < q$ and derive a contradiction.

By the definition of $\operatorname{red.rank}_R(B, E)$, there exists $e \in E$ such that $e \neq 0$ and $\mathfrak{A}_{\nu+1}(B) e = 0$. Consequently, by Lemma 1, some power of $\mathfrak{A}_{\mu}(A)$ annihilates $\mathfrak{A}_{\nu}(B) e$. But $0:_E \mathfrak{A}_{\mu}(A) = 0$ because A is stable relative to E, and thus we see that $\mathfrak{A}_{\nu}(B) e = 0$. However, this is a contradiction since, by the definition of reduced rank, $\mathfrak{A}_{\nu}(B)$ does not annihilate any non-zero element of E.

4.2 Stable homomorphisms

We shall now restate some of the results of section (4.1) in a form which is more convenient for dealing with finite free resolutions.

To this end let $\phi \colon F \to G$ be a homomorphism, where F and G are finite free modules. For the moment it will be assumed that $F \neq 0$ and $G \neq 0$. Select bases $x_1, x_2, \ldots, x_p$ for F and $y_1, y_2, \ldots, y_q$ for G, and put

$$\phi(x_j) = \sum_{k=1}^{q} a_{jk} y_k \quad (1 \leqslant j \leqslant p), \qquad (4.2.1)$$

where $a_{jk} \in R$. The matrix $A = \|a_{jk}\|$ will be referred to as the *matrix of ϕ relative to the bases* $x_1, x_2, \ldots, x_p$ and $y_1, y_2, \ldots, y_q$. If we choose other bases for F and G the effect on A is simply to replace it by an equivalent matrix A' say. However, by Chapter 1 Theorem 4, $\mathfrak{A}_{\nu}(A) = \mathfrak{A}_{\nu}(A')$ for all $\nu \geqslant 0$. This allows us to put

$$\mathfrak{A}_{\nu}(\phi) = \mathfrak{A}_{\nu}(A). \qquad (4.2.2)$$

It then follows that

$$\mathfrak{A}_{\nu}(\phi) = 0 \quad \text{for} \quad \nu > \min\left[\operatorname{rank}_R(F), \operatorname{rank}_R(G)\right] \quad (4.2.3)$$

and
$$R = \mathfrak{A}_0(\phi) \supseteq \mathfrak{A}_1(\phi) \supseteq \mathfrak{A}_2(\phi) \supseteq \ldots \qquad (4.2.4)$$

by (1.4.1) and (1.4.2) respectively.

For an R-module $E \neq 0$ set

$$\operatorname{rank}_R(\phi, E) = \text{largest } \nu \text{ such that } \mathfrak{A}_{\nu}(\phi) E \neq 0, \qquad (4.2.5)$$

and

$$\text{red.rank}_R(\phi, E) = \text{largest } \nu \text{ such that } 0:_E \mathfrak{A}_\nu(\phi) = 0. \quad (4.2.6)$$

This ensures that

$$\text{rank}_R(\phi, E) = \text{rank}_R(A, E), \quad (4.2.7)$$

$$\text{red.rank}_R(\phi, E) = \text{red.rank}_R(A, E), \quad (4.2.8)$$

and, by (3.2.5),

$$0 \leqslant \text{red.rank}_R(\phi, E) \leqslant \text{rank}_R(\phi, E) \leqslant \text{rank}_R(\phi, R)$$
$$\leqslant \min[\text{rank}_R(F), \text{rank}_R(G)]. \quad (4.2.9)$$

Naturally $\phi: F \to G$ is said to be *stable relative to E* if

$$\text{red.rank}_R(\phi, E)$$

is equal to $\text{rank}_R(\phi, E)$, and this happens precisely when A is stable relative to E. In the special case where ϕ is stable relative to R (equivalently when A is stable) we simplify the language and say merely that $\phi: F \to G$ is *stable*. Since this is an important case we also simplify the notation for dealing with this situation and put

$$\text{rank}_R(\phi) = \text{rank}_R(\phi, R). \quad (4.2.10)$$

It is now time to get rid of the assumption that neither F nor G is the zero module. Suppose then that either $F = 0$ or $G = 0$ (or both). In this case we put $\mathfrak{A}_0(\phi) = R$ and $\mathfrak{A}_\nu(\phi) = 0$ for all $\nu > 0$. Then (4.2.3) and (4.2.4) hold without exceptions, (4.2.5) and (4.2.6) are just definitions with slightly wider scopes, and the inequalities (4.2.9) remain valid. Indeed we can regard (4.2.2), (4.2.7) and (4.2.8) as holding as well provided that for a matrix A with an empty set of rows or columns we put

$$\mathfrak{A}_\nu(A) = \begin{cases} R & \text{for } \nu = 0, \\ 0 & \text{for } \nu > 0, \end{cases}$$

and define $\text{rank}_R(A, E)$ and $\text{red.rank}_R(A, E)$ exactly as in the non-degenerate case. For future reference we note the following: *every null homomorphism $\phi: F \to G$ between an* ARBITRARY *pair of finite free modules is stable of rank zero relative to any non-zero R-module whatsoever.*

Once again suppose momentarily that $F \neq 0$ and $G \neq 0$ and let S be a multiplicatively closed subset of R such that $E_S \neq 0$. Select a base $x_1, x_2, \ldots, x_p$ for F and a base $y_1, y_2, \ldots, y_q$ for G. Then, by Chapter 2 Theorem 9, $x_1/1, x_2/1, \ldots, x_p/1$ and $y_1/1, y_2/1, \ldots, y_q/1$ are bases for the free R_S-modules F_S and G_S. Now if the matrix of $\phi \colon F \to G$ with respect to the chosen bases is A, the matrix of the R_S-homomorphism $\phi_S \colon F_S \to G_S$ will be A_S with respect to the corresponding bases of F_S and G_S. It therefore follows, from (3.2.11) and (3.2.12), that

$$\operatorname{rank}_{R_S}(\phi_S, E_S) \leqslant \operatorname{rank}_R(\phi, E) \qquad (4.2.11)$$

and

$$\operatorname{red.rank}_R(\phi, E) \leqslant \operatorname{red.rank}_{R_S}(\phi_S, E_S), \qquad (4.2.12)$$

and it is clear that (4.2.11) and (4.2.12) continue to hold even if the conditions $F \neq 0$ and $G \neq 0$ are dropped. Hence *if $\phi \colon F \to G$ is a homomorphism between arbitrary finite free modules which is stable relative to E, and S is a multiplicatively closed subset of R such that $E_S \neq 0$, then $\phi_S \colon F_S \to G_S$ is stable relative to the R_S-module E_S and $\operatorname{rank}_{R_S}(\phi_S, E_S) = \operatorname{rank}_R(\phi, E)$.*

Let us suppose once more that $F \neq 0$, $G \neq 0$ and let

$$x_1, x_2, \ldots, x_p, A \text{ and } y_1, y_2, \ldots, y_q$$

have the same meanings as in the last paragraph. There exists an isomorphism

$$F \otimes_R E \approx E^p \qquad (4.2.13)$$

of R-modules in which the element $\sum_{i=1}^{p} (x_i \otimes e_i)$ of $F \otimes_R E$ is matched with $(e_1, e_2, \ldots, e_p)$ of E^p. We shall use (4.2.13) to identify $F \otimes_R E$ with E^p. Now $\phi \colon F \to G$ induces a homomorphism

$$\phi \otimes E \colon F \otimes_R E \to G \otimes_R E \qquad (4.2.14)$$

in which $x \otimes e$ in $F \otimes_R E$ is mapped into $\phi(x) \otimes e$ in $G \otimes_R E$. Hence if we identify $F \otimes_R E$ with E^p and $G \otimes_R E$ with E^q, then (4.2.14) becomes the homomorphism $E^p \to E^q$ previously encountered in (4.1.2).

T H E O R E M 4. *Suppose that $\phi \colon F > G$ is a homomorphism, where F, G are arbitrary finite free R-modules, and let $E \neq 0$ be an R-module. Then the sequence*

$$0 \to F \otimes_R E \xrightarrow{\phi \otimes E} G \otimes_R E$$

is exact if and only if both (i) ϕ *is stable relative to* E, *and* (ii) $\mathrm{rank}_R(\phi, E) = \mathrm{rank}_R(F)$.

Proof. We can suppose that ϕ is not a null homomorphism because otherwise everything follows trivially. In particular we may assume that $F \neq 0$ and $G \neq 0$.

Choose bases for F and G, let $A = \|a_{jk}\|$ be the matrix of ϕ with respect to these bases, and identify $F \otimes_R E$ and $G \otimes_R E$ with E^p and E^q respectively, where $p = \mathrm{rank}_R(F)$ and $q = \mathrm{rank}_R(G)$. Then $\phi \otimes E$ corresponds to a certain homomorphism† $E^p \to E^q$. It follows that $\phi \otimes E$ is a monomorphism if and only if the equations

$$\sum_{j=1}^{p} e_j a_{jk} = 0 \quad (1 \leqslant k \leqslant q)$$

have only the trivial solution in E. Now by Chapter 3 Theorem 8 for this to happen it is necessary and sufficient that $0:_E \mathfrak{A}_p(A) = 0$. However $0:_E \mathfrak{A}_p(A) = 0$ implies that

$$\mathrm{red}.\mathrm{rank}_R(A, E) = \mathrm{rank}_R(A, E) = p$$

and, of course, conversely. The theorem follows.

Let $\phi\colon F \to G$ and $\psi\colon G \to H$ be homomorphisms, where F, G, H are finite free modules, and let us choose bases $\{x_1, x_2, ..., x_p\}$, $\{y_1, y_2, ..., y_q\}$ and $\{z_1, z_2, ..., z_t\}$ for F, G and H respectively. Denote the matrix of ϕ with respect to $\{x_1, x_2, ..., x_p\}$ and $\{y_1, y_2, ..., y_q\}$ by A and that of ψ with respect to $\{y_1, y_2, ..., y_q\}$ and $\{z_1, z_2, ..., z_t\}$ by B. Then AB is the matrix of the composite homomorphism

$$F \xrightarrow{\phi} G \xrightarrow{\psi} H$$

with respect to $\{x_1, x_2, ..., x_p\}$ and $\{z_1, z_2, ..., z_t\}$. Accordingly $AB = 0$ if and only if $\psi\phi = 0$. Now let E be an R-module. It is easily verified that $AB = 0$ *exactly on* E *if and only if* $\psi\phi = 0$ *and the sequence*

$$F \otimes_R E \xrightarrow{\phi \otimes E} G \otimes_R E \xrightarrow{\psi \otimes E} H \otimes_R E$$

is exact.

We are now ready to reformulate and slightly extend Theorem 3.

<hr>

† See (4.1.2).

THEOREM 5. *Let $\phi\colon F \to G$ and $\psi\colon G \to H$ be homomorphisms such that $\psi\phi = 0$, where F, G, H are finite free modules. Further let $E \neq 0$ be an R-module. If now ϕ is stable relative to E and the sequence*

$$F \otimes_R E \xrightarrow{\phi \otimes E} G \otimes_R E \xrightarrow{\psi \otimes E} H \otimes_R E$$

is exact, then ψ is also stable relative to E and

$$\operatorname{rank}_R(\phi, E) + \operatorname{rank}_R(\psi, E) = \operatorname{rank}_R(G).$$

Proof. If $F \neq 0$, $G \neq 0$ and $H \neq 0$, then Theorem 5 follows from Theorem 3 on taking bases for F, G, H and representing ϕ and ψ by means of matrices. When $F = 0$ Theorem 5 is a consequence of Theorem 4 and if $G = 0$ then ϕ, ψ are null homomorphisms and our conventions immediately yield the desired conclusion.

Finally suppose that $F \neq 0$, $G \neq 0$ but $H = 0$. Then $\phi \otimes E$ is an epimorphism and we have only to prove that

$$\operatorname{rank}_R(\phi, E) = \operatorname{rank}_R(G).$$

Let H' be a non-zero free R-module and $\psi'\colon G \to H'$ the null homomorphism. This ensures that $\psi'\phi = 0$ and that the sequence

$$F \otimes_R E \xrightarrow{\phi \otimes E} G \otimes_R E \xrightarrow{\psi' \otimes E} H' \otimes_R E$$

is exact. Consequently $\operatorname{rank}_R(\phi, E) = \operatorname{rank}_R(G)$ by the case that was first considered.

There is an important special case of Theorem 5 that it is worth-while recording separately. We recall that if C is an R-module, then $C \otimes_R R$ and C are isomorphic R-modules under an isomorphism which matches the element $c \otimes r$ of the former with the element rc of the latter. We use this isomorphism to identify $C \otimes_R R$ and C. Suppose now that we have a homomorphism $f\colon C \to D$. This induces a homomorphism

$$f \otimes R\colon C \otimes_R R \to D \otimes_R R.$$

The point to be noted is that if we identify $C \otimes_R R$ with C and $D \otimes_R R$ with D, then $f \otimes R$ reverts to the original homomorphism $f\colon C \to D$.

In view of these observations, the next theorem arises from Theorem 5 by taking $E = R$.

THEOREM 6. *Let* $F \xrightarrow{\phi} G \xrightarrow{\psi} H$ *be an exact sequence of R-modules, where F, G, H are free modules with finite bases. If now ϕ is stable, then ψ is also stable and*

$$\operatorname{rank}_R(\phi) + \operatorname{rank}_R(\psi) = \operatorname{rank}_R(G).$$

There is one final remark to be made before we leave this section and start to apply our results. Suppose that

$$F \xrightarrow{\phi} G \to M \to 0$$

is an exact sequence, where F and G are finite free modules, and let $q = \operatorname{rank}_R(G)$. *We claim that*, with our previous notation for Fitting invariants,

$$\mathfrak{F}_\mu(M) = \begin{cases} \mathfrak{A}_{q-\mu}(\phi) & \text{for } 0 \leqslant \mu \leqslant q, \\ R & \text{for } \mu > q. \end{cases} \qquad (4.2.15)$$

Indeed if ϕ is a null homomorphism this is obvious because of our conventions and if ϕ is non-null, then $F \neq 0$, $G \neq 0$ and the result follows from (3.1.13) on introducing bases.

EXERCISE 3. *Let $\psi: F \to G$ be an epimorphism, where F and G are finite free modules. Show that ψ is stable and that*

$$\operatorname{rank}_R(\psi) = \operatorname{rank}_R(G).$$

4.3 Further results on Euler characteristics

A sequence

$$\ldots \to C_{n+1} \xrightarrow{d_{n+1}} C_n \xrightarrow{d_n} C_{n-1} \xrightarrow{d_{n-1}} C_{n-2} \to \ldots \qquad (4.3.1)$$

composed of R-modules and R-homomorphisms is called a *complex* if the product of any two consecutive homomorphisms is zero. Thus (4.3.1) is a complex if $d_n d_{n+1} = 0$ for all relevant values of n. According to some definitions a complex is required to extend indefinitely in both directions, but on the present occasion we shall allow complexes to be of finite length. If the sequence (4.3.1) is exact, then we shall say that it constitutes an *exact complex*.

Let (4.3.1) be a complex and let us denote it by $\mathbb{C}$. If now E is an R-module, then by $\mathbb{C} \otimes_R E$ is meant the complex

$$\ldots \to C_{n+1} \otimes_R E \to C_n \otimes_R E \to C_{n-1} \otimes_R E \to C_{n-2} \otimes_R E \to \ldots,$$

where $C_n \otimes_R E \to C_{n-1} \otimes_R E$ is the homomorphism $d_n \otimes E$. Naturally we may identify $\mathbb{C} \otimes_R R$ with $\mathbb{C}$ itself.

THEOREM 7. *Suppose that*

$$0 \to F_n \xrightarrow{\phi_n} F_{n-1} \to \ldots \to F_1 \xrightarrow{\phi_1} F_0 \xrightarrow{\epsilon} M \to 0,$$

where $n \geqslant 1$, is a complex, that $F_0, F_1, \ldots, F_n$ are finite free modules, and that the sequence

$$F_1 \xrightarrow{\phi_1} F_0 \xrightarrow{\epsilon} M \to 0$$

is exact. Suppose further that $E \neq 0$ is an R-module and that

$$0 \to F_n \otimes_R E \to F_{n-1} \otimes_R E \to \ldots \to F_1 \otimes_R E$$
$$\to F_0 \otimes_R E \to M \otimes_R E \to 0$$

is exact. Then

(a) $\phi_1, \phi_2, \ldots, \phi_n$ *are all stable relative to E;*

(b) $\mathrm{rank}_R(F_n) = \mathrm{rank}_R(\phi_n, E)$;

(c) $\mathrm{rank}_R(F_m) = \mathrm{rank}_R(\phi_{m+1}, E) + \mathrm{rank}_R(\phi_m, E)$ *for* $m = 1, 2, \ldots, n-1$.

Proof. By Theorem 4, ϕ_n is stable relative to E and

$$\mathrm{rank}_R(\phi_n, E) = \mathrm{rank}_R(F_n).$$

Now suppose that $n \geqslant m > 1$ and that we have shown that ϕ_m is stable relative to E. Since

$$F_m \otimes_R E \to F_{m-1} \otimes_R E \to F_{m-2} \otimes_R E$$

is exact, it follows, from Theorem 5, that ϕ_{m-1} is stable relative to E and that

$$\mathrm{rank}_R(\phi_m, E) + \mathrm{rank}_R(\phi_{m-1}, E) = \mathrm{rank}_R(F_{m-1}).$$

This establishes the theorem.

COROLLARY 1. *Let the hypotheses be as in Theorem 7. Then*

$$\sum_{\nu=0}^{n} (-1)^{\nu} \mathrm{rank}_R(F_\nu) = \mathrm{rank}_R(F_0) - \mathrm{rank}_R(\phi_1, E) \geqslant 0.$$

Proof. That $\sum_{\nu=0}^{n} (-1)^{\nu} \mathrm{rank}_R(F_\nu)$ has the value stated follows from (*b*) and (*c*) of the theorem. Since

$$\operatorname{rank}_R(\phi_1, E) \leqslant \min\left[\operatorname{rank}_R(F_0), \operatorname{rank}_R(F_1)\right]$$

by (4.2.9), the other assertion is immediate as well.

COROLLARY 2. *Let the hypotheses be as in Theorem* 7 *and put*

$$k = \sum_{\nu=0}^{n} (-1)^{\nu} \operatorname{rank}_R(F_\nu)$$

so that $k \geqslant 0$ *by Corollary* 1. *Then, with the usual notation for Fitting invariants,*

$$\mathfrak{F}_\mu(M)E = 0 \quad for \quad 0 \leqslant \mu < k$$

and $\mathfrak{F}_k(M)E \neq 0$. *Indeed* $\mathfrak{F}_k(M)$ *has a stronger property namely* $0:_E \mathfrak{F}_k(M) = 0$.

Proof. Put $q = \operatorname{rank}_R(F_0)$ so that, by Corollary 1,

$$\operatorname{rank}_R(\phi_1, E) = q - k$$

and suppose that $0 \leqslant \mu \leqslant k$. Then $\mu \leqslant q$ and, by (4.2.15), $\mathfrak{F}_\mu(M) = \mathfrak{A}_{q-\mu}(\phi_1)$. Thus

$$\mathfrak{F}_\mu(M)E = \mathfrak{A}_{q-\mu}(\phi_1)E$$

and this, by virtue of the definition of $\operatorname{rank}_R(\phi_1, E)$, will be zero when $\mu < k$ but non-zero when $\mu = k$. Finally, since ϕ_1 is stable relative to E and its rank relative to E is $q - k$, we have

$$0:_E \mathfrak{A}_{q-k}(\phi_1) = 0$$

that is to say $0:_E \mathfrak{F}_k(M) = 0$.

We recall that $\mathfrak{F}(M)$ denotes the invariant $\mathfrak{F}_0(M)$.

COROLLARY 3. *Let the assumptions be as in Theorem* 7. *Then the following two statements are equivalent:*

(i) $\displaystyle\sum_{\nu=0}^{n} (-1)^{\nu} \operatorname{rank}_R(F_\nu) = 0$;

(ii) $\mathfrak{F}(M)E \neq 0$.

Further when (i) *and* (ii) *hold we have* $0:_E \mathfrak{F}(M) = 0$.

This follows at once from Corollary 2.

Our next result is obtained by taking $E = R$ in the last theorem and its corollaries.

THEOREM 8. *Suppose that $n \geqslant 1$ and that*

$$0 \to F_n \xrightarrow{\phi_n} F_{n-1} \to \dots \to F_1 \xrightarrow{\phi_1} F_0 \to M \to 0$$

is a finite free resolution of the R-module M. Then

(a) $\phi_1, \phi_2, \dots, \phi_n$ *are all stable;*

(b) $\mathrm{rank}_R(F_n) = \mathrm{rank}_R(\phi_n)$;

(c) $\mathrm{rank}_R(F_m) = \mathrm{rank}_R(\phi_{m+1}) + \mathrm{rank}_R(\phi_m)$ *for $1 \leqslant m < n$;*

(d) $\mathrm{Char}_R(M) = \mathrm{rank}_R(F_0) - \mathrm{rank}_R(\phi_1) \geqslant 0$.

Moreover if $k = \mathrm{Char}_R(M)$, then

$$\mathfrak{F}_\mu(M) = \begin{cases} 0 & for \quad 0 \leqslant \mu < k, \\ \neq 0 & for \quad \mu = k, \end{cases}$$

and $0 :_R \mathfrak{F}_k(M) = 0$.

COROLLARY. *Let the R-module M have a finite free resolution of finite length. Then the following two statements are equivalent:*

(i) $\mathrm{Char}_R(M) = 0$;

(ii) $\mathfrak{F}(M) \neq 0$.

Moreover when (i) *and* (ii) *hold we have $0 :_R \mathfrak{F}(M) = 0$.*

EXERCISE 4. *The R-module M can be generated by q elements and it has a finite free resolution of finite length. Show that $\mathrm{Char}_R(M) \leqslant q$.*

Suppose that the R-module M has a finite free resolution of finite length. It was a matter of conjecture for some time that $\mathrm{Char}_R(M) = 0$ if and only if $\mathrm{Ann}_R(M) \neq 0$. This is in fact true as has been shown by W. V. Vasconcelos (44). The purpose of the apparent digression which now follows is to provide the auxiliary theorems needed to prove this result. However the information which we shall obtain in the process will also prove useful in other contexts at a later stage.

EXERCISE 5. *Let $\{P_i\}_{i \in I}$ be a non-empty family of prime ideals of R with the property that if $i_1, i_2 \in I$ then either $P_{i_1} \subseteq P_{i_2}$ or $P_{i_2} \subseteq P_{i_1}$. Show that the intersection of the family is itself a prime ideal.*

Let $\mathfrak{A}$ be an ideal of R and P a prime ideal.

DEFINITION. *We say that P is a 'minimal prime ideal of $\mathfrak{A}$' if* (i) $\mathfrak{A} \subseteq P$ *and* (ii) *there is no prime ideal strictly contained in P which also contains $\mathfrak{A}$.*

THEOREM 9. *Let $\mathfrak{A}$ be an ideal of R and P a prime ideal containing $\mathfrak{A}$. Then there exists a minimal prime ideal P' of $\mathfrak{A}$ such that $\mathfrak{A} \subseteq P' \subseteq P$.*

Proof. Let Σ consist of all the prime ideals of R that contain $\mathfrak{A}$ and are contained in P. Then $P \in \Sigma$ and therefore Σ is not empty. We now introduce a partial order $\leqslant$ on Σ as follows: if $P_1, P_2 \in \Sigma$ then $P_1 \leqslant P_2$ precisely when $P_2 \subseteq P_1$. (Thus our partial order is the reverse of inclusion.) It then follows, from Exercise 5, that Σ is an inductive system, and hence, by Zorn's Lemma, it possesses a maximal element P' say. Since $P' \in \Sigma$, we have $\mathfrak{A} \subseteq P' \subseteq P$ and, since P' is maximal in Σ, it must be a minimal prime ideal of $\mathfrak{A}$. Thus the theorem is proved.

Let $\mathfrak{A}$ be an ideal of R and put

$$\operatorname{Rad} \mathfrak{A} = \{r \mid r \in R \text{ and } r^m \in \mathfrak{A} \text{ for some } m > 0\}. \quad (4.3.2)$$

It is usual to call $\operatorname{Rad} \mathfrak{A}$ the *radical* of $\mathfrak{A}$. The next theorem shows, among other things, that $\operatorname{Rad} \mathfrak{A}$ is an ideal. This, of course, can easily be verified directly.

THEOREM 10. *Let $\mathfrak{A}$ be an ideal of R. Then $\operatorname{Rad} \mathfrak{A}$ is the intersection of all the minimal prime ideals of $\mathfrak{A}$.*

Remark. If $\mathfrak{A} = R$, then, of course, $\mathfrak{A}$ has no minimal prime ideals. We deal with this situation by adopting the convention that the intersection of any empty set of ideals (of R) is R itself.

Proof. If $\alpha \in \operatorname{Rad} \mathfrak{A}$, then there exists an integer m such that $\alpha^m \in \mathfrak{A}$. Hence α^m belongs to every minimal prime ideal of $\mathfrak{A}$ and therefore α itself has this property.

Now suppose that $\beta \in R$ and $\beta \notin \operatorname{Rad} \mathfrak{A}$. Then $\{\beta^n\}_{n \geqslant 0}$ is a multiplicatively closed subset of R not meeting $\mathfrak{A}$ and therefore, by Chapter 1 Theorem 10, there exists a prime ideal P such that $\mathfrak{A} \subseteq P$ and $\beta \notin P$. It follows, by Theorem 9, that there exists a minimal prime ideal P', of $\mathfrak{A}$, such that $P' \subseteq P$. Thus $\beta \notin P'$ and the theorem follows.

EXERICSE 6. *Let $\mathfrak{A}_1, \mathfrak{A}_2, ..., \mathfrak{A}_n$ be ideals of R. Show that*

$$\operatorname{Rad} (\mathfrak{A}_1 \mathfrak{A}_2 ... \mathfrak{A}_n) = \operatorname{Rad} (\mathfrak{A}_1 \cap \mathfrak{A}_2 \cap ... \cap \mathfrak{A}_n)$$
$$= \operatorname{Rad} \mathfrak{A}_1 \cap \operatorname{Rad} \mathfrak{A}_2 \cap ... \cap \operatorname{Rad} \mathfrak{A}_n.$$

Let $\mathfrak{A}$ and $\mathfrak{B}$ be ideals of R. If $\mathfrak{A} \subseteq \mathfrak{B}$, then, of course,

$$\mathrm{Rad}\,\mathfrak{A} \subseteq \mathrm{Rad}\,\mathfrak{B}.$$

Again, by Exercise 6, if $m > 0$ is an integer then

$$\mathrm{Rad}\,(\mathfrak{A}^m) = \mathrm{Rad}\,\mathfrak{A}.$$

Now suppose that E is an R-module that can be generated by q elements. We know, from Chapter 3 Theorem 5, that

$$[\mathrm{Ann}_R\,(E)]^q \subseteq \mathfrak{F}(E) \subseteq \mathrm{Ann}_R\,(E).$$

It follows that

$$\mathrm{Rad}\,(\mathrm{Ann}_R\,(E)) = \mathrm{Rad}\,(\mathfrak{F}(E)). \qquad (4.3.3)$$

EXERCISE 7. *Let $\mathfrak{A}$ be an ideal of R and $\mathfrak{B}$ a finitely generated ideal contained in $\mathrm{Rad}\,\mathfrak{A}$. Show that there exists a positive integer m such that $\mathfrak{B}^m \subseteq \mathfrak{A}$.*

EXERCISE 8. *Let R be a quasi-local ring with maximal ideal P and let M be a finitely generated R-module. Show that there exists an epimorphism $\omega \colon F \to M$, where F is a finite free module and $\mathrm{Ker}\,\omega \subseteq PF$.*

THEOREM 11. *Let R be a quasi-local ring with maximal ideal P and suppose that for every finitely generated ideal $I \subseteq P$ we have $0 :_R I \neq 0$. If now M is an R-module with a finite free resolution of finite length, then M is free.*

Proof. Put $k = \mathrm{Pd}_R^*\,(M)$. If $k \leqslant 0$, then M is projective and hence free because R is a quasi-local ring. We shall therefore suppose that $k \geqslant 1$ and derive a contradiction.

Assume, for the moment, that $k > 1$ and construct an exact sequence $0 \to M' \to F \to M \to 0$, where F is a finite free module. Then $\mathrm{Pd}_R^*\,(M') = k - 1$ and we can look for a contradiction using M' rather than M. The outcome of this observation is that we may add the assumption that $\mathrm{Pd}_R^*\,(M) = 1$.

By Exercise 8, we can construct an exact sequence

$$0 \to G \xrightarrow{\ \phi\ } F_0 \xrightarrow{\ \omega\ } M \to 0,$$

where F_0 is a finite free module and $\mathrm{Ker}\,\omega \subseteq PF_0$. Since

$$\mathrm{Pd}_R^*\,(M) = 1,$$

G is a supplementable projective module and therefore, since R is a quasi-local ring, a finite free module. Note that, because $\mathrm{Pd}_R^*(M) = 1$, $G \neq 0$ and therefore, if $p = \mathrm{rank}_R(G)$, we have $p > 0$.

Now, by Theorem 4, ϕ is stable and $\mathrm{rank}_R(\phi) = p$. Thus $0 :_R \mathfrak{A}_p(\phi) = 0$. However, $\phi(G) \subseteq PF_0$ and therefore if we take bases for G and F_0, the matrix representing ϕ will have all its entries in P. Thus $\mathfrak{A}_p(\phi)$ is a finitely generated ideal contained in P and hence $0 :_R \mathfrak{A}_p(\phi) \neq 0$ by hypothesis. This provides the contradiction that we were seeking.

We now have all we need to prove Vasconcelos's theorem.

THEOREM 12. *Let the R-module M have a finite free resolution of finite length. Then the following statements are equivalent:*

(a) $\mathrm{Char}_R(M) = 0$;

(b) $\mathrm{Ann}_R(M) \neq 0$.

Proof. Theorem 8 Cor. shows that (a) implies (b). In what follows we shall suppose that $\mathrm{Ann}_R(M) \neq 0$ and $\mathrm{Char}_R(M) \neq 0$. From these assumptions we shall derive a contradiction and this will prove the theorem.

Choose a finitely generated ideal $\mathfrak{A} \neq 0$ such that

$$\mathfrak{A} \subseteq \mathrm{Ann}_R(M).$$

Put $\mathfrak{B} = \mathrm{Ann}_R(\mathfrak{A})$. Then $\mathfrak{B} \neq R$. Theorem 9 now shows that $\mathfrak{B}$ has at least one minimal prime ideal. Let P be such a minimal prime ideal. Next, because $\mathfrak{A}$ is finitely generated, we have

$$\mathrm{Ann}_{R_P}(\mathfrak{A}R_P) = \mathfrak{B}R_P \subseteq PR_P$$

and therefore $\mathfrak{A}R_P \neq 0$.

The connection† between the prime ideals of R and those of R_P shows that PR_P is the only prime ideal of R_P containing $\mathfrak{B}R_P$. Hence, by Theorem 10, $\mathrm{Rad}\{\mathfrak{B}R_P\} = PR_P$.

Let I be a finitely generated ideal of R_P that is contained in PR_P. Then, by Exercise 7, there exists a positive integer m such that $I^m \subseteq \mathfrak{B}R_P$ and therefore I^m annihilates $\mathfrak{A}R_P$. But $\mathfrak{A}R_P$ is a non-zero ideal of R_P. Thus the quasi-local ring R_P satisfies the first hypothesis of Theorem 11. Further the R_P-module M_P has a

† See Chapter 1 Exercise 6.

finite free resolution of finite length. Consequently, by Theorem 11, M_P is free. But

$$\text{Char}_{R_P}(M_P) = \text{Char}_R(M)$$

and thus is non-zero. It follows that $M_P \neq 0$ and therefore, since M_P is R_P-free, $\text{Ann}_{R_P}(M_P) = 0$. However, M is finitely generated. Accordingly, by Chapter 2 Theorem 12,

$$\text{Ann}_R(M)\, R_P = \text{Ann}_{R_P}(M_P) = 0$$

and hence, *a fortiori*, $\mathfrak{A}R_P = 0$. This provides the contradiction we have been seeking and so the proof is complete.

Suppose that the module M has a finite free resolution of finite length and that $\text{Char}_R(M) = 0$. Since $\mathfrak{F}(M) \subseteq \text{Ann}_R(M)$, it follows, from Theorem 8 Cor., that $0 :_R \text{Ann}_R(M) = 0$. Now in the case where R is a Noetherian ring, this is enough to ensure that $\text{Ann}_R(M)$ contains a non-zerodivisor. For some time it was thought possible that this conclusion might hold even without the Noetherian condition. However we shall now give an example to show that this is not the case. In common with Theorem 12, the example is due to W. V. Vasconcelos (**44**). Later we shall be able to fill in the gap which arises in this way by means of a modified version of the theory of grade.

AN EXAMPLE. A construction will now be given which produces an ideal $\mathfrak{A}$ in a commutative ring R having the following properties:

(1) $\mathfrak{A}$ *is generated by two elements and is neither zero nor the whole ring;*

(2) *every element of $\mathfrak{A}$ is a zerodivisor in R;*

(3) *there is an exact sequence*

$$0 \to R \to R^2 \to R \to R/\mathfrak{A} \to 0;$$

(4) $0 :_R \mathfrak{A} = 0$.

If therefore $M = R/\mathfrak{A}$, then M has a finite free resolution of length two, $\text{Char}_R(M) = 0$, and $\text{Ann}_R(R/\mathfrak{A}) = \mathfrak{A}$ is composed of zerodivisors. Thus the example illustrates the point previously mentioned. Later the same example will be useful for illustrating another point of interest.

Before beginning the construction, let us observe that if Γ is a commutative ring with an identity element and E is a Γ-module, then $\Gamma \oplus E$ can be given the structure of a commutative ring (with an identity element) by means of the following device:†‡ if $\gamma_1, \gamma_2 \in \Gamma$ and $e_1, e_2 \in E$, then the product of (γ_1, e_1) and (γ_2, e_2) is taken to be

$$(\gamma_1, e_1)(\gamma_2, e_2) = (\gamma_1 \gamma_2, \gamma_1 e_2 + \gamma_2 e_1).$$

Now suppose that Ω is a field and that x, y are distinct indeterminates. Then $\Omega[x, y]$ is a unique factorization domain. In what follows P will denote a typical prime ideal generated by an irreducible element.

If P is as above, then $\Omega[x, y]/P$ is an integral domain. Let K_P be its quotient field. Evidently K_P has a natural structure as an $\Omega[x, y]$-module. Suppose that $\phi \in \Omega[x, y]$ and consider the mapping $K_P \to K_P$ induced by multiplication by ϕ. The point to be noted is that if $\phi \in P$, then the mapping is the null homomorphism; however if $\phi \notin P$, then it is an automorphism.

Define an $\Omega[x, y]$-module E by

$$E = \sum_P K_P \quad \text{(direct sum)},$$

put $R = \Omega[x, y] \oplus E$, and turn R into a commutative ring in the manner already explained. We use the following notation: if $\alpha \in R$, then $\alpha = (a, \sum_P a_P)$ where $a \in \Omega[x, y]$ and the family $\{a_P\}$ belongs to the direct sum $\sum_P K_P$.

We are now ready to define the ideal $\mathfrak{A}$. To this end put $\bar{x} = (x, 0)$, $\bar{y} = (y, 0)$, these being elements of R, and let $\mathfrak{A} = R\bar{x} + R\bar{y}$. Then certainly (1) is satisfied. Next suppose that $\alpha = (a, \sum_P a_P)$ belongs to $\mathfrak{A}$. It is clear that if $a = 0$, then α annihilates any element of the form $(0, \sum_P b_P)$. On the other hand if $a \neq 0$, then $a \in x\Omega[x, y] + y\Omega[x, y]$ and therefore it has at least one irreducible factor. Let Π be the prime ideal generated by such a factor. If now b_Π is any element of K_Π, then α annihilates $(0, b_\Pi)$. This shows that, in any event, α is a zerodivisor and we have verified (2).

---

† This is the idealization principle of M. Nagata. See [(**34**) Chapter 1, §1].

Consider the sequence

$$0 \to R \xrightarrow{\;\lambda\;} R^2 \xrightarrow{\;\mu\;} \mathfrak{A} \to 0, \qquad (4.3.4)$$

where $\lambda(\alpha) = \alpha(\bar{y}, -\bar{x})$ and $\mu(\beta, \gamma) = \beta\bar{x} + \gamma\bar{y}$. Certainly (4.3.4) is a complex and μ is an epimorphism. Suppose that $\mu(\beta, \gamma) = 0$. *We claim that there is a unique $\alpha \in R$ such that $(\beta, \gamma) = \alpha(\bar{y}, -\bar{x})$.* Note that if we establish this claim it will show that (4.3.4) is exact and (3) will follow.

Let $\beta = (b, \sum_P b_P)$, $\gamma = (c, \sum_P c_P)$. Then, since $\mu(\beta, \gamma) = 0$, we have

$$xb + yc = 0 \qquad (4.3.5)$$

and

$$xb_P + yc_P = 0 \quad (\text{all } P). \qquad (4.3.6)$$

It follows that there exists a unique $a \in \Omega[x, y]$ such that $(b, c) = a(y, -x)$. Again, if we select a particular prime ideal P, then one at least of x, y will induce (by multiplication) an automorphism of K_P. Hence there is a unique $a_P \in K_P$ such that $(b_P, c_P) = a_P(y, -x)$. Now a_P will be zero for all but a finite number of choices of P. Thus we see that there is a *unique* $\alpha = (a, \sum_P a_P)$ in R for which $(\beta, \gamma) = \alpha(\bar{y}, -\bar{x})$ and we have justified our claim. At the same time we see that if $\alpha \in (0 :_R \mathfrak{A})$ so that $\alpha(\bar{y}, -\bar{x}) = (0, 0)$, then $\alpha = 0$ by the uniqueness property. Thus not only does (3) hold, but also (4) holds as well. Accordingly our example exhibits all the features which we undertook to provide.

We return to the main discussion.

THEOREM 13. *If the R-module E has a finite free resolution of finite length, then the following two statements are equivalent:*

(a) *E is a free module;*

(b) *E can be generated by $\mathrm{Char}_R(E)$ elements.*

Proof. Obviously (a) implies (b). Now assume (b) and put $\mathrm{Char}_R(E) = q$. We can then construct an exact sequence

$$0 \to F_n \to F_{n-1} \to \ldots \to F_1 \to F_0 \to E \to 0,$$

where each F_j is a finite free R-module and $\mathrm{rank}_R(F_0) = q$. Hence if $K = \mathrm{Ker}(F_0 \to E)$, then K has a finite free resolution and

$\mathrm{Char}_R(K) = 0$. It therefore follows, by Theorem 8 Cor., that $0 :_R \mathrm{Ann}_R(K) = 0$.

Let $x_1, x_2, \ldots, x_q$ be a base for F_0 and suppose that $x \in K$. Then $x = r_1 x_1 + r_2 x_2 + \ldots + r_q x_q$ for suitable elements $r_1, r_2, \ldots, r_q$ in R. Since x is annihilated by $\mathrm{Ann}_R(K)$, we have $r_j \mathrm{Ann}_R(K) = 0$ for $j = 1, 2, \ldots, q$. It follows that $r_j = 0$ $(1 \leqslant j \leqslant q)$ and hence that $x = 0$. Thus $K = 0$ and therefore E is free.

EXERCISE 9. *Let $\mathfrak{A} \neq 0$ be an ideal of R which has a finite free resolution of finite length. Show that* $\mathrm{Char}_R(\mathfrak{A}) = 1$.

4.4 The rôle of idempotents

Let E be an R-module and suppose that we have an exact sequence

$$0 \to \Pi_n \to \Pi_{n-1} \to \ldots \to \Pi_1 \to \Pi_0 \to E \to 0, \qquad (4.4.1)$$

where each Π_j is a finitely generated *projective* R-module. If now P is a prime ideal of R and we localize at P, then (4.4.1) will give rise to a finite *free* resolution of the R_P-module E_P of length n. In particular we see that $\mathrm{Char}_{R_P}(E_P)$ is defined for each prime ideal P. However, it may happen that as P varies we get several values for the Euler characteristic. (Of course, if E has a finite free resolution of finite length this kind of behaviour cannot occur.) We therefore have a phenomenon which can be regarded as a kind of instability. This will now be investigated. To some extent the investigation constitutes a digression from the main development since the results obtained in section (4.4) will not be used in the sequel.

At this point it is convenient to introduce some topological ideas. The set of prime ideals of R is customarily called the *spectrum* of R and it is denoted by $\mathrm{Spec}\,(R)$. Let $\mathfrak{A}$ be an ideal of R and let $V(\mathfrak{A})$ be the subset of $\mathrm{Spec}\,(R)$ defined by

$$V(\mathfrak{A}) = \{P \,|\, P \in \mathrm{Spec}\,(R) \text{ and } \mathfrak{A} \subseteq P\}. \qquad (4.4.2)$$

This subset is known as the *variety* of $\mathfrak{A}$.

It is clear that a prime ideal contains $\mathfrak{A}$ if and only if it contains $\mathrm{Rad}\,\mathfrak{A}$. Hence

$$V(\mathfrak{A}) = V(\mathrm{Rad}\,\mathfrak{A}). \qquad (4.4.3)$$

Again, by Theorems 9 and 10, $\mathrm{Rad}\,\mathfrak{A}$ is the intersection of all the prime ideals containing $\mathfrak{A}$. Accordingly

$$\mathrm{Rad}\,\mathfrak{A} = \bigcap_{P \in V(\mathfrak{A})} P. \tag{4.4.4}$$

Hence, by combining (4.4.3) and (4.4.4), we obtain

Theorem 14. *Let $\mathfrak{A}$ and $\mathfrak{B}$ be ideals of R. Then $V(\mathfrak{A}) = V(\mathfrak{B})$ if and only if $\mathrm{Rad}\,\mathfrak{A} = \mathrm{Rad}\,\mathfrak{B}$.*

Corollary. *Let E be a finitely generated R-module. Then*

$$\mathrm{Supp}_R(E) = V(\mathrm{Ann}_R(E)) = V(\mathfrak{F}(E)).$$

Proof. The equation $\mathrm{Supp}_R(E) = V(\mathrm{Ann}_R(E))$ is a restatement of Chapter 2 Theorem 13, and we have $V(\mathrm{Ann}_R(E)) = V(\mathfrak{F}(E))$ by virtue of (4.3.3) and the theorem just proved.

It is clear that on the one hand $V(0) = \mathrm{Spec}\,(R)$ and on the other that $V(R)$ is the empty subset of $\mathrm{Spec}\,(R)$. Next if $\mathfrak{A}$, $\mathfrak{B}$ are ideals, then

$$V(\mathfrak{A}\mathfrak{B}) = V(\mathfrak{A}) \cup V(\mathfrak{B}). \tag{4.4.5}$$

Finally if $\{\mathfrak{A}_i\}_{i \in I}$ is a family of ideals of R, then

$$V\!\left(\sum_{i \in I} \mathfrak{A}_i\right) = \bigcap_{i \in I} V(\mathfrak{A}_i). \tag{4.4.6}$$

These simple observations amount to a proof of

Theorem 15. *There is a topology on $\mathrm{Spec}\,(R)$ in which the closed subsets are the varieties $V(\mathfrak{A})$, where $\mathfrak{A}$ ranges over all the ideals of R.*

The topology described in this theorem is known as the *Zariski topology*. From now on in any reference to a topology on $\mathrm{Spec}\,(R)$ it is to be understood that it is the Zariski topology that we have in mind.

Corollary. *If E is a finitely generated R-module, then $\mathrm{Supp}_R(E)$ is a closed subset of $\mathrm{Spec}\,(R)$.*

This follows from the corollary to Theorem 14.

Theorem 16. *Let η be an idempotent and put $\eta' = 1 - \eta$. Then $\eta\eta' = 0$, η' is also an idempotent, and $V(R\eta)$ and $V(R\eta')$ are complementary subsets of $\mathrm{Spec}\,(R)$. Consequently $V(R\eta)$ is both open and closed. Furthermore, if η_1 and η_2 are idempotents and*

$$V(R\eta_1) = V(R\eta_2),$$

then $\eta_1 = \eta_2$.

Proof. It is obvious that η' is an idempotent. Since $\eta + \eta' = 1$, no prime ideal can contain both η and η' and therefore $V(R\eta)$ and $V(R\eta')$ do not intersect. On the other hand

$$\eta\eta' = \eta - \eta^2 = 0.$$

If therefore P is a prime ideal, then either $\eta \in P$ or $\eta' \in P$. This shows that $V(R\eta)$ and $V(R\eta')$ are complementary subsets of Spec (R).

Now assume that $V(R\eta_1) = V(R\eta_2)$. By Theorem 14,

$$\eta_1 \in \operatorname{Rad}(R\eta_1) = \operatorname{Rad}(R\eta_2)$$

and so some power of η_1 and therefore η_1 itself belongs to $R\eta_2$. But, because $\eta_2^2 = \eta_2$, this implies that $\eta_1\eta_2 = \eta_1$. Since $\eta_1\eta_2 = \eta_2$ for similar reasons, we have $\eta_1 = \eta_2$ as required.

THEOREM 17. *There is a natural bijection between the idempotents of R and the subsets of* Spec (R) *that are both open and closed. In this bijection the idempotent η is matched with the subset $V(R\eta)$.*

Proof. Let $\mathfrak{A}$ be an ideal with the property that $V(\mathfrak{A})$ is an open as well as a closed subset of Spec (R). Then there exists an ideal $\mathfrak{B}$ such that

$$V(\mathfrak{A}) \cup V(\mathfrak{B}) = \operatorname{Spec}(R)$$

and $V(\mathfrak{A}) \cap V(\mathfrak{B})$ is empty. From the latter property we see that there is no prime ideal which contains $\mathfrak{B}$ as well as $\mathfrak{A}$. Consequently $\mathfrak{A} + \mathfrak{B} = R$ and therefore we can find $a \in \mathfrak{A}$ and $b \in \mathfrak{B}$ such that $a + b = 1$. Next, by (4.4.5),

$$V(\mathfrak{A}\mathfrak{B}) = V(\mathfrak{A}) \cup V(\mathfrak{B}) = \operatorname{Spec}(R) = V(0)$$

and therefore, by Theorem 14,

$$\mathfrak{A}\mathfrak{B} \subseteq \operatorname{Rad}(\mathfrak{A}\mathfrak{B}) = \operatorname{Rad}(0).$$

Accordingly we can find an integer m such that $a^m b^m = 0$. Now, because $a + b = 1$, no prime ideal can contain both a^m and b^m. Thus $Ra^m + Rb^m = R$. It is therefore possible to find $\eta \in Ra^m$ and $\eta' \in Rb^m$ so that $\eta + \eta' = 1$ and then, since $\eta\eta' = 0$, η and η' will be idempotents.

By construction $R\eta \subseteq \mathfrak{A}$ and therefore $\operatorname{Rad}(R\eta) \subseteq \operatorname{Rad}\mathfrak{A}$. Let

P be a prime ideal containing $R\eta$. Then P does not contain η' and therefore it does not contain $\mathfrak{B}$. Accordingly $P \notin V(\mathfrak{B})$ which means therefore that $P \in V(\mathfrak{A})$. Hence, by (4.4.4),

$$\operatorname{Rad} \mathfrak{A} = \bigcap_{P' \in V(\mathfrak{A})} P' \subseteq \bigcap_{R\eta \subseteq P} P = \operatorname{Rad}(R\eta).$$

This shows that $\operatorname{Rad} \mathfrak{A} = \operatorname{Rad}(R\eta)$ and now it follows, from Theorem 14, that $V(\mathfrak{A}) = V(R\eta)$. The remaining assertions of the theorem are now seen to be consequences of Theorem 16.

COROLLARY. *The ring R contains no non-trivial idempotents if and only if the space* Spec (R) *is connected.*

Our next result will enable us to relate the results of this section to those obtained in the rest of the chapter.

THEOREM 18. *Let $\phi: F \to G$ be a homomorphism, where F and G are finite free modules. Then the following three statements are equivalent:*

(a) Coker ϕ *is a projective module;*
(b) *there is a homomorphism $\psi: G \to F$ such that $\phi\psi\phi = \phi$;*
(c) *for each $\nu \geqslant 0$, $\mathfrak{A}_\nu(\phi)$ is generated by an idempotent.*

Proof. If ϕ is a null homomorphism, then all of (a), (b) and (c) hold anyway. It follows that we may restrict our attention to the case where $F \neq 0$ and $G \neq 0$. On this understanding we choose bases for F and G and let A be the matrix of ϕ with respect to these bases.

We know, from Chapter 2 Exercise 7, that (a) and (b) are equivalent. *Assume* (b). Then there is a matrix Ω such that $A\Omega A = A$. Consequently, by Chapter 1 Lemma 1, $\mathfrak{A}_\nu(A)$, which is the same as $\mathfrak{A}_\nu(\phi)$, is generated by an idempotent. Thus (b) implies (c).

Finally *assume* (c) and let P be a prime ideal. If η is an idempotent of R, then $\eta/1$ is an idempotent of R_P and therefore,† in R_P, $\eta/1$ is either zero or it is the identity element. Since $\mathfrak{A}_\nu(A_P) = \mathfrak{A}_\nu(A) R_P$ and since $\mathfrak{A}_\nu(A)$ is generated by an idempotent, we may conclude that $\mathfrak{A}_\nu(A_P)$ is either 0 or R_P. Hence, by Chapter 1 Theorem 6, we can find an R_P-matrix Ω say such that $A_P \Omega A_P = A_P$ and therefore there exists an R_P-homo-

† See Chapter 1 Theorem 11.

morphism $\omega\colon G_P \to F_P$ with the property that $\phi_P \omega \phi_P = \phi_P$. Accordingly, by Chapter 2 Exercise 7, $\operatorname{Coker} \phi_P$, which may be identified with $(\operatorname{Coker} \phi)_P$, is a projective and hence free R_P-module. Thus $\operatorname{Coker} \phi$ is a finitely presented R-module all of whose localizations are free. Consequently, by Chapter 2 Theorem 14, $\operatorname{Coker} \phi$ is a projective R-module. This completes the proof.

COROLLARY. *Let E be a finitely presented R-module. Then E is projective if and only if each of its Fitting invariants is generated by an idempotent.*

This follows from the equivalence of (a) and (c) together with (4.2.15).

EXERCISE 10. *Let K and N be submodules of an R-module E and suppose that, for every maximal ideal P, K_P and N_P coincide as submodules of E_P. Show that $K = N$. Deduce that if $\mathfrak{A}$ and $\mathfrak{B}$ are ideals of R and $\mathfrak{A}R_P = \mathfrak{B}R_P$ for all maximal ideals, then $\mathfrak{A} = \mathfrak{B}$.*

THEOREM 19. *Let E be a projective R-module generated by q elements, where q is finite. Then, for each prime ideal P, E_P is a finite free R_P-module and $0 \leqslant \operatorname{rank}_{R_P}(E_P) \leqslant q$. Moreover, every Fitting invariant $\mathfrak{F}_\mu(E)$ is generated by an idempotent and if $k \geqslant 0$ is an integer, then the set*

$$\{P \,|\, P \in \operatorname{Spec}(R) \ and \ \operatorname{rank}_{R_P}(E_P) = k\}$$

is both open and closed in $\operatorname{Spec}(R)$.

Proof. First E_P is a projective R_P-module and therefore a free R_P-module because R_P is a quasi-local ring. Next E_P can be generated by q elements and therefore $\mathfrak{F}_q(E_P) = R_P$. The inequality $\operatorname{rank}_{R_P}(E_P) \leqslant q$ now follows because $\mathfrak{F}_\mu(E_P) = 0$ when $\mu < \operatorname{rank}_{R_P}(E_P)$ as may be seen from the solution to Exercise 1 of Chapter 3. Again it is evident that E is finitely presented as well as projective. Consequently Theorem 18 Cor. shows that every one of its Fitting invariants is generated by an idempotent.

Let $\mathfrak{F}_\mu(E) = R\eta_\mu$, where η_μ is an idempotent, and let $m \geqslant 0$ be an integer. Since the statements

(i) $\operatorname{rank}_{R_P}(E_P) \leqslant m$,
(ii) $\mathfrak{F}_m(E_P) = R_P$,

(iii) $\eta_m \notin P$,

(iv) $P \notin V(R\eta_m)$,

are equivalent, Theorem 16 shows that the set X_m (say) of prime ideals for which $\operatorname{rank}_{R_P}(E_P) \leqslant m$ is both open and closed. Now, provided we interpret X_{-1} as being the empty subset of $\operatorname{Spec}(R)$, we have

$$\{P|P \in \operatorname{Spec}(R) \text{ and } \operatorname{rank}_{R_P}(E_P) = k\} = X_k \backslash X_{k-1}.$$

Consequently this set is open and closed as well.

COROLLARY. *Let E be a finitely generated projective R-module. Then $\operatorname{Ann}_R(E) = \mathfrak{F}(E)$. Also $\operatorname{Supp}_R(E)$ is both open and closed.*

Proof. Let P be a prime ideal. Then, by Chapter 2 Theorem 12,

$$\operatorname{Ann}_{R_P}(E_P) = \operatorname{Ann}_R(E)R_P$$

and, by Chapter 3 Theorem 3,

$$\mathfrak{F}(E_P) = \mathfrak{F}(E)R_P.$$

Now E_P is a free R_P-module. Consequently either $\operatorname{Ann}_{R_P}(E_P) = 0$ or $\operatorname{Ann}_{R_P}(E_P) = R_P$. It therefore follows, by Chapter 3 Theorem 5, that $\operatorname{Ann}_{R_P}(E_P) = \mathfrak{F}(E_P)$. Thus $\operatorname{Ann}_R(E)R_P = \mathfrak{F}(E)R_P$ for all prime ideals P and therefore $\operatorname{Ann}_R(E) = \mathfrak{F}(E)$ by Exercise 10. Finally $\operatorname{Supp}_R(E) = V(\mathfrak{F}(E))$ and now, since we know that $\mathfrak{F}(E)$ is generated by an idempotent, Theorem 16 shows that $\operatorname{Supp}_R(E)$ is both open and closed.

We are now ready to consider the phenomenon to which our attention was drawn at the beginning of section (4.4).

THEOREM 20. *Suppose that*

$$0 \to \Pi_n \to \Pi_{n-1} \to \ldots \to \Pi_1 \to \Pi_0 \to E \to 0$$

is an exact sequence, where each Π_j is a finitely generated projective module. Then for each prime ideal P the R_P-module E_P has a finite free resolution of finite length. Moreover if k is an integer, then the set

$$\{P|P \in \operatorname{Spec}(R) \text{ and } \operatorname{Char}_{R_P}(E_P) = k\} \qquad (4.4.7)$$

is both open and closed in $\operatorname{Spec}(R)$.

Proof. If $0 \leqslant \mu \leqslant n$, then, by Theorem 19, $(\Pi_\mu)_P$ is a finite free R_P-module and there are only a finite number of possibilities for its rank; moreover the subset of Spec (R) on which its rank has any prescribed value is both open and closed. Consequently, since

$$\text{Char}_{R_P}(E_P) = \sum_{\mu=0}^{n} (-1)^\mu \text{rank}_{R_P}((\Pi_\mu)_P),$$

the set occurring in (4.4.7) is the union of a finite number of sets each of which is both open and closed. The theorem follows.

EXERCISE 11. *Let $\mathfrak{A}$ be an ideal of R. Show that the following two statements are equivalent:*
  (1) *$\mathfrak{A}$ is generated by an idempotent;*
  (2) *$\mathfrak{A}$ is finitely generated and $\mathfrak{A}^2 = \mathfrak{A}$.*

THEOREM 21. *Suppose that*

$$0 \to \Pi_n \to \Pi_{n-1} \to \ldots \to \Pi_1 \to \Pi_0 \to E \to 0$$

is an exact sequence, where each Π_j is a finitely generated projective R-module. Then

$$\text{Ann}_R(\text{Ann}_R(E)) = \text{Ann}_R(\mathfrak{F}(E))$$

and this ideal is generated by an idempotent.

Remark. If the module E has a *finite free* resolution of finite length, then Theorem 8 Cor. shows that the ideal here in question is either the zero ideal or it is the whole ring.

Proof. It is easy to see that E is finitely presented. Consequently $\mathfrak{F}(E)$ is finitely generated. Put $\mathfrak{B} = \text{Ann}_R(\mathfrak{F}(E))$ and let P be a prime ideal. Then, by Chapter 2 Theorem 12 and Chapter 3 Theorem 3, $\mathfrak{B}R_P = \text{Ann}_{R_P}(\mathfrak{F}(E_P))$.

If $\mathfrak{F}(E_P) \neq 0$, then, by Theorem 8 Cor.,

$$\mathfrak{B}R_P = 0 \quad \text{and} \quad \text{Char}_{R_P}(E_P) = 0.$$

On the other hand if $\mathfrak{F}(E_P) = 0$, then

$$\mathfrak{B}R_P = R_P \quad \text{and} \quad \text{Char}_{R_P}(E_P) > 0.$$

Thus $\qquad \mathfrak{B} \subseteq P \Leftrightarrow \mathfrak{B}R_P = 0 \Leftrightarrow \text{Char}_{R_P}(E_P) = 0 \qquad (4.4.8)$

and

$\mathfrak{B} \not\subseteq P \Leftrightarrow \mathfrak{B}R_P = R_P \Leftrightarrow \text{Char}_{R_P}(E_P) > 0 \Leftrightarrow \text{Ann}_R(E)R_P = 0,$
$$(4.4.9)$$

where the final equivalence in (4.4.9) follows from Theorem 12. Next (4.4.8) and Theorem 20 show that

$$V(\mathfrak{B}) = \{P \mid P \in \mathrm{Spec}\,(R) \text{ and } \mathrm{Char}_{R_P}(E_P) = 0\}$$

is an open and closed subset of $\mathrm{Spec}\,(R)$. Hence $V(\mathfrak{B}) = V(R\eta)$ for some idempotent η.

We next observe that if $\eta \in P$, that is if $\mathfrak{B} \subseteq P$, then $(R\eta)\,R_P = 0 = \mathfrak{B}R_P$. On the other hand if $\eta \notin P$, i.e. if $\mathfrak{B} \nsubseteq P$, then $(R\eta)\,R_P = R_P = \mathfrak{B}R_P$. Hence $(R\eta)\,R_P = \mathfrak{B}R_P$ for all prime ideals P and therefore $\mathfrak{B} = R\eta$ by Exercise 10.

Now consider the ideal $\eta\,\mathrm{Ann}_R(E)$. If $\eta \in P$, then

$$(\eta\,\mathrm{Ann}_R(E))\,R_P = 0$$

because $\eta/1 = 0$ in R_P. If however $\eta \notin P$, then $\mathfrak{B} \nsubseteq P$ and therefore, as we saw above, $\mathrm{Ann}_R(E)\,R_P = 0$. Thus we see that $(\eta\,\mathrm{Ann}_R(E))\,R_P = 0$ for every prime ideal P and so it follows, by Exercise 10, that $\eta\,\mathrm{Ann}_R(E) = 0$. This shows that

$$\mathfrak{B} = R\eta \subseteq \mathrm{Ann}_R(\mathrm{Ann}_R(E)).$$

However, by Chapter 3 Theorem 5, $\mathfrak{F}(E) \subseteq \mathrm{Ann}_R(E)$ whence $\mathrm{Ann}_R(\mathrm{Ann}_R(E)) \subseteq \mathfrak{B}$. Thus $\mathfrak{B} = \mathrm{Ann}_R(\mathrm{Ann}_R(E))$ and the proof is complete.

Solutions to the Exercises on Chapter 4

EXERCISE 1. *Let E be an R-module and let $\mathfrak{A}$ be the ideal generated by the finite set of elements $\alpha_1, \alpha_2, ..., \alpha_p$. If now for each $i(1 \leqslant i \leqslant p)$ some power of α_i annihilates E, show that there exists a positive integer m such that $\mathfrak{A}^m E = 0$.*

Solution. Choose an integer $\nu > 0$ so that $\alpha_i^{\nu} E = 0$ for $i = 1, 2, ..., p$ and put $m = p\nu$. If now $\mu_1, \mu_2, ..., \mu_p$ are non-negative integers such that $\mu_1 + \mu_2 + ... + \mu_p = m$, then $\mu_i \geqslant \nu$ for at least one value of i and therefore $\alpha_1^{\mu_1}\alpha_2^{\mu_2} ... \alpha_p^{\mu_p} E = 0$. Since $\mathfrak{A}^m$ is generated by elements of the form $\alpha_1^{\mu_1}\alpha_2^{\mu_2} ... \alpha_p^{\mu_p}$, it follows that $\mathfrak{A}^m E = 0$.

EXERCISE 2. *Suppose that $AB = 0$ exactly on the R-module E and that S is a multiplicatively closed subset of R not containing zero. Show that $A_S B_S = 0$ exactly on the R_S-module E_S.*

Solution. Let A have p rows and q columns and let B have q rows and t columns. It is clear that $A_S B_S = 0$. Suppose now that $\eta^* \in (E_S)^q$ and $\eta^* B_S = 0$. Then η^* can be expressed in the form

$$\eta^* = (\eta_1/s, \eta_2/s, \ldots, \eta_q/s),$$

where $\eta = (\eta_1, \eta_2, \ldots, \eta_q) \in E^q$ and $s \in S$. Next, in view of the fact that $\eta^* B_S = 0$, there must exist $\sigma \in S$ such that $(\sigma \eta) B = 0$. Accordingly we can find $\xi = (\xi_1, \xi_2, \ldots, \xi_p) \in E^p$ having the property that $\xi A = \sigma \eta$. If therefore

$$\xi^* = (\xi_1/s\sigma, \xi_2/s\sigma, \ldots, \xi_p/s\sigma),$$

then $\xi^* \in (E_S)^p$ and $\xi^* A_S = \eta^*$. Thus $A_S B_S = 0$ exactly on E_S.

EXERCISE 3. *Let $\psi: F \to G$ be an epimorphism, where F and G are finite free modules. Show that ψ is stable and that*

$$\operatorname{rank}_R (\psi) = \operatorname{rank}_R (G).$$

Solution. Let $q = \operatorname{rank}_R (G)$. By (4.2.15)

$$\mathfrak{F}_0(\operatorname{Coker} \psi) = \mathfrak{A}_q(\psi)$$

and therefore $\mathfrak{A}_q(\psi) = R$ because $\operatorname{Coker} \psi = 0$. Thus both $\operatorname{rank}_R (\psi)$ and $\operatorname{red.rank}_R (\psi)$ are equal to q.

EXERCISE 4. *The R-module M can be generated by q elements and it has a finite free resolution of finite length. Show that*

$$\operatorname{Char}_R (M) \leqslant q.$$

Solution. By Chapter 3 Theorem 2, $\mathfrak{F}_q(M) = R$ and hence, in particular, $\mathfrak{F}_q(M) \neq 0$. That $\operatorname{Char}_R (M) \leqslant q$ now follows from Theorem 8.

EXERCISE 5. *Let $\{P_i\}_{i \in I}$ be a non-empty family of prime ideals of R with the property that if $i_1, i_2 \in I$, then either $P_{i_1} \subseteq P_{i_2}$ or $P_{i_2} \subseteq P_{i_1}$. Show that the intersection of the family is itself a prime ideal.*

Solution. Let P be the intersection of the family $\{P_i\}_{i \in I}$. Then, since I is not empty, $P \neq R$. Suppose that $\alpha, \beta \in R$, that $\alpha\beta \in P$ and $\alpha \notin P$. Then there exists $j \in I$ such that $\alpha \notin P_j$ in which case $\beta \in P_j$.

Take any i in I. By hypothesis either $P_j \subseteq P_i$ or $P_i \subseteq P_j$. If $P_j \subseteq P_i$, then certainly $\beta \in P_i$. Suppose that $P_i \subseteq P_j$. In this case $\alpha \notin P_i$, so again $\beta \in P_i$. Thus in either event $\beta \in P_i$. This shows that $\beta \in P$ and establishes that P is a prime ideal.

EXERCISE 6. *Let* $\mathfrak{A}_1, \mathfrak{A}_2, \ldots, \mathfrak{A}_n$ *be ideals of* R. *Show that*
$$\mathrm{Rad}\,(\mathfrak{A}_1 \mathfrak{A}_2 \ldots \mathfrak{A}_n) = \mathrm{Rad}\,(\mathfrak{A}_1 \cap \mathfrak{A}_2 \cap \ldots \cap \mathfrak{A}_n)$$
$$= \mathrm{Rad}\,\mathfrak{A}_1 \cap \mathrm{Rad}\,\mathfrak{A}_2 \cap \ldots \cap \mathrm{Rad}\,\mathfrak{A}_n.$$

Solution. It is clear that
$$\mathrm{Rad}\,(\mathfrak{A}_1 \mathfrak{A}_2 \ldots \mathfrak{A}_n) \subseteq \mathrm{Rad}\,(\mathfrak{A}_1 \cap \mathfrak{A}_2 \cap \ldots \cap \mathfrak{A}_n)$$
$$\subseteq \mathrm{Rad}\,\mathfrak{A}_1 \cap \mathrm{Rad}\,\mathfrak{A}_2 \cap \ldots \cap \mathrm{Rad}\,\mathfrak{A}_n.$$

Now suppose that $\alpha \in \mathrm{Rad}\,\mathfrak{A}_i$ for $i = 1, 2, \ldots, n$. We can choose a positive integer k so that $\alpha^k \in \mathfrak{A}_i$ $(1 \leqslant i \leqslant n)$ and this will secure that
$$\alpha^{kn} \in \mathfrak{A}_1 \mathfrak{A}_2 \ldots \mathfrak{A}_n.$$

Accordingly $\alpha \in \mathrm{Rad}\,(\mathfrak{A}_1 \mathfrak{A}_2 \ldots \mathfrak{A}_n)$ and with this the solution is complete.

EXERCISE 7. *Let* $\mathfrak{A}$ *be an ideal of* R *and* $\mathfrak{B}$ *a finitely generated ideal contained in* $\mathrm{Rad}\,\mathfrak{A}$. *Show that there exists a positive integer* m *such that* $\mathfrak{B}^m \subseteq \mathfrak{A}$.

Solution. Let $\mathfrak{B} = R\beta_1 + B\beta_2 + \ldots + R\beta_p$ and suppose that $1 \leqslant i \leqslant p$. Then some power of β_i is in $\mathfrak{A}$ and this power will annihilate $R/\mathfrak{A}$. It follows, from Exercise 1, that there exists a positive integer m such that $\mathfrak{B}^m(R/\mathfrak{A}) = 0$. We now have $\mathfrak{B}^m \subseteq \mathfrak{A}$.

EXERCISE 8. *Let* R *be a quasi-local ring with maximal ideal* P *and let* M *be a finitely generated* R-*module. Show that there exists an epimorphism* $\omega \colon F \to M$, *where* F *is a finite free module and* $\mathrm{Ker}\,\omega \subseteq PF$.

Solution. We may suppose that $M \neq 0$. Let
$$M = Ru_1 + Ru_2 + \ldots + Ru_q,$$
where q is minimal, and let F be a free R-module with a base $x_1, x_2, \ldots, x_q$ of q elements. We can then construct an epimorphism $\omega \colon F \to M$ with $\omega(x_i) = u_i$. Put $K = \mathrm{Ker}\,\omega$.

Let $x \in K$ say $x = r_1 x_1 + r_2 x_2 + \ldots + r_q x_q$, where $r_j \in R$. *We claim that* $r_1 \in P$. For suppose that r_1 is not in P. Then, since $\omega(x) = 0$, we have $r_1 u_1 + r_2 u_2 + \ldots + r_q u_q = 0$ and r_1 is a unit. However, this implies that $M = Ru_2 + Ru_3 + \ldots + Ru_q$ contradicting the minimal property of q. This establishes our claim. Similar considerations show that $r_j \in P$ for $j = 1, 2, \ldots, q$. Thus $x \in PF$ and therefore $K \subseteq PF$.

EXERCISE 9. *Let* $\mathfrak{A} \neq 0$ *be an ideal of* R *which has a finite free resolution of finite length. Show that* $\mathrm{Char}_R(\mathfrak{A}) = 1$.

Solution. Suppose that

$$0 \to F_n \to F_{n-1} \to \ldots \to F_1 \to F_0 \to \mathfrak{A} \to 0$$

is a finite free resolution of $\mathfrak{A}$. This gives rise to an exact sequence

$$0 \to F_n \to F_{n-1} \to \ldots \to F_1 \to F_0 \to R \to R/\mathfrak{A} \to 0.$$

But $\mathrm{Ann}_R(R/\mathfrak{A}) = \mathfrak{A}$ and this is not zero. Consequently, by Theorem 12, $\mathrm{Char}_R(R/\mathfrak{A}) = 0$ and therefore $\mathrm{Char}_R(\mathfrak{A}) = 1$.

EXERCISE 10. *Let* K *and* N *be submodules of an* R-*module* E *and suppose that, for every maximal ideal* P, K_P *and* N_P *coincide as submodules of* E_P. *Show that* $K = N$. *Deduce that if* $\mathfrak{A}$ *and* $\mathfrak{B}$ *are ideals of* R *and* $\mathfrak{A}R_P = \mathfrak{B}R_P$ *for all maximal ideals, then* $\mathfrak{A} = \mathfrak{B}$.

Solution. Let $x \in K$ and let P be a maximal ideal. Then, in E_P, $x/1 = n/s$ for some n in N and s in $R \backslash P$. Thus there exists σ in $R \backslash P$ such that $\sigma x \in N$. Accordingly $N :_R (Rx + N)$ meets $R \backslash P$ and therefore $N :_R (Rx + N)$ is not contained in any maximal ideal. This shows that $N :_R (Rx + N) = R$ whence $x \in N$. We now have $K \subseteq N$ and a similar argument yields $N \subseteq K$. This proves that $K = N$. The other assertion follows by taking $E = R$, $K = \mathfrak{A}$ and $N = \mathfrak{B}$.

EXERCISE 11. *Let* $\mathfrak{A}$ *be an ideal of* R. *Show that the following two statements are equivalent:*
(1) $\mathfrak{A}$ *is generated by an idempotent;*
(2) $\mathfrak{A}$ *is finitely generated and* $\mathfrak{A}^2 = \mathfrak{A}$.

Solution. Obviously (1) implies (2). *Assume* (2). Then

$$\mathfrak{A} = Ra_1 + Ra_2 + \ldots + Ra_m$$

say and therefore $\mathfrak{A} = \mathfrak{A}a_1 + \mathfrak{A}a_2 + \ldots + \mathfrak{A}a_m$. We therefore have relations

$$a_i = c_{i1}a_1 + c_{i2}a_2 + \ldots + c_{im}a_m \quad (1 \leqslant i \leqslant m)$$

or

$$\sum_{j=1}^{m} (\delta_{ij} - c_{ij})\, a_j = 0 \quad (1 \leqslant i \leqslant m),$$

where δ_{ij} is the Kronecker symbol and $c_{ij} \in \mathfrak{A}$. It follows that if D is the determinant of the $m \times m$ matrix $\|\delta_{ij} - c_{ij}\|$, then $D\mathfrak{A} = 0$. But $D = 1 - \alpha$ for some $\alpha \in \mathfrak{A}$. Accordingly $(1 - \alpha)\alpha = 0$ and therefore α is an idempotent. Finally from $R\alpha \subseteq \mathfrak{A}$ and $(1 - \alpha)\mathfrak{A} = 0$ we deduce that $\mathfrak{A} = R\alpha$ and with this the solution is complete.

5. Latent non-zerodivisors

General remarks

In this chapter we shall give an account of the theory of *grade* because this theory is needed for the further study of finite free resolutions. The concept of grade develops out of that of an *R-sequence* which is defined as follows: if E is an R-module, then a sequence $\alpha_1, \alpha_2, ..., \alpha_s$ of elements of R is called *an R-sequence on E* provided that for each i ($1 \leqslant i \leqslant s$) the element α_i annihilates no non-zero element of the factor module

$$E/(\alpha_1 E + \alpha_2 E + ... + \alpha_{i-1} E).$$

Thus we shall be very much concerned with questions which have to do with the existence (or non-existence) of non-zerodivisors on a module.

The theory surrounding this subject was first created for modules over Noetherian rings because at one time it seemed that the Noetherian condition was needed to ensure the existence of non-zerodivisors under appropriate conditions. In the general, i.e. non-Noetherian, case it can happen that the conditions in question may be satisfied without the necessary non-zerodivisors being present, and indeed we shall give an example at a later stage† to illustrate this phenomenon. However, in these circumstances it turns out that, by adjoining indeterminates, a non-zerodivisor can be made to appear almost as if by magic. This state of affairs may be described by saying that a *latent non-zerodivisor* was present in the original situation.

The notion of a latent non-zerodivisor leads on to that of a latent R-sequence although in fact we shall not use this term. What is important is that this line of thought will enable us to

† See section (5.3).

develop the theory of grade in a form which dispenses completely with Noetherian conditions† and which is sufficiently rich in results for our various purposes.

As usual, R will denote a non-trivial commutative ring with an identity element. We shall also reserve the symbols $x, x_1, x_2, x_3, \ldots$ to denote indeterminates.

5.1   R-sequences

Let E be an R-module and let α belong to R. We recall that α is said to be a *zerodivisor* on E if there exists $e \in E$ such that $e \neq 0$ but $\alpha e = 0$. Thus α is a zerodivisor on E if and only if $0 :_E R\alpha \neq 0$.

Let K be a submodule of E and put

$$K :_E \alpha = \{e \,|\, e \in E \text{ and } \alpha e \in K\}$$

so that in fact $K :_E \alpha$ is the same as $K :_E R\alpha$.‡ Then α *is a non-zerodivisor on E/K if and only if* $K :_E \alpha = K$.

Let $\alpha_1, \alpha_2, \ldots, \alpha_s$ be a sequence of elements of R. As already observed $\alpha_1, \alpha_2, \ldots, \alpha_s$ is called an *R-sequence on E* if, for each $i (1 \leqslant i \leqslant s)$, α_i is a non-zerodivisor on

$$E/(\alpha_1 E + \alpha_2 E + \ldots + \alpha_{i-1} E).$$

Thus $\alpha_1, \alpha_2, \ldots, \alpha_s$ is an R-sequence on E if and only if

$$(\alpha_1 E + \alpha_2 E + \ldots + \alpha_{i-1} E) :_E \alpha_i = \alpha_1 E + \alpha_2 E + \ldots + \alpha_{i-1} E \quad (5.1.1)$$

for $i = 1, 2, \ldots, s$. Note that a single element α, of R, forms by itself an R-sequence on E when and only when $0 :_E \alpha = 0$. Again, to take care of certain extreme situations, we introduce the convention that an *empty sequence*, that is a sequence $\alpha_1, \alpha_2, \ldots, \alpha_s$ with $s = 0$, is an R-sequence on every R-module. Finally, *if we say simply that $\alpha_1, \alpha_2, \ldots, \alpha_s$ is an R-sequence, and do not mention a module, then it is to be understood that we mean that it is an R-sequence on R and therefore*

$$(R\alpha_1 + R\alpha_2 + \ldots + R\alpha_{i-1}) :_R \alpha_i = R\alpha_1 + R\alpha_2 + \ldots + R\alpha_{i-1}$$

for $i = 1, 2, \ldots, s$.

---

† The realization that indeterminates can be used in this way is due to M. Hochster. See [(**25**) §1].

‡ See (2.4.6).

Suppose that $0 \leqslant p < j \leqslant s$ and put

$$E' = E/(\alpha_1 E + \alpha_2 E + \ldots + \alpha_p E).$$

Then in view of the isomorphism

$$E/(\alpha_1 E + \ldots + \alpha_p E + \ldots + \alpha_{j-1} E) \approx E'/(\alpha_{p+1} E' + \ldots + \alpha_{j-1} E')$$

we at once obtain

LEMMA 1. *Let E be an R-module, let $\alpha_1, \alpha_2, \ldots, \alpha_s$ be elements of R, and suppose that $0 \leqslant p \leqslant s$. Put*

$$E' = E/(\alpha_1 E + \alpha_2 E + \ldots + \alpha_p E).$$

Then $\alpha_1, \alpha_2, \ldots, \alpha_s$ is an R-sequence on E if and only if $\alpha_1, \alpha_2, \ldots, \alpha_p$ is an R-sequence on E and $\alpha_{p+1}, \alpha_{p+2}, \ldots, \alpha_s$ is an R-sequence on E'.

THEOREM 1. *Let $\alpha_1, \alpha_2, \ldots, \alpha_s$ be an R-sequence on the R-module E and suppose that $1 \leqslant p \leqslant s$. Then $\alpha_1, \alpha_{p+1}, \alpha_{p+2}, \ldots, \alpha_s$ is an R-sequence on $E/(\alpha_2 E + \alpha_3 E + \ldots + \alpha_p E)$.*

Proof. The case $p = 1$ is trivial. Now suppose that $p = 2$ and put $K = E/\alpha_2 E$. To show that $\alpha_1, \alpha_3, \alpha_4, \ldots, \alpha_s$ is an R-sequence on K it is enough, in view of Lemma 1, to prove that α_1 is not a zerodivisor on K and that $\alpha_3, \alpha_4, \ldots, \alpha_s$ is an R-sequence on $K/\alpha_1 K$. However $K/\alpha_1 K$ is isomorphic to $E/(\alpha_1 E + \alpha_2 E)$ and therefore, again by Lemma 1, the latter point is clear.

To establish that α_1 is not a zerodivisor on K it suffices to prove that

$$\alpha_2 E :_E \alpha_1 = \alpha_2 E.$$

Assume therefore that $\alpha_1 e \in \alpha_2 E$, where $e \in E$. Then $\alpha_1 e = \alpha_2 e'$ for some $e' \in E$. Then, because $\alpha_1 E :_E \alpha_2 = \alpha_1 E$, it follows that $e' \in \alpha_1 E$ say $e' = \alpha_1 e''$. Thus $\alpha_1 e = \alpha_1 \alpha_2 e''$ whence $e = \alpha_2 e''$ because α_1 is not a zerodivisor on E. Accordingly we have shown that

$$\alpha_2 E :_E \alpha_1 \subseteq \alpha_2 E$$

and, of course, the opposite inclusion is obvious.

We have now established that $\alpha_1, \alpha_3, \alpha_4, \ldots, \alpha_s$ is an R-sequence on $K = E/\alpha_2 E$. If we apply the result just proved to K and $\alpha_1, \alpha_3, \alpha_4, \ldots, \alpha_s$ we find that $\alpha_1, \alpha_4, \alpha_5, \ldots, \alpha_s$ is an R-sequence on $K/\alpha_3 K$. But $K/\alpha_3 K$ is isomorphic to $E/(\alpha_2 E + \alpha_3 E)$. Consequently $\alpha_1, \alpha_4, \alpha_5, \ldots, \alpha_s$ is an R-sequence on $E/(\alpha_2 E + \alpha_3 E)$, i.e.

the theorem is proved for the case $p = 3$. It is clear how the argument continues.

LEMMA 2. *Let E be an R-module and let α, β, r belong to R. Then the following statements are equivalent:*

(a) *α, r and β, r are R-sequences on E;*

(b) *$\alpha\beta$, r is an R-sequence on E.*

Proof. Assume (a). Then each of α, β is a non-zerodivisor on E. Consequently $\alpha\beta$ is a non-zerodivisor on E. Now suppose that e belongs to $\alpha\beta E:_E r$. Then $re = \alpha\beta e_1$ for some $e_1 \in E$ and, moreover, e belongs to $\beta E:_E r = \beta E$. Let $e = \beta e'$. We now have $r\beta e' = \alpha\beta e_1$ whence $re' = \alpha e_1$ because β is not a zerodivisor on E. It follows that e' is in $\alpha E:_E r = \alpha E$ say $e' = \alpha e''$. Accordingly $e = \beta\alpha e'' \in \alpha\beta E$. Thus $\alpha\beta E:_E r \subseteq \alpha\beta E$ and therefore

$$\alpha\beta E:_E r = \alpha\beta E.$$

Thus (a) implies (b).

Assume (b). Since $\alpha\beta$ is a non-zerodivisor on E, it follows that each of α, β is a non-zerodivisor on E. Let e belong to $\alpha E:_E r$. Then βe is in $\alpha\beta E:_E r = \alpha\beta E$ say $\beta e = \alpha\beta e^*$. But β is not a zerodivisor on E. Consequently $e = \alpha e^*$. We conclude therefore that $\alpha E:_E r = \alpha E$ and now we have shown that α, r is an R-sequence on E. The proof that β, r is an R-sequence on E is similar.

In the next theorem $r_1, \ldots, r_{p-1}$, $r_{p+1}, \ldots, r_q$ and α, β denote elements of R.

THEOREM 2. *If E is an R-module, then the following statements are equivalent:*

(a) *each of $r_1, \ldots, r_{p-1}, \alpha, r_{p+1}, \ldots, r_q$ and $r_1, \ldots, r_{p-1}, \beta, r_{p+1}, \ldots, r_q$ is an R-sequence on E;*

(b) *$r_1, \ldots, r_{p-1}, \alpha\beta, r_{p+1}, \ldots, r_q$ is an R-sequence on E.*

Proof. Assume (a). Then, for $1 \leqslant i \leqslant p-1$, r_i is a non-zerodivisor on $E/(r_1 E + \ldots + r_{i-1}E)$. Also, since both α and β are non-zerodivisors on $E/(r_1 E + \ldots + r_{p-1}E)$, $\alpha\beta$ is a non-zerodivisor on $E/(r_1 E + \ldots + r_{p-1}E)$.

Next suppose that $1 \leqslant j \leqslant q-p$. Since $\alpha, r_{p+1}, \ldots, r_q$ is an R-sequence on $E/(r_1 E + \ldots + r_{p-1}E)$, it follows, from Theorem 1, that $\alpha, r_{p+j}, \ldots, r_q$ is an R-sequence on

$$E/(r_1 E + \ldots + r_{p-1}E + r_{p+1}E + \ldots + r_{p+j-1}E) = K$$

say. Thus α, r_{p+j} and likewise β, r_{p+j} are R-sequences on K and therefore, by Lemma 2, $\alpha\beta, r_{p+j}$ is an R-sequence on K as well. It follows that r_{p+j} is a non-zerodivisor on $K/\alpha\beta K$. But $K/\alpha\beta K$ is isomorphic to

$$E/(r_1 E + \ldots + r_{p-1} E + \alpha\beta E + r_{p+1} E + \ldots + r_{p+j-1} E)$$

and with this we have shown that (a) implies (b).

Assume (b). Then, for $1 \leqslant i \leqslant p-1$, r_i is a non-zerodivisor on $E/(r_1 E + \ldots + r_{i-1} E)$. Furthermore, since $\alpha\beta$ is a non-zerodivisor on $E/(r_1 E + \ldots + r_{p-1} E)$, α itself is not a zerodivisor on this module. Suppose next that $1 \leqslant j \leqslant q - p$. Then, because $\alpha\beta, r_{p+1}, \ldots, r_q$ is an R-sequence on $E/(r_1 E + \ldots + r_{p-1} E)$, Theorem 1 shows that $\alpha\beta, r_{p+j}$ is an R-sequence on

$$E/(r_1 E + \ldots + r_{p-1} E + r_{p+1} E + \ldots + r_{p+j-1} E) = K$$

say, Thus, by Lemma 2, r_{p+j} is a non-zerodivisor on $K/\alpha K$ and therefore it is a non-zerodivisor on

$$E/(r_1 E + \ldots + r_{p-1} E + \alpha E + r_{p+1} E + \ldots + r_{p+j-1} E).$$

We have therefore proved that $r_1, \ldots, r_{p-1}, \alpha, r_{p+1}, \ldots, r_q$ is an R-sequence on E and, by symmetry, the same is true of $r_1, \ldots, r_{p-1}, \beta, r_{p+1}, \ldots, r_q$. The proof is now complete.

THEOREM 3. *Let $\alpha_1, \alpha_2, \ldots, \alpha_p$ be elements of R, let E be an R-module, and let $\nu_1, \nu_2, \ldots, \nu_p$ be positive integers. Then $\alpha_1, \alpha_2, \ldots, \alpha_p$ is an R-sequence on E if and only if $\alpha_1^{\nu_1}, \alpha_2^{\nu_2}, \ldots, \alpha_p^{\nu_p}$ is an R-sequence on E.*

This is a straightforward application of Theorem 2.

Usually if $\alpha_1, \alpha_2, \ldots, \alpha_p$ is an R-sequence on an R-module E, then this property is destroyed if the terms of the sequence are rearranged in any way. However we do have

LEMMA 3. *Let $\alpha_1, \ldots, \alpha_p, \alpha_{p+1}, \ldots, \alpha_q$ be an R-sequence on the R-module E. Then*

$$(\alpha_1 E + \ldots + \alpha_{p-1} E + \alpha_{p+1} E):_E \alpha_p = \alpha_1 E + \ldots + \alpha_{p-1} E + \alpha_{p+1} E.$$

Hence $\alpha_1, \ldots, \alpha_{p-1}, \alpha_{p+1}, \alpha_p, \alpha_{p+2}, \ldots, \alpha_q$ is an R-sequence on E if and only if

$$(\alpha_1 E + \ldots + \alpha_{p-1} E):_E \alpha_{p+1} = \alpha_1 E + \alpha_2 E + \ldots + \alpha_{p-1} E.$$

Proof. Since $\alpha_p, \alpha_{p+1}, ..., \alpha_q$ is an R-sequence on

$$E/(\alpha_1 E + ... + \alpha_{p-1} E) = K$$

say, it follows, from Theorem 1, that $\alpha_p, \alpha_{p+2}, ..., \alpha_q$ is an R-sequence on $K/\alpha_{p+1}K$. In particular, α_p is not a zerodivisor on $K/\alpha_{p+1}K$ and therefore it is not a zerodivisor on

$$E/(\alpha_1 E + ... + \alpha_{p-1} E + \alpha_{p+1} E).$$

Accordingly

$$(\alpha_1 E + ... + \alpha_{p-1} E + \alpha_{p+1} E) :_E \alpha_p = \alpha_1 E + ... + \alpha_{p-1} E + \alpha_{p+1} E$$

and the proof is complete.

EXERCISE 1. *Give an example of a commutative ring R, an R-module E, and elements α, β (of R) such that α, β is an R-sequence on E, but β, α is not.*

The next theorem will be needed later to extend the theory of R-sequences.

THEOREM 4. *Let $\mathfrak{A}$ be an ideal of R and α, β elements of $\mathfrak{A}$ which are non-zerodivisors on the R-module E. Then $\alpha E :_E \mathfrak{A}$ and $\beta E :_E \mathfrak{A}$ are isomorphic modules under an isomorphism which matches the submodule αE of the former with the submodule βE of the latter. Consequently $(\alpha E :_E \mathfrak{A})/\alpha E$ and $(\beta E :_E \mathfrak{A})/\beta E$ are isomorphic.*

Proof. Let $u \in (\alpha E :_E \mathfrak{A})$. Since $\beta \in \mathfrak{A}$, we have $\beta u \in \alpha E$. Consequently, because α is not a zerodivisor on E, there is a *unique* $e \in E$ such that $\beta u = \alpha e$. Further, if $a \in \mathfrak{A}$, then $\alpha a e = \beta a u \in \beta \alpha E$, whence $a e \in \beta E$ because α is a non-zerodivisor on E.

This argument shows that $e \in (\beta E :_E \mathfrak{A})$. Accordingly we have a mapping
$$f: (\alpha E :_E \mathfrak{A}) \to (\beta E :_E \mathfrak{A})$$

with the property that $\beta u = \alpha f(u)$ for all u in $\alpha E :_E \mathfrak{A}$. This is clearly a homomorphism. Similarly we have a homomorphism

$$g: (\beta E :_E \mathfrak{A}) \to (\alpha E :_E \mathfrak{A})$$

in which $\alpha v = \beta g(v)$ for each v in $\beta E :_E \mathfrak{A}$. An easy verification now shows that $gf(u) = u$ and $fg(v) = v$. Consequently f and g are inverse isomorphisms.

Finally we have $\alpha f(\alpha E) = \alpha(\beta E)$ whence $f(\alpha E) = \beta E$ because α is not a zerodivisor on E. This completes the proof.

EXERCISE 2. *Let E be an R-module and S a multiplicatively closed subset of R not containing the zero element. Show that if $\alpha_1, \alpha_2, ..., \alpha_p$ is an R-sequence on E, then $\alpha_1/1, \alpha_2/1, ..., \alpha_p/1$ is an R_S-sequence on E_S.*

5.2 Polynomials

We must now leave the subject of R-sequences for the moment in order to establish certain facts about polynomials. To this end let $x_1, x_2, x_3, ...$ be an infinite sequence of distinct indeterminates and let m be a positive integer. Then the polynomials in $x_1, x_2, ..., x_m$ with coefficients in R form a commutative ring which is denoted by $R[x_1, x_2, ..., x_m]$. Of course we can identify an element of R with the corresponding 'constant' polynomial. In this way R becomes a subring of $R[x_1, x_2, ..., x_m]$.

The typical element of $R[x_1, x_2, ..., x_m]$ is a *finite* formal sum

$$\Sigma a_{\mu_1 \mu_2 ... \mu_m} x_1^{\mu_1} x_2^{\mu_2} ... x_m^{\mu_m},$$

where $\mu_1, \mu_2, ..., \mu_m$ is a sequence of non-negative integers and the coefficient $a_{\mu_1 \mu_2 ... \mu_m}$ of the power product $x_1^{\mu_1} x_2^{\mu_2} ... x_m^{\mu_m}$ belongs to R. Note that if $\mathfrak{A}$ is an ideal of R, then it forms a subset of

$$R[x_1, x_2, ..., x_m]$$

and therefore it generates an ideal in the polynomial ring. This ideal, which is denoted by $\mathfrak{A}R[x_1, x_2, ..., x_m]$, consists of all those polynomials whose coefficients belong to $\mathfrak{A}$.

Let E be an R-module. Then besides considering polynomials in $x_1, x_2, ..., x_m$ with coefficients in R we may also consider polynomials in $x_1, x_2, ..., x_m$ with coefficients in E. Such a polynomial is simply a finite formal sum

$$\Sigma e_{\nu_1 \nu_2 ... \nu_m} x_1^{\nu_1} x_2^{\nu_2} ... x_m^{\nu_m},$$

where the ν_i are non-negative integers and $e_{\nu_1 \nu_2 ... \nu_m} \in E$. Let us denote the aggregate of these new polynomials by $E[x_1, x_2, ..., x_m]$. It is clear how we can define addition in $E[x_1, x_2, ..., x_m]$ and that with respect to this obvious addition $E[x_1, x_2, ..., x_m]$ is an abelian group. However we can do more because $E[x_1, x_2, ..., x_m]$ has a natural structure as a module over the ring $R[x_1, x_2, ..., x_m]$. In this structure if $\alpha \in R$ and $e \in E$, then the product of $\alpha x_1^{\mu_1} x_2^{\mu_2} \ldots x_m^{\mu_m}$ and $e x_1^{\nu_1} x_2^{\nu_2} \ldots x_m^{\nu_m}$ is given by

$$(\alpha x_1^{\mu_1} x_2^{\mu_2} \ldots x_m^{\mu_m})(e x_1^{\nu_1} x_2^{\nu_2} \ldots x_m^{\nu_m}) = (\alpha e) x_1^{\mu_1+\nu_1} x_2^{\mu_2+\nu_2} \ldots x_m^{\mu_m+\nu_m};$$

furthermore the product of a general member of $R[x_1, x_2, \ldots, x_m]$ with a general member of $E[x_1, x_2, \ldots, x_m]$ is obtained from this special case by using the requirement that multiplication has to be bilinear.

Now suppose that $f: E \to E'$ is a homomorphism of R-modules and define a mapping

$$f^*: E[x_1, x_2, \ldots, x_m] \to E'[x_1, x_2, \ldots, x_m]$$

by means of the formula

$$f^*(\Sigma e_{\nu_1 \nu_2 \ldots \nu_m} x_1^{\nu_1} x_2^{\nu_2} \ldots x_m^{\nu_m}) = \Sigma f(e_{\nu_1 \nu_2 \ldots \nu_m}) x_1^{\nu_1} x_2^{\nu_2} \ldots x_m^{\nu_m}.$$

An easy verification shows that f^* is a homomorphism of $R[x_1, x_2, \ldots, x_m]$-modules. Obviously if $i: E \to E$ is the identity mapping of E, then $i^*: E[x_1, x_2, \ldots, x_m] \to E[x_1, x_2, \ldots, x_m]$ is also an identity mapping. Again if $f: E \to E'$ and $g: E' \to E''$ are homomorphisms of R-modules, then

$$(gf)^* = g^*f^*.$$

These remarks may be summarized, by using the language of Category Theory, as follows: *forming polynomials in m variables produces a covariant functor from modules over R to modules over $R[x_1, x_2, \ldots, x_m]$.* Our next theorem shows that the process of forming polynominals preserves exact sequences.

THEOREM 5. *Let $E' \to E \to E''$ be an exact sequence of R-modules. Then the derived sequence*

$$E'[x_1, x_2, \ldots, x_m] \to E[x_1, x_2, \ldots, x_m] \to E''[x_1, x_2, \ldots, x_m]$$

of $R[x_1, x_2, \ldots, x_m]$-modules is also exact.

Although Theorem 5 states a very important fact its proof is no more than a trivial verification. We shall therefore omit the details.

Note that $E[x_1, x_2, \ldots, x_m] = 0$ if and only if $E = 0$.

EXERCISE 3. *Let E and E' be R-modules and $m > 0$ an integer. Show that E and E' are isomorphic if and only if $E[x_1, x_2, \ldots, x_m]$ and $E'[x_1, x_2, \ldots, x_m]$ are isomorphic as $R[x_1, x_2, \ldots, x_m]$-modules.*

We shall now derive a number of useful formulae. In these $\mathfrak{A}$ will denote an ideal of R and K a submodule of an R-module E. In order to prevent our expressions from becoming too cumbersome we shall, when convenient, use the abbreviations

$$R[x; m] \quad \text{for} \quad R[x_1, x_2, ..., x_m] \qquad (5.2.1)$$

and $\qquad E[x; m] \quad \text{for} \quad E[x_1, x_2, ..., x_m]. \qquad (5.2.2)$

Let $j \colon K \to E$ be the inclusion mapping. Then the exact sequence

$$0 \to K \xrightarrow{\;j\;} E$$

gives rise to the sequence

$$0 \to K[x; m] \xrightarrow{\;j^*\;} E[x; m]$$

of $R[x; m]$-modules and this, by Theorem 5, is also exact. Thus j^* is an injection of $K[x; m]$ into $E[x; m]$ and it enables us to regard the former as a submodule of the latter. Note that *when $K[x_1, x_2, ..., x_m]$ is regarded as a submodule of $E[x_1, x_2, ..., x_m]$ it consists of all the members of $E[x_1, x_2, ..., x_m]$ whose coefficients belong to K.*

LEMMA 4. *Let K' and K'' be submodules of an R-module E, let $m > 0$ be an integer, and suppose that*

$$K'[x_1, x_2, ..., x_m] = K''[x_1, x_2, ..., x_m]$$

in $E[x_1, x_2, ..., x_m]$. Then $K' = K''$.

This is clear as soon as one compares the constant polynomials in $K'[x_1, x_2, ..., x_m]$ with those in $K''[x_1, x_2, ..., x_m]$.

Let $\mathfrak{A}$, K, E be as explained above, and let us apply Theorem 5 to the exact sequence

$$0 \to K \xrightarrow{\;j\;} E \to E/K \to 0.$$

This yields the exact sequence

$$0 \to K[x; m] \xrightarrow{\;j^*\;} E[x; m] \to (E/K)[x; m] \to 0$$

and now, if we regard $K[x; m]$ as a submodule of $E[x; m]$, then we obtain an isomorphism

$$E[x_1, ..., x_m]/K[x_1, ..., x_m] \approx (E/K)[x_1, ..., x_m] \qquad (5.2.3)$$

of $R[x_1, x_2, ..., x_m]$-modules.

Consider next the submodule

$$(\mathfrak{A}R[x_1, x_2, ..., x_m]) E[x_1, x_2, ..., x_m]$$

of the $R[x_1, x_2, ..., x_m]$-module $E[x_1, x_2, ..., x_m]$. Evidently it is composed of all members of $E[x_1, x_2, ..., x_m]$ whose coefficients belong to $\mathfrak{A}E$. Consequently

$$(\mathfrak{A}R[x; m]) E[x; m] = \mathfrak{A}(E[x; m]) = (\mathfrak{A}E)[x; m] \quad (5.2.4)$$

and therefore the expression $\mathfrak{A}E[x_1, x_2, ..., x_m]$ may be used without risk of ambiguity.

Again if
$$\phi = \Sigma e_{\nu_1 \nu_2 ... \nu_m} x_1^{\nu_1} x_2^{\nu_2} ... x_m^{\nu_m}$$

belongs to $E[x_1, x_2, ..., x_m]$, then ϕ belongs to

$$K[x; m]:_{E[x; m]} \mathfrak{A}R[x; m]$$

if and only if $\mathfrak{A}\phi \subseteq K[x; m]$. However, this is the case if and only if $\mathfrak{A}e_{\nu_1 \nu_2 ... \nu_m} \subseteq K$ for every coefficient $e_{\nu_1 \nu_2 ... \nu_m}$ and this in turn happens when and only when $\phi \in (K:_E \mathfrak{A})[x; m]$. Accordingly

$$K[x; m]:_{E[x; m]} \mathfrak{A}R[x; m] = (K:_E \mathfrak{A})[x; m]. \quad (5.2.5)$$

Let us apply (5.2.5) to the case where $\mathfrak{A}$ is a principal ideal $R\alpha$. We then find that

$$K[x; m]:_{E[x; m]} \alpha = (K:_E \alpha)[x; m]. \quad (5.2.6)$$

Consequently $\quad K[x; m]:_{E[x; m]} \alpha = K[x; m]$

when and only when $(K:_E \alpha)[x; m] = K[x; m]$. But, by Lemma 4, this occurs precisely when $K:_E \alpha = K$. Thus, to sum up, α *is a non-zerodivisor on* E/K *when and only when* α *is a non-zerodivisor on* $E[x_1, x_2, ..., x_m]/K[x_1, x_2, ..., x_m]$.

LEMMA 5. *Let* $\alpha_1, \alpha_2, ..., \alpha_s$ *belong to* R, *let* E *be an* R-module, *and let* $m > 0$ *be an integer. Then* $\alpha_1, \alpha_2, ..., \alpha_s$ *is an* R-sequence on E *if and only if it is an* $R[x_1, x_2, ..., x_m]$-sequence on $E[x_1, x_2, ..., x_m]$.

Proof. Suppose that $1 \leqslant i \leqslant s$ and put

$$N = E/(\alpha_1 E + \alpha_2 E + ... + \alpha_{i-1} E)$$

so that $N = E/\mathfrak{A}E$, where $\mathfrak{A} = R\alpha_1 + R\alpha_2 + ... + R\alpha_{i-1}$. By (5.2.3) and (5.2.4), $N[x_1, x_2, ..., x_m]$ and

$$E[x_1, x_2, ..., x_m]/\mathfrak{A}E[x_1, x_2, ..., x_m]$$

are isomorphic as $R[x_1, x_2, ..., x_m]$-modules. The observation made just before Lemma 5 now shows that α_i is a non-zerodivisor on

$N = E/(\alpha_1 E + \alpha_2 E + ... + \alpha_{i-1} E)$ if and only if it is a non-zero-divisor on

$$E[x; m]/\mathfrak{A}E[x; m]$$
$$= E[x; m]/(\alpha_1 E[x; m] + \alpha_2 E[x; m] + ... + \alpha_{i-1} E[x; m]).$$

This, of course, implies the truth of the lemma.

We conclude this section by noting what happens if we first adjoin a set of indeterminates and then afterwards adjoin some new ones. To this end suppose that $m < n$ and, as before, let E be an R-module. Put $R^* = R[x_1, x_2, ..., x_m]$ and $E^* = E[x_1, x_2, ..., x_m]$ so that E^* is an R^*-module. If further indeterminates

$$x_{m+1}, x_{m+2}, ..., x_n$$

are introduced, then we can form both

$$R^*[x_{m+1}, x_{m+2}, ..., x_n] \quad \text{and} \quad E^*[x_{m+1}, x_{m+2}, ..., x_n]$$

and, of course, the latter is a module with respect to the former.

Consider $R[x_1, ..., x_m, x_{m+1}, ..., x_n]$. Its members are polynomials in $x_1, ..., x_m, x_{m+1}, ..., x_n$. Now such a polynomial can be written in a *unique* way as a polynomial in $x_{m+1}, x_{m+2}, ..., x_n$ with coefficients which are polynomials in $x_1, x_2, ..., x_m$. In fact we have a bijection between

$$R[x_1, ..., x_m, x_{m+1}, ..., x_n] \quad \text{and} \quad R^*[x_{m+1}, ..., x_n]$$

and on the basis of this we write

$$R[x_1, ..., x_m, x_{m+1}, ..., x_n] = R^*[x_{m+1}, ..., x_n] \qquad (5.2.7)$$

because, as is clear, the bijection is a ring-isomorphism.

For similar reasons we can write

$$E[x_1, ..., x_m, x_{m+1}, ..., x_n] = E^*[x_{m+1}, ..., x_n]. \qquad (5.2.8)$$

The bijection involved in (5.2.8) is certainly an isomorphism of abelian groups. However more is true. Let us identify

$$R[x_1, ..., x_m, x_{m+1}, ..., x_n]$$

with $R^*[x_{m+1}, ..., x_n]$ by means of (5.2.7) and

$$E[x_1, ..., x_m, x_{m+1}, ..., x_n]$$

with $E^*[x_{m+1}, ..., x_n]$ by means of (5.2.8). The important fact to be noted is the following: *the structure of $E[x_1, ..., x_m, x_{m+1}, ..., x_n]$*

as a module over $R[x_1, ..., x_m, x_{m+1}, ..., x_n]$ is the same as the structure of $E^*[x_{m+1}, ..., x_n]$ as a module over $R^*[x_{m+1}, ..., x_n]$. It will be convenient to refer to this observation as the *principle of successive adjunction of indeterminates*.

EXERCISE 4. *Show that an idempotent in the polynomial ring* $R[x_1, x_2, ..., x_m]$ *is necessarily a constant.*

EXERCISE 5. *Let F be a free R-module and m a positive integer. Show that $F[x; m]$ is a free $R[x; m]$-module and that*

$$\mathrm{rank}_R(F) = \mathrm{rank}_{R[x; m]}(F[x; m]).$$

5.3   Zerodivisors in a polynomial ring

Before we come to the main results of this section, it will be convenient to make some further general observations concerning polynomials.

To this end let m be a positive integer and denote by Ω the set of all sequences $(\mu_1, \mu_2, ..., \mu_m)$, where each μ_i is a non-negative integer. If $(\mu_1, \mu_2, ..., \mu_m)$ and $(\nu_1, \nu_2, ..., \nu_m)$ both belong to Ω, we shall put

$$(\mu_1, \mu_2, ..., \mu_m) < (\nu_1, \nu_2, ..., \nu_m) \tag{5.3.1}$$

if there exists j $(1 \leqslant j \leqslant m)$ such that

$$\mu_1 = \nu_1, \quad \mu_2 = \nu_2, ..., \mu_{j-1} = \nu_{j-1}$$

and $\mu_j < \nu_j.$

It is easy to see that this provides a total order on Ω. Indeed any non-empty subset of Ω has a first member, so Ω is *well-ordered* by our relation. The order provided by (5.3.1) is known as the *lexicographical* order. Note that if

$$(\mu_1, \mu_2, ..., \mu_m) < (\nu_1, \nu_2, ..., \nu_m)$$

and $\sigma_1, \sigma_2, ..., \sigma_m$ are arbitrary non-negative integers, then

$$(\mu_1 + \sigma_1, \mu_2 + \sigma_2, ..., \mu_m + \sigma_m) < (\nu_1 + \sigma_1, \nu_2 + \sigma_2, ..., \nu_m + \sigma_m).$$

The next result is crucial for our theory.

THEOREM 6. *Let E be an R-module, let $f \in R[x_1, x_2, ..., x_m]$ and suppose that f is a zerodivisor on $E[x_1, x_2, ..., x_m]$. Then there exists $e \in E$, $e \neq 0$ such that $fe = 0$.*

Proof. By hypothesis, there exists $\phi \in E[x_1, x_2, ..., x_m]$ such that $\phi \neq 0$ and $f\phi = 0$. We choose ϕ so that it has the smallest possible number of non-zero coefficients.

We may assume that $f \neq 0$. Let us write

$$f = ax_1^{\mu_1} x_2^{\mu_2} ... x_m^{\mu_m} + bx_1^{\nu_1} x_2^{\nu_2} ... x_m^{\nu_m} + ... + rx_1^{\sigma_1} x_2^{\sigma_2} ... x_m^{\sigma_m},$$

where $a, b, ..., r$ are non-zero elements of R and

$$(\mu_1, \mu_2, ..., \mu_m) > (\nu_1, \nu_2, ..., \nu_m) > ... > (\sigma_1, \sigma_2, ..., \sigma_m)$$

in the lexicographical order. Similarly we let

$$\phi = ex_1^{p_1} x_2^{p_2} ... x_m^{p_m} + e'x_1^{q_1} x_2^{q_2} ... x_m^{q_m} + ... + e^* x_1^{t_1} x_2^{t_2} ... x_m^{t_m},$$

where $e, e', ..., e^*$ are non-zero elements of E and

$$(p_1, p_2, ..., p_m) > (q_1, q_2, ..., q_m) > ... > (t_1, t_2, ..., t_m).$$

Since $f\phi = 0$, it follows from consideration of the coefficient of $x_1^{\mu_1+p_1} x_2^{\mu_2+p_2} ... x_m^{\mu_m+p_m}$ that $ae = 0$. Now $a\phi$ has fewer non-zero terms than ϕ and $f(a\phi) = 0$. Accordingly our choice of ϕ shows that $a\phi = 0$ and therefore that

$$(f - ax_1^{\mu_1} x_2^{\mu_2} ... x_m^{\mu_m}) \phi = 0.$$

Next consideration of the coefficient of $x_1^{\nu_1+p_1} x_2^{\nu_2+p_2} ... x_m^{\nu_m+p_m}$ in this new relation shows that $be = 0$. Accordingly $b\phi$ has fewer non-zero terms than ϕ and $f(b\phi) = 0$. Consequently $b\phi = 0$ and therefore

$$(f - ax_1^{\mu_1} x_2^{\mu_2} ... x_m^{\mu_m} - bx_1^{\nu_1} x_2^{\nu_2} ... x_m^{\nu_m}) \phi = 0.$$

Proceeding in this way we discover that

$$ae = be = ... = re = 0.$$

Thus $fe = 0$ and, since $e \neq 0$, the proof is complete.

We shall now restate this result in a somewhat different form.

THEOREM 7. *Let E be an R-module, let $f \in R[x_1, x_2, ..., x_m]$, and let $\mathfrak{A}$ be the ideal of R generated by the coefficients of f. Then $0:_E \mathfrak{A} = 0$ if and only if f is a non-zerodivisor on $E[x_1, x_2, ..., x_m]$.*

Proof. If $0:_E \mathfrak{A} \neq 0$, then there exists $e \in E$, $e \neq 0$ such that $fe = 0$. Consequently f is a zerodivisor on $E[x_1, x_2, ..., x_m]$. On the other hand if f is a zerodivisor on $E[x_1, x_2, ..., x_m]$, then, by

Theorem 6, there exists $e' \in E$, $e' \neq 0$ such that $fe' = 0$. But then $\mathfrak{A}e' = 0$ and therefore $0:_E \mathfrak{A} \neq 0$. This completes the proof.

THEOREM 8. *Let $\mathfrak{A}$ be a finitely generated ideal of R. Then the following statements are equivalent:*

(i) $0:_E \mathfrak{A} = 0$;

(ii) *for all $m > 0$, $\mathfrak{A}R[x_1, x_2, ..., x_m]$ contains a non-zerodivisor on $E[x_1, x_2, ..., x_m]$;*

(iii) *for some $m > 0$, $\mathfrak{A}R[x_1, x_2, ..., x_m]$ contains a non-zerodivisor on $E[x_1, x_2, ..., x_m]$.*

Proof. Let $\mathfrak{A} = Ra_1 + Ra_2 + ... + Ra_p$. By Theorem 7 if $0:_E \mathfrak{A} = 0$, then $a_1 x_1 + a_2 x_2 + ... + a_p x_p$ is a non-zerodivisor on $E[x_1, x_2, ..., x_p]$. Now suppose that $n > 0$ and f in $\mathfrak{A}R[x_1, x_2, ..., x_n]$ is a non-zerodivisor on $E[x_1, x_2, ..., x_n]$. Let e belong to $0:_E \mathfrak{A}$. Since the coefficients of f belong to $\mathfrak{A}$ we have $fe = 0$ and therefore $e = 0$. Thus (i) and (iii) are equivalent. To complete the proof we establish

LEMMA 6. *Let $\mathfrak{B}$ be an ideal of R and E an R-module. Then the following statements are equivalent:*

(a) *for some $m > 0$, $\mathfrak{B}R[x_1, x_2, ..., x_m]$ contains a non-zerodivisor on $E[x_1, x_2, ..., x_m]$;*

(b) *for all $m > 0$, $\mathfrak{B}R[x_1, x_2, ..., x_m]$ contains a non-zerodivisor on $E[x_1, x_2, ..., x_m]$.*

Proof. Suppose that $p > 0$ and that $\mathfrak{B}R[x_1, x_2, ..., x_p]$ contains a polynomial f which is not a zerodivisor on $E[x_1, x_2, ..., x_p]$. Now let m be an arbitrary positive integer.

Denote by $\mathfrak{A}$ the ideal generated by the coefficients of f. Then $\mathfrak{A} \subseteq \mathfrak{B}$ and, by Theorem 7, $0:_E \mathfrak{A} = 0$. Clearly we can find ϕ in $R[x_1, x_2, ..., x_m]$ with the property that its coefficients also generate $\mathfrak{A}$. Then $\phi \in \mathfrak{B}R[x_1, x_2, ..., x_m]$ and, again by Theorem 7, ϕ is not a zerodivisor on $E[x_1, x_2, ..., x_m]$. Thus (a) implies (b). The converse, of course, is trivial.

DEFINITION. *The ideal $\mathfrak{A}$, of R, is said to contain a 'latent non-zerodivisor' on E if $\mathfrak{A}R[x]$ contains a non-zerodivisor on $E[x]$.*

In accordance with our usual practice, if $\mathfrak{A}$ contains a latent non-zerodivisor on R, then we shall say simply that it contains a latent non-zerodivisor. Note that if $\mathfrak{A}$ is *finitely generated*, then, by Theorem 8, $\mathfrak{A}$ contains a latent non-zerodivisor on E if and only if $0:_E \mathfrak{A} = 0$.

AN EXAMPLE. We note that it is possible to have an ideal $\mathfrak{A}$ in a ring R such that (i) $\mathfrak{A}$ is generated by two elements, (ii) $\mathfrak{A}$ is composed of zerodivisors, and (iii) $\mathfrak{A}$ contains a latent non-zerodivisor. Indeed the example following the proof of Theorem 12 in Chapter 4 certainly has properties (i) and (ii). Since it is finitely generated and $0:_E \mathfrak{A} = 0$, it has property (iii) as well.

EXERCISE 6. *Let the R-module E have a finite free resolution of finite length. Show that the following statements are equivalent:*
(a) $\mathrm{Char}_R (E) = 0$;
(b) $\mathrm{Ann}_R (E)$ *contains a latent non-zerodivisor.*

5.4 Further properties of R-sequences

In the lemma which follows $\mathfrak{A} = Ra_0 + Ra_1 + \ldots + Ra_m$ is an ideal of R and (as usual) x denotes an indeterminate. We put

$$f = a_0 + a_1 x + a_2 x^2 + \ldots + a_m x^m. \qquad (5.4.1)$$

LEMMA 7. *Let $\mathfrak{A}$ and f be as above, let E be an R-module, and let $\beta_1, \beta_2, \ldots, \beta_s$ be an R-sequence on E contained in $\mathfrak{A}$. Then, for $1 \leqslant i \leqslant s$, the sequence $\beta_1, \ldots, \beta_{i-1}, f, \beta_i, \ldots, \beta_{s-1}$ is an $R[x]$-sequence on $E[x]$ contained in $\mathfrak{A}R[x]$.*

Proof. Certainly $\beta_1, \ldots, \beta_{i-1}, f, \beta_i, \ldots, \beta_{s-1}$ are all contained in $\mathfrak{A}R[x]$. Suppose that $1 \leqslant i \leqslant s$ and put $K = E/(\beta_1 E + \ldots + \beta_{i-1}E)$. Then β_i is a non-zerodivisor on K and $\beta_i \in \mathfrak{A}$. Consequently $0:_K \mathfrak{A} = 0$ and therefore, by Theorem 7, f is a non-zerodivisor on

$$K[x] = E[x]/(\beta_1 E[x] + \ldots + \beta_{i-1}E[x]).$$

Accordingly

$$(\beta_1 E[x] + \ldots + \beta_{i-1}E[x]):_{E[x]} f = \beta_1 E[x] + \ldots + \beta_{i-1}E[x] \quad (5.4.2)$$

for $i = 1, 2, \ldots, s$. Since, by Lemma 5, $\beta_1, \beta_2, \ldots, \beta_{s-1}$ is an $R[x]$-sequence on $E[x]$, it follows that $\beta_1, \ldots, \beta_{s-1}, f$ is an $R[x]$-sequence on $E[x]$. Finally in view of (5.4.2) we can make repeated applications of Lemma 3 to show that $\beta_1, \ldots, \beta_{i-1}, f, \beta_i, \ldots, \beta_{s-1}$ is an $R[x]$-sequence on $E[x]$.

THEOREM 9. *Let* $\mathfrak{A}$ *be an ideal generated by* n *elements and* E *an R-module such that* $\mathfrak{A}E \neq E$. *If now* $\beta_1, \beta_2, \ldots, \beta_s$ *is an R-sequence on* E *composed of elements of* $\mathfrak{A}$, *then* $s \leqslant n$.

Proof. We shall use induction on n. When $n = 0$ the assertion is trivial. It will therefore be assumed that the inequality in question has been proved for the case $n = m$ $(m \geqslant 0)$ and we now examine the situation where $n = m+1$, say

$$\mathfrak{A} = Ra_0 + Ra_1 + \ldots + Ra_m.$$

Evidently from here on we may suppose that $s \geqslant 1$.

Define f as in (5.4.1). Then

$$\mathfrak{A}R[x] = R[x]f + R[x]a_1 + \ldots + R[x]a_m$$

and, by Lemma 7, $f, \beta_1, \beta_2, \ldots, \beta_{s-1}$ is an $R[x]$-sequence on $E[x]$. Furthermore

$$fE[x] + a_1E[x] + \ldots + a_mE[x] = \mathfrak{A}E[x] \neq E[x].$$

Put $K = E[x]/fE[x]$, $\bar{R} = R[x]/fR[x]$, $\overline{\mathfrak{A}} = \mathfrak{A}R[x]/fR[x]$ and denote by $\bar{\beta}_i$ the natural image of β_i in $\bar{R}$. Then K is an $\bar{R}$-module, $\overline{\mathfrak{A}}$ can be generated by $m = n-1$ elements, $\overline{\mathfrak{A}}K \neq K$, and $\bar{\beta}_1, \bar{\beta}_2, \ldots, \bar{\beta}_{s-1}$ is an $\bar{R}$-sequence on K composed of elements of $\overline{\mathfrak{A}}$. The inductive hypothesis therefore shows that $s-1 \leqslant n-1$. Accordingly $s \leqslant n$ and the theorem is proved.

In preparation for the next result we note that if $\mathfrak{B}$ is an ideal of R, and K, N are submodules of an R-module E with $N \subseteq K$, then

$$(K :_E \mathfrak{B})/N = (K/N) :_{E/N} \mathfrak{B}$$

and therefore we have an isomorphism

$$(K :_E \mathfrak{B})/K \approx ((K/N) :_{E/N} \mathfrak{B})/(K/N). \qquad (5.4.3)$$

THEOREM 10. *Let* $\mathfrak{B}$ *be an ideal of* R, *let* E *be an R-module, and let* $\alpha_1, \alpha_2, \ldots, \alpha_s$ *and* $\gamma_1, \gamma_2, \ldots, \gamma_s$ *be R-sequences on* E *each composed of* s *elements of* $\mathfrak{B}$. *In these circumstances there exists an isomorphism*

$$\frac{(\alpha_1 E + \alpha_2 E + \ldots + \alpha_s E) :_E \mathfrak{B}}{\alpha_1 E + \alpha_2 E + \ldots + \alpha_s E} \approx \frac{(\gamma_1 E + \gamma_2 E + \ldots + \gamma_s E) :_E \mathfrak{B}}{\gamma_1 E + \gamma_2 E + \ldots + \gamma_s E}$$

of R-modules.

Proof. We use induction on s. The case $s = 0$ is trivial and the case $s = 1$ follows from Theorem 4. We now suppose that $s > 1$ and that the theorem has been proved for R-sequences of length $s - 1$.

Choose $a_0, a_1, ..., a_m$ in $\mathfrak{B}$ so that the ideal

$$\mathfrak{A} = Ra_0 + Ra_1 + ... + Ra_m$$

contains all the α_i and γ_i and define f as in (5.4.1). By Lemmas 5 and 7, $\alpha_1, \alpha_2, ..., \alpha_{s-1}, \alpha_s$ and $\alpha_1, \alpha_2, ..., \alpha_{s-1}, f$ are $R[x]$-sequences on $E[x]$ contained in $\mathfrak{B}R[x]$. Hence α_s and f are non-zerodivisors on

$$E[x]/(\alpha_1 E[x] + ... + \alpha_{s-1} E[x]) = E' \quad \text{(say)}$$

and they are contained in $\mathfrak{B}R[x]$. It therefore follows, from Theorem 4, that we have an isomorphism

$$\frac{f E' :_{E'} \mathfrak{B}R[x]}{f E'} \approx \frac{\alpha_s E' :_{E'} \mathfrak{B}R[x]}{\alpha_s E'}$$

which in view of (5.4.3) shows that

$$\frac{(\alpha_1 E[x] + ... + \alpha_{s-1} E[x] + f E[x]) :_{E[x]} \mathfrak{B}R[x]}{\alpha_1 E[x] + ... + \alpha_{s-1} E[x] + f E[x]} \tag{5.4.4}$$

is isomorphic to

$$\frac{(\alpha_1 E[x] + ... + \alpha_{s-1} E[x] + \alpha_s E[x]) :_{E[x]} \mathfrak{B}R[x]}{\alpha_1 E[x] + ... + \alpha_{s-1} E[x] + \alpha_s E[x]}$$

and this in turn is isomorphic to

$$\left[\frac{(\alpha_1 E + \alpha_2 E + ... + \alpha_s E) :_E \mathfrak{B}}{\alpha_1 E + \alpha_2 E + ... + \alpha_s E} \right] [x]. \tag{5.4.5}$$

Similarly we can show that

$$\frac{(\gamma_1 E[x] + ... + \gamma_{s-1} E[x] + f E[x]) :_{E[x]} \mathfrak{B}R[x]}{\gamma_1 E[x] + ... + \gamma_{s-1} E[x] + f E[x]} \tag{5.4.6}$$

is isomorphic to

$$\left[\frac{(\gamma_1 E + \gamma_2 E + ... + \gamma_s E) :_E \mathfrak{B}}{\gamma_1 E + \gamma_2 E + ... + \gamma_s E} \right] [x]. \tag{5.4.7}$$

Now, in view of Exercise 3, it will suffice to prove that the modules in (5.4.5) and (5.4.7) are isomorphic and this will follow if we show that this is so for the modules in (5.4.4) and (5.4.6). Let us next observe that these latter modules do not depend on the order of the terms in the sequences $\alpha_1, ..., \alpha_{s-1}, f$ and $\gamma_1, ..., \gamma_{s-1}, f$. Furthermore, by Lemma 7, $f, \alpha_1, \alpha_2, ..., \alpha_{s-1}$ and $f, \gamma_1, \gamma_2, ..., \gamma_{s-1}$ are $R[x]$-sequences on $E[x]$. Thus we have reduced the problem to the case where the two sequences have the same initial term. In other words *it will suffice to prove the assertion of the theorem under the additional assumption that $\alpha_1 = \gamma_1$*.

Assume therefore that we have this situation. Then $\alpha_2, \alpha_3, ..., \alpha_s$ and $\gamma_2, \gamma_3, ..., \gamma_s$ are R-sequences on

$$E/\alpha_1 E = E/\gamma_1 E = M \quad \text{(say)}$$

composed of elements of $\mathfrak{B}$. Accordingly, by the induction hypothesis, we have an isomorphism

$$\frac{(\alpha_2 M + ... + \alpha_s M):_M \mathfrak{B}}{\alpha_2 M + \alpha_3 M + ... + \alpha_s M} \approx \frac{(\gamma_2 M + ... + \gamma_s M):_M \mathfrak{B}}{\gamma_2 M + \gamma_3 M + ... + \gamma_s M}.$$

However, in view of (5.4.3), this is equivalent to

$$\frac{(\alpha_1 E + \alpha_2 E + ... + \alpha_s E):_E \mathfrak{B}}{\alpha_1 E + \alpha_2 E + ... + \alpha_s E} \approx \frac{(\gamma_1 E + \gamma_2 E + ... + \gamma_s E):_E \mathfrak{B}}{\gamma_1 E + \gamma_2 E + ... + \gamma_s E},$$

which is what we were aiming to prove.

5.5 Classical grade and true grade

Let $\mathfrak{B}$ be an ideal of R and E an R-module. We consider (finite) R-sequences on E composed of elements of $\mathfrak{B}$. The upper bound of the lengths of all such R-sequences will be called the *classical grade* of $\mathfrak{B}$ on E and it will be denoted by $\mathrm{gr}_R\{\mathfrak{B}; E\}$. Since the sequence of zero length is, by our conventions, an R-sequence on E, we have

$$0 \leqslant \mathrm{gr}_R\{\mathfrak{B}; E\} \leqslant \infty.$$

As before let $x_1, x_2, x_3, ...$ be distinct indeterminates and let us use the abbreviations introduced in (5.2.1) and (5.2.2). It follows, from Lemma 5 and the principle of the successive adjunction of indeterminates, that

$$\mathrm{gr}_{R[x;\,m]}\{\mathfrak{B}R[x;\,m]; E[x;\,m]\}$$

is an increasing function of m. We put

$$\mathrm{Gr}_R\{\mathfrak{B}; E\} = \lim_{m\to\infty} \mathrm{gr}_{R[x;m]}\{\mathfrak{B}R[x; m]; E[x; m]\}$$

and refer to $\mathrm{Gr}_R\{\mathfrak{B}; E\}$ as the *true grade* or *polynomial grade* of $\mathfrak{B}$ on E.

The special case $E = R$ is important and, in accordance with our usual practice, we often simplify both the language and the notation in this situation. This we do by putting

$$\mathrm{Gr}_R\{\mathfrak{B}\} = \mathrm{Gr}_R\{\mathfrak{B}; R\} \qquad (5.5.1)$$

and by referring to $\mathrm{Gr}_R\{\mathfrak{B}\}$ as the *true grade* or *polynomial grade* of the ideal $\mathfrak{B}$.

Let us return to the general situation. Here we note that as an immediate consequence of the definition

$$\mathrm{gr}_R\{\mathfrak{B}; E\} \leqslant \mathrm{Gr}_R\{\mathfrak{B}; E\} \qquad (5.5.2)$$

and also

$$\mathrm{Gr}_R\{\mathfrak{B}R[x; m]; E[x; m]\} = \mathrm{Gr}_R\{\mathfrak{B}; E\} \qquad (5.5.3)$$

for all $m > 0$. Observe too that

$$\mathfrak{B}_1 \subseteq \mathfrak{B}_2 \text{ implies } \mathrm{Gr}_R\{\mathfrak{B}_1; E\} \leqslant \mathrm{Gr}_R\{\mathfrak{B}_2; E\}. \qquad (5.5.4)$$

Furthermore, in view of Lemma 6 and Theorem 8, we have

LEMMA 8. *In order that* $\mathrm{Gr}_R\{\mathfrak{B}; E\}$ *should be strictly positive it is necessary and sufficient that* $\mathfrak{B}$ *should contain a latent non-zerodivisor on* E. *Moreover if* $\mathfrak{B}$ *is finitely generated this occurs when and only when* $0:_E \mathfrak{B} = 0$.

Let us return for a moment to the example immediately preceding Exercise 6. This provides an ideal $\mathfrak{A}$, generated by two elements, with the property that $\mathrm{gr}_R\{\mathfrak{A}; R\} = 0$ but $\mathrm{Gr}_R\{\mathfrak{A}; R\} > 0$. Thus it is possible for the classical grade and the true grade to be different.

THEOREM 11. *Let* $\mathfrak{B}$ *be an ideal of* R *and* E *an* R-module. *Then*

$$\mathrm{Gr}_R\{\mathfrak{B}; E\} = \sup_{\mathfrak{A}} \mathrm{Gr}_R\{\mathfrak{A}; E\},$$

where $\mathfrak{A}$ *ranges over all finitely generated ideals contained in* $\mathfrak{B}$.

Proof. Suppose that $m > 0$ and that $\phi_1, \phi_2, \ldots, \phi_s$ is an $R[x; m]$-sequence on $E[x; m]$ contained in $\mathfrak{B}R[x; m]$. There exists a finitely generated ideal $\mathfrak{A}$ contained in $\mathfrak{B}$ and containing all the coefficients of $\phi_1, \phi_2, ..., \phi_s$. Then $\phi_i \in \mathfrak{A}R[x; m]$ and therefore

$$s \leqslant \mathrm{gr}_{R[x;\, m]}\{\mathfrak{A}R[x;\, m];\, E[x;\, m]\} \leqslant \mathrm{Gr}_R\{\mathfrak{A};\, E\}.$$

This shows that

$$\mathrm{Gr}_R\{\mathfrak{B};\, E\} \leqslant \sup_{\mathfrak{A}} \mathrm{Gr}_R\{\mathfrak{A};\, E\}$$

and now the theorem follows because the opposite inequality is trivial.

THEOREM 12. *Let $\mathfrak{B}$ be an ideal of R and E an R-module. Then*

$$\mathrm{Gr}_R\{\mathfrak{B};\, E\} = \mathrm{Gr}_R\{\mathrm{Rad}\,\mathfrak{B};\, E\}.$$

Proof. Suppose that $m > 0$ and let $\phi_1, \phi_2, ..., \phi_s$ be an $R[x;\, m]$-sequence on $E[x;\, m]$ contained in $(\mathrm{Rad}\,\mathfrak{B})\,R[x;\, m]$. Then each coefficient of ϕ_j belongs to $\mathrm{Rad}\,\mathfrak{B}$ and therefore some power of the coefficient is in $\mathfrak{B}$ itself. Accordingly the individual terms of ϕ_j are contained in $\mathrm{Rad}\,(\mathfrak{B}R[x;\, m])$ and therefore ϕ_j too will belong to this ideal. It follows that if k is large, then all of $\phi_1^k, \phi_2^k, ..., \phi_s^k$ are in $\mathfrak{B}R[x;\, m]$. Moreover, by Theorem 3, these will form an $R[x;\, m]$-sequence on $E[x;\, m]$. Thus

$$s \leqslant \mathrm{gr}_{R[x;\, m]}\{\mathfrak{B}R[x;\, m];\, E[x;\, m]\} \leqslant \mathrm{Gr}_R\{\mathfrak{B};\, E\},$$

whence $\qquad\qquad \mathrm{Gr}_R\{\mathrm{Rad}\,\mathfrak{B};\, E\} \leqslant \mathrm{Gr}_R\{\mathfrak{B};\, E\}.$

The theorem now follows because $\mathfrak{B}$ is contained in $\mathrm{Rad}\,\mathfrak{B}$.

EXERCISE 7. *Let $\mathfrak{A}$ be an ideal of R and put $\mathfrak{B} = \mathrm{Rad}\,\mathfrak{A}$. Show that*
$$\mathrm{Rad}\,\{\mathfrak{A}R[x_1, x_2, ..., x_m]\} = \mathfrak{B}R[x_1, x_2, ..., x_m]$$
for all $m > 0$.

THEOREM 13. *Let $\mathfrak{A}$ be an ideal of R which can be generated by n elements and let E be an R-module such that $\mathfrak{A}E \neq E$. Then $\mathrm{Gr}_R\{\mathfrak{A};\, E\} \leqslant n$.*

Proof. Theorem 9 applied to the ideal $\mathfrak{A}R[x_1, x_2, ..., x_m]$ and the $R[x_1, x_2, ..., x_m]$-module $E[x_1, x_2, ..., x_m]$ shows that

$$\mathrm{gr}_{R[x;\, m]}\{\mathfrak{A}R[x;\, m];\, E[x;\, m]\} \leqslant n.$$

The theorem follows on letting m tend to infinity.

THEOREM 14. *Let $\mathfrak{B}$ be an ideal of R, E an R-module, and $\beta_1, \beta_2, ..., \beta_s$ ($s \geqslant 0$) an R-sequence on E composed of elements of $\mathfrak{B}$. If now $\mathrm{Gr}_R\{\mathfrak{B}; E\} > s$, then there exists $f \in \mathfrak{B}R[x]$ such that $\beta_1, \beta_2, ..., \beta_s, f$ is an $R[x]$-sequence on $E[x]$ composed of elements of $\mathfrak{B}R[x]$.*

Proof. By Theorem 11 there exists a finitely generated ideal $\mathfrak{A} \subseteq \mathfrak{B}$ such that all the β_i are in $\mathfrak{A}$ and $\mathrm{Gr}_R\{\mathfrak{A}; E\} > s$. Now choose $m > 0$ sufficiently large to ensure that

$$\mathrm{gr}_{R[x;\,m]}\{\mathfrak{A}R[x;\,m]; E[x;\,m]\} > s.$$

Then there exist elements $u_1, u_2, ..., u_{s+1}$ in $\mathfrak{A}R[x;\,m]$ which form an $R[x;\,m]$-sequence on $E[x;\,m]$. Since $\beta_1, \beta_2, ..., \beta_s$ are also in $\mathfrak{A}R[x;\,m]$ and form an $R[x;\,m]$-sequence on $E[x;\,m]$, it follows from Theorem 10 that we have an isomorphism between the modules

$$\frac{(\beta_1 E[x;\,m] + ... + \beta_s E[x;\,m]) :_{E[x;\,m]} \mathfrak{A}R[x;\,m]}{\beta_1 E[x;\,m] + \beta_2 E[x;\,m] + ... + \beta_s E[x;\,m]}$$

and

$$\frac{(u_1 E[x;\,m] + ... + u_s E[x;\,m]) :_{E[x;\,m]} \mathfrak{A}R[x;\,m]}{u_1 E[x;\,m] + u_2 E[x;\,m] + ... + u_s E[x;\,m]}. \tag{5.5.5}$$

But $u_{s+1} \in \mathfrak{A}R[x;\,m]$ and it is not a zerodivisor on

$$E[x;\,m]/(u_1 E[x;\,m] + ... + u_s E[x;\,m]).$$

Consequently (5.5.5) is a null module and therefore

$$(\beta_1 E[x;\,m] + ... + \beta_s E[x;\,m]) :_{E[x;\,m]} \mathfrak{A}R[x;\,m]$$
$$= \beta_1 E[x;\,m] + ... + \beta_s E[x;\,m].$$

Accordingly, by Lemma 4,

$$(\beta_1 E + \beta_2 E + ... + \beta_s E) :_E \mathfrak{A} = \beta_1 E + \beta_2 E + ... + \beta_s E$$

and therefore $\mathfrak{A}$ contains a latent non-zerodivisor on

$$E/(\beta_1 E + ... + \beta_s E).$$

In other words there exists $f \in \mathfrak{A}R[x] \subseteq \mathfrak{B}R[x]$ such that f is a non-zerodivisor on $E[x]/(\beta_1 E[x] + ... + \beta_s E[x])$. Thus $\beta_1, \beta_2, ..., \beta_s, f$ is an $R[x]$-sequence on $E[x]$ and it is composed of elements of $\mathfrak{B}R[x]$.

COROLLARY. *Let $\mathfrak{B}$ be an ideal of R, E an R-module, and $\beta_1, \beta_2, ..., \beta_s$ ($s \geqslant 0$) an R-sequence on E composed of elements of $\mathfrak{B}$.*

If now m > 0 is an integer and $\mathrm{Gr}_R\{\mathfrak{B}; E\} \geqslant s + m$, *then there exist* $f_1, f_2, \ldots, f_m$ *in* $\mathfrak{B}R[x; m]$ *such that* $\beta_1, \beta_2, \ldots, \beta_s, f_1, f_2, \ldots, f_m$ *is an* $R[x; m]$*-sequence on* $E[x; m]$.

This follows by successive applications of the theorem.

THEOREM 15. *Let* $\mathfrak{B}$ *be an ideal of* R *and* $\beta_1, \beta_2, \ldots, \beta_s$ *an* R*-sequence on* E *contained in* $\mathfrak{B}$. *Then*

$$\mathrm{Gr}_R\{\mathfrak{B}; E\} = \mathrm{Gr}_R\{\mathfrak{B}; E/(\beta_1 E + \ldots + \beta_s E)\} + s.$$

This follows immediately from the corollary to Theorem 14.

EXERCISE 8. *Let* $\mathfrak{B}_1, \mathfrak{B}_2$ *be ideals of* R *and let* E *be an* R*-module. Show that*

$$\mathrm{Gr}_R\{\mathfrak{B}_1 \cap \mathfrak{B}_2; E\} = \min\left[\mathrm{Gr}_R\{\mathfrak{B}_1; E\}, \mathrm{Gr}_R\{\mathfrak{B}_2; E\}\right].$$

EXERCISE 9. *Let* $\mathfrak{A} = Ra_0 + Ra_1 + \ldots + Ra_m$ *be an ideal of* R, *and let* E *be an* R*-module. Put*

$$f_i = a_0 + a_1 x_i + a_2 x_i^2 + \ldots + a_m x_i^m \quad (i = 1, 2, \ldots).$$

Show that if $s \geqslant 0$, *then the following two statements are equivalent:*

(a) $\mathrm{Gr}_R\{\mathfrak{A}; E\} \geqslant s$;

(b) $f_1, f_2, \ldots, f_s$ *is an* $R[x_1, x_2, \ldots, x_s]$*-sequence on* $E[x_1, x_2, \ldots, x_s]$.

The next theorem guarantees the existence of prime ideals with certain properties in relation to grade.

THEOREM 16.† *Let* $\mathfrak{B} \neq R$ *be an ideal and* E *an* R*-module. Then there exists a prime ideal* P *such that*

$$\mathfrak{B} \subseteq P \quad \text{and} \quad \mathrm{Gr}_R\{P; E\} = \mathrm{Gr}_R\{\mathfrak{B}; E\}.$$

Proof. We may assume that $\mathrm{Gr}_R\{\mathfrak{B}; E\} = k$, where $0 \leqslant k < \infty$. First assume that $\mathrm{Gr}_R\{\mathfrak{B}; E\} = 0$ and let Σ consist of all ideals I such that $\mathfrak{B} \subseteq I$ and $\mathrm{Gr}_R\{I; E\} = 0$. Then Σ is not empty and we can partially order it by means of the inclusion relation. This, we claim, turns Σ into an inductive system. For let $\{I_\lambda\}_{\lambda \in \Lambda}$ be a non-empty totally ordered subset of Σ and let I be their union. Then I is an ideal and $\mathfrak{B} \subseteq I$. Suppose that $\mathfrak{A}$ is a finitely generated ideal contained in I. Then $\mathfrak{A} \subseteq I_\lambda$ for some $\lambda \in \Lambda$ and therefore $\mathrm{Gr}_R\{\mathfrak{A}; E\} = 0$. It follows, from Theorem 11, that

† Note that this is a stronger result than Theorem 12.

$\mathrm{Gr}_R\{I; E\} = 0$ and with this our claim that Σ is an inductive system is established.

By Zorn's Lemma, Σ contains a maximal member P say. Certainly $\mathfrak{B} \subseteq P$, $\mathrm{Gr}_R\{P; E\} = 0$ and $P \neq R$. *We claim that P is a prime ideal.* For suppose that α, β are elements of R and $\alpha \notin P$, $\beta \notin P$. Then $P + R\alpha$ and $P + R\beta$ strictly contain P and therefore they are not in Σ. Thus

$$\mathrm{Gr}_R\{P + R\alpha; E\} > 0 \quad \text{and} \quad \mathrm{Gr}_R\{P + R\beta; E\} > 0.$$

It follows, by Exercise 8, that

$$\mathrm{Gr}_R\{(P + R\alpha) \cap (P + R\beta); E\} > 0.$$

Now, by Chapter 4 Exercise 6,

$$(P + R\alpha) \cap (P + R\beta) \quad \text{and} \quad (P + R\alpha)(P + R\beta)$$

have the same radical. Consequently, by Theorem 12,

$$\mathrm{Gr}_R\{(P + R\alpha)(P + R\beta); E\} > 0$$

whence, *a fortiori,* $\quad \mathrm{Gr}_R\{P + R\alpha\beta; E\} > 0.$

This shows that $\alpha\beta \notin P$ and establishes the claim that P is a prime ideal. Thus the proof is complete so far as the case $k = 0$ is concerned.

Suppose next that $k > 0$ and assume for the moment that $\mathfrak{B}$ contains an R-sequence $\beta_1, \beta_2, \ldots, \beta_k$ on E of length k. Put $E' = E/(\beta_1 E + \beta_2 E + \ldots + \beta_k E)$. Then, by Theorem 15,

$$\mathrm{Gr}_R\{\mathfrak{B}; E'\} = 0.$$

Hence, by what has already been proved, there exists a prime ideal P such that $\mathfrak{B} \subseteq P$ and $\mathrm{Gr}_R\{P; E'\} = 0$. But then, again by Theorem 15, $\mathrm{Gr}_R\{P; E\} = k$.

Finally suppose simply that $k > 0$. Then there exists $m > 0$ such that $\mathfrak{B}R[x; m]$ contains an $R[x; m]$-sequence on $E[x; m]$ of length k. Accordingly, by the last paragraph, there exists a prime ideal Π of $R[x; m]$ such that $\mathfrak{B}R[x; m] \subseteq \Pi$ and

$$\begin{aligned}
\mathrm{Gr}_{R[x; m]}\{\Pi; E[x; m]\} &= \mathrm{Gr}_{R[x; m]}\{\mathfrak{B}R[x; m]; E[x; m]\} \\
&= \mathrm{Gr}_R\{\mathfrak{B}; E\} \\
&= k.
\end{aligned}$$

Put $P = \Pi \cap R$. Then P is a prime ideal of R and $\mathfrak{B} \subseteq P$. Furthermore

$$
\begin{aligned}
k \leqslant \mathrm{Gr}_R\{P; E\} &= \mathrm{Gr}_{R[x;\,m]}\{PR[x; m]; E[x; m]\} \\
&\leqslant \mathrm{Gr}_{R[x;\,m]}\{\Pi; E[x; m]\} \\
&= k.
\end{aligned}
$$

The theorem follows.

We add a few remarks about grade and direct sums. To this end let $\{E_i\}_{i \in I}$ be an arbitrary family of R-modules and put

$$
E = \sum_{i \in I} E_i \quad \text{(direct sum).} \tag{5.5.6}
$$

If now $\mathfrak{B}$ is an ideal of R, then

$$
\mathfrak{B}E = \sum_{i \in I} \mathfrak{B}E_i \quad \text{(direct sum),}
$$

and we have an isomorphism

$$
E/\mathfrak{B}E \approx \sum_{i \in I} (E_i/\mathfrak{B}E_i) \quad \text{(direct sum).}
$$

Furthermore, if $\beta \in R$ then β is a non-zerodivisor on E if and only if it is a non-zerodivisor on each E_i. These remarks combine to show that if $\beta_1, \beta_2, \ldots, \beta_s$ belong to R, then $\beta_1, \beta_2, \ldots, \beta_s$ *is an R-sequence on E if and only if it is an R-sequence on every* E_i.

Suppose that (5.5.6) holds. Then

$$
E[x_1, x_2, \ldots, x_m] = \sum_{i \in I} E_i[x_1, x_2, \ldots, x_m]
$$

this being a direct sum of $R[x_1, x_2, \ldots, x_m]$-modules. It therefore follows that

$$
\begin{aligned}
\mathrm{gr}_{R[x;\,m]}\{\mathfrak{B}R[x; m]; E[x; m]\} &\leqslant \min_{i \in I} \mathrm{gr}_{R[x;\,m]}\{\mathfrak{B}R[x; m]; E_i[x; m]\} \\
&\leqslant \min_{i \in I} \mathrm{Gr}_R\{\mathfrak{B}; E_i\},
\end{aligned}
$$

whence, on letting m tend to infinity, we deduce that

$$
\mathrm{Gr}_R\{\mathfrak{B}; E\} \leqslant \min_{i \in I} \mathrm{Gr}_R\{\mathfrak{B}; E_i\}. \tag{5.5.7}
$$

THEOREM 17. *Let $\mathfrak{A}$ be a finitely generated ideal of R and $\{E_i\}_{i \in I}$ an arbitary family of R-modules. Put $E = \sum\limits_{i \in I} E_i$ (direct sum).*

Then $\qquad\qquad \mathrm{Gr}_R\{\mathfrak{A}; E\} = \min\limits_{i \in I} \mathrm{Gr}_R\{\mathfrak{A}; E_i\}.$

Proof. If $\mathrm{Gr}_R\{\mathfrak{A}; E_i\} = 0$ for some i, then $\mathrm{Gr}_R\{\mathfrak{A}; E\} = 0$ by (5.5.7). Now suppose that $\mathrm{Gr}_R\{\mathfrak{A}; E_i\} \geqslant s > 0$ for all i. Let $\mathfrak{A} = Ra_0 + Ra_1 + \dots + Ra_m$ and put

$$f_j = a_0 + a_1 x_j + a_2 x_j^2 + \dots + a_m x_j^m.$$

Then, by Exercise 9, $f_1, f_2, \dots, f_s$ is an $R[x; s]$-sequence on $E_i[x; s]$ and therefore it is an $R[x; s]$-sequence on $E[x; s]$. Accordingly $\mathrm{Gr}_R\{\mathfrak{A}; E\} \geqslant s$. By letting s approach the value

$$\min_{i \in I} \mathrm{Gr}_R\{\mathfrak{A}; E_i\}$$

from below, we are now able to deduce that

$$\min_{i \in I} \mathrm{Gr}_R\{\mathfrak{A}; E_i\} \leqslant \mathrm{Gr}_R\{\mathfrak{A}; E\}.$$

This proves the theorem because, as is shown by (5.5.7), the opposite inequality holds almost trivially.

THEOREM 18. *Let $\mathfrak{B}$ be an arbitrary ideal of R, and let $E_1, E_2, \dots, E_m$ be R-modules. Put $E = E_1 \oplus E_2 \oplus \dots \oplus E_m$. Then*

$$\mathrm{Gr}_R\{\mathfrak{B}; E\} = \min_{1 \leqslant i \leqslant m} \mathrm{Gr}_R\{\mathfrak{B}; E_i\}.$$

Proof. Suppose that $\mathrm{Gr}_R\{\mathfrak{B}; E_i\} \geqslant q$ for $i = 1, 2, \dots, m$. By Theorem 11, we can find a finitely generated ideal $\mathfrak{A}_i \subseteq \mathfrak{B}$ so that $\mathrm{Gr}_R\{\mathfrak{A}_i; E_i\} \geqslant q$. Put $\mathfrak{A} = \mathfrak{A}_1 + \mathfrak{A}_2 + \dots + \mathfrak{A}_m$. Then $\mathfrak{A}$ is finitely generated, $\mathfrak{A} \subseteq \mathfrak{B}$ and $\mathrm{Gr}_R\{\mathfrak{A}; E_i\} \geqslant q$ for all i. Consequently, by Theorem 17, $\mathrm{Gr}_R\{\mathfrak{A}; E\} \geqslant q$. This shows that $\mathrm{Gr}_R\{\mathfrak{B}; E\} \geqslant q$ and thus we may conclude that

$$\mathrm{Gr}_R\{\mathfrak{B}; E\} \geqslant \min_{1 \leqslant i \leqslant m} \mathrm{Gr}_R\{\mathfrak{B}; E_i\}.$$

The theorem now follows because the opposite inequality holds by virtue of (5.5.7).

COROLLARY. *Let $\mathfrak{B}$ be an ideal of R and $E_1, E_2, \dots, E_m$ R-modules. Suppose that $\mathrm{Gr}_R\{\mathfrak{B}; E_i\} > 0$ for $i = 1, 2, \dots, m$. Then there exists $f \in \mathfrak{B}R[x]$ such that f is not a zerodivisor on any of $E_1[x], E_2[x], \dots, E_m[x]$.*

Proof. Put $E = E_1 \oplus E_2 \oplus \dots \oplus E_m$. Then $\mathrm{Gr}_R\{\mathfrak{B}; E\} > 0$ by

Theorem 18. Consequently there exists $f \in \mathfrak{B}R[x]$ such that f is not a zerodivisor on

$$E[x] = E_1[x] \oplus E_2[x] \oplus \ldots \oplus E_m[x].$$

This has the required property.

We end this section with some results which show how grade is affected by certain operations involving a change of ring. To this end let Σ be a non-trivial commutative ring with an identity element, and $\phi: R \to \Sigma$ a ring-homomorphism.† Suppose that E is a Σ-module. Then E may be regarded as an R-module by putting $re = \phi(r)\,e$, where r and e denote elements of R and E respectively.

Suppose now that $\beta_1, \beta_2, \ldots, \beta_s$ are elements of R. Clearly $\beta_1, \beta_2, \ldots, \beta_s$ *is an R-sequence on E if and only if*

$$\phi(\beta_1), \phi(\beta_2), \ldots, \phi(\beta_s)$$

is a Σ-sequence on E.

After these preliminaries, let $m > 0$ be an integer. The ring-homomorphism ϕ gives rise to a ring-homomorphism

$$\phi_m: R[x_1, x_2, \ldots, x_m] \to \Sigma[x_1, x_2, \ldots, x_m].$$

Here when $f \in R[x_1, x_2, \ldots, x_m]$ we obtain $\phi_m(f)$ by applying ϕ to all the coefficients of f. Naturally $E[x_1, x_2, \ldots, x_m]$ is a $\Sigma[x_1, x_2, \ldots, x_m]$-module and we can convert it into an $R[x_1, x_2, \ldots, x_m]$-module with the aid of the homomorphism ϕ_m. On the other hand we can equally well regard E as an R-module and then proceed directly to give $E[x_1, x_2, \ldots, x_m]$ the structure of an $R[x_1, x_2, \ldots, x_m]$-module. It should be noted that these two procedures produce identical results.

Now let $\mathfrak{A} = Ra_0 + Ra_1 + \ldots + Ra_q$ be a finitely generated ideal of R. Then
$$\mathfrak{A}\Sigma = \phi(a_0)\,\Sigma + \phi(a_1)\,\Sigma + \ldots + \phi(a_q)\,\Sigma.$$

Put
$$f_i = a_0 + a_1 x_i + a_2 x_i^2 + \ldots + a_q x_i^q$$

and observe that for $1 \leqslant i \leqslant m$ we have

$$\phi_m(f_i) = \phi(a_0) + \phi(a_1)\,x_i + \phi(a_2)\,x_i^2 + \ldots + \phi(a_q)\,x_i^q.$$

† It is understood that a ring-homomorphism takes identity element into identity element.

Our earlier remarks now show that $f_1, f_2, ..., f_m$ is an $R[x; m]$-sequence on $E[x; m]$ if and only if $\phi_m(f_1), \phi_m(f_2), ..., \phi_m(f_m)$ is a $\Sigma[x; m]$-sequence on $E[x; m]$. It therefore follows, from Exercise 9, that $\mathrm{Gr}_R\{\mathfrak{A}; E\} \geqslant m$ if and only if $\mathrm{Gr}_\Sigma\{\mathfrak{A}\Sigma; E\} \geqslant m$. Thus when $\mathfrak{A}$ is a *finitely generated* ideal of R and E is a Σ-module we have

$$\mathrm{Gr}_R\{\mathfrak{A}; E\} = \mathrm{Gr}_\Sigma\{\mathfrak{A}\Sigma; E\}. \tag{5.5.8}$$

This result is extended in

THEOREM 19. *Let Σ be a non-trivial commutative ring and $\phi: R \to \Sigma$ a ring-homomorphism. If now $\mathfrak{B}$ is any ideal of R and E is a Σ-module, then*

$$\mathrm{Gr}_R\{\mathfrak{B}; E\} = \mathrm{Gr}_\Sigma\{\mathfrak{B}\Sigma; E\}.$$

Proof. Let $\mathfrak{A} \subseteq \mathfrak{B}$ be a finitely generated ideal of R. Then, by (5.5.8),

$$\mathrm{Gr}_R\{\mathfrak{A}; E\} = \mathrm{Gr}_\Sigma\{\mathfrak{A}\Sigma; E\} \leqslant \mathrm{Gr}_\Sigma\{\mathfrak{B}\Sigma; E\}$$

whence, taking upper bounds as $\mathfrak{A}$ varies and using Theorem 11, we obtain

$$\mathrm{Gr}_R\{\mathfrak{B}; E\} \leqslant \mathrm{Gr}_\Sigma\{\mathfrak{B}\Sigma; E\}.$$

On the other hand, if I is any finitely generated ideal of Σ contained in $\mathfrak{B}\Sigma$, it is easy to see that we can find a finitely generated ideal $\mathfrak{A}$, of R, contained in $\mathfrak{B}$ and such that $I \subseteq \mathfrak{A}\Sigma$. It then follows that

$$\mathrm{Gr}_\Sigma\{I; E\} \leqslant \mathrm{Gr}_\Sigma\{\mathfrak{A}\Sigma; E\} = \mathrm{Gr}_R\{\mathfrak{A}; E\} \leqslant \mathrm{Gr}_R\{\mathfrak{B}; E\}$$

and this yields $\mathrm{Gr}_\Sigma\{\mathfrak{B}\Sigma; E\} \leqslant \mathrm{Gr}_R\{\mathfrak{B}; E\}$ by allowing I to vary. The theorem follows.

The final observations of this section concern the behaviour of grade in relation to the process of forming rings and modules of fractions. To this end let S be a multiplicatively closed subset of R not containing the zero element and let E be an R-module. Then S is also a multiplicatively closed subset of $R[x_1, x_2, ..., x_m]$ and we obtain a bijection

$$(R[x_1, x_2, ..., x_m])_S \xrightarrow{\sim} R_S[x_1, x_2, ..., x_m] \tag{5.5.9}$$

by mapping the element

$$\frac{\Sigma a_{\nu_1 \nu_2 \ldots \nu_m} x_1^{\nu_1} x_2^{\nu_2} \ldots x_m^{\nu_m}}{s}$$

(where $a_{\nu_1 \nu_2 \ldots \nu_m} \in R$ and $s \in S$) of the first ring into the element

$$\sum \frac{a_{\nu_1 \nu_2 \ldots \nu_m}}{s} x_1^{\nu_1} x_2^{\nu_2} \ldots x_m^{\nu_m}$$

of the second. Indeed this bijection is a ring-homomorphism and it enables us to identify $(R[x; m])_S$ with $R_S[x; m]$. In a similar manner we can construct a bijection

$$(E[x_1, x_2, \ldots, x_m])_S \xrightarrow{\sim} E_S[x_1, x_2, \ldots, x_m]. \qquad (5.5.10)$$

A straightforward verification now shows that the bijection (5.5.10) is actually an isomorphism between two *modules* over the ring $(R[x; m])_S = R_S[x; m]$. This fact is relevant to the next exercise which shows that grade increases on forming fractions.

EXERCISE 10. *Let $\mathfrak{B}$ be an ideal of R, E an R-module, and S a multiplicatively closed subset of R not containing zero. Show that*

$$\mathrm{Gr}_R\{\mathfrak{B}; E\} \leqslant \mathrm{Gr}_{R_S}\{\mathfrak{B}R_S; E_S\}.$$

5.6 Grade and exact sequences

Let K, M be R-modules, $\phi \colon K \to M$ a homomorphism, and α an element of R. Then $\phi(0 :_K \alpha) \subseteq 0 :_M \alpha$ and $\phi(\alpha K) \subseteq \alpha M$. Thus ϕ induces homomorphisms

$$\tilde{\phi} \colon (0 :_K \alpha) \to (0 :_M \alpha)$$

and $\qquad\qquad \hat{\phi} \colon (K/\alpha K) \to (M/\alpha M).$

Now assume that we have an exact sequence

$$0 \to E' \xrightarrow{\,f\,} E \xrightarrow{\,g\,} E'' \to 0 \qquad (5.6.1)$$

and, as before, let $\alpha \in R$. The observations of the last paragraph enable us to construct sequences

$$0 \to (0 :_{E'} \alpha) \xrightarrow{\tilde{f}} (0 :_E \alpha) \xrightarrow{\tilde{g}} (0 :_{E''} \alpha) \qquad (5.6.2)$$

and $\qquad E'/\alpha E' \xrightarrow{\hat{f}} E/\alpha E \xrightarrow{\hat{g}} E''/\alpha E'' \to 0. \qquad (5.6.3)$

Indeed it may be verified without difficulty that both (5.6.2) and (5.6.3) are exact. The interesting fact is that these two sequences may be combined into a single long exact sequence. The details of the connecting homomorphism are explained below.

Suppose that $e'' \in (0:_{E''} \alpha)$. Since g is an epimorphism, we can find $e \in E$ such that $g(e) = e''$. Then $g(\alpha e) = \alpha e'' = 0$ and therefore there exists $e' \in E'$ such that $f(e') = \alpha e$. Denote by $[e']$ the natural image of e' in $E'/\alpha E'$ and put

$$\Delta(e'') = [e']. \tag{5.6.4}$$

We pause at this point to make a few observations. In the above construction there is a certain freedom of choice in the selection of the element e. However if we choose differently, then the effect on e' is simply to add to it an element of $\alpha E'$ and therefore $[e']$ itself will remain unchanged. Thus (5.6.4) provides us with a well-defined mapping

$$\Delta \colon (0:_{E''} \alpha) \to E'/\alpha E' \tag{5.6.5}$$

and a trivial verification shows that, in fact, Δ is a homomorphism of R-modules. This is the *connecting homomorphism* referred to above.

LEMMA 9.† *Let* $0 \to E' \xrightarrow{f} E \xrightarrow{g} E'' \to 0$ *be an exact sequence of R-modules and let* $\alpha \in R$. *Then the resulting sequence*

$$0 \to (0:_{E'} \alpha) \xrightarrow{\tilde{f}} (0:_E \alpha) \xrightarrow{\tilde{g}} (0:_{E''} \alpha)$$
$$\xrightarrow{\Delta} (E'/\alpha E') \xrightarrow{\hat{f}} (E/\alpha E) \xrightarrow{\hat{g}} (E''/\alpha E'') \to 0$$

is also exact.

Proof. In view of our previous observations it is enough to show that the sequences

$$(0:_E \alpha) \xrightarrow{\tilde{g}} (0:_{E''} \alpha) \xrightarrow{\Delta} (E'/\alpha E') \tag{5.6.6}$$

and $\quad (0:_{E''} \alpha) \xrightarrow{\Delta} (E'/\alpha E') \xrightarrow{\hat{f}} (E/\alpha E) \tag{5.6.7}$

are exact. There are four items to be checked.

(1) Suppose that $e \in (0:_E \alpha)$. Then $\Delta(g(e)) = [e']$, where $e' \in E'$

---

† This is a special case of the well known *Ker–Coker sequence*. See for example [**(38)** §3.4, p. 62].

and is chosen to satisfy $f(e') = \alpha e = 0$. But f is a monomorphism and therefore $e' = 0$. Thus $\Delta(g(e)) = 0$ and we have shown that $\Delta\tilde{g} = 0$.

(2) Assume that $e'' \in (0:_{E''} \alpha)$ and that $\Delta(e'') = 0$. Then there exist $e \in E$ and $e' \in E'$ such that $g(e) = e''$, $f(e') = \alpha e$, and $[e'] = 0$. Thus $e' = \alpha e'_1$ for some $e'_1 \in E'$ and therefore

$$\alpha(e - f(e'_1)) = f(e') - f(\alpha e'_1) = 0.$$

Accordingly $e - f(e'_1)$ belongs to $0:_E \alpha$ and

$$\tilde{g}(e - f(e'_1)) = g(e) - gf(e'_1) = g(e) = e''.$$

This shows that $\operatorname{Ker}\Delta \subseteq \operatorname{Im}\tilde{g}$. The exactness of the sequence (5.6.6) has thus been established.

(3) We now consider (5.6.7). To this end suppose that $e'' \in (0:_{E''} \alpha)$. Then $\Delta(e'') = [e']$, where $e' \in E'$ and is obtained in the manner explained when the connecting homomorphism was defined. It follows that $f(e') \in \alpha E$ and hence $\hat{f}([e']) = 0$. This shows that $\hat{f}\Delta = 0$.

(4) Finally assume that $e' \in E'$ and $\hat{f}([e']) = 0$. Then $f(e') \in \alpha E$, say $f(e') = \alpha e$, and
$$\alpha g(e) = gf(e') = 0.$$

Consequently $g(e) \in (0:_E \alpha)$. Furthermore $\Delta(g(e)) = [e']$ and therefore $[e'] \in \operatorname{Im}\Delta$. This proves that $\operatorname{Ker}\hat{f} \subseteq \operatorname{Im}\Delta$ and now the lemma is established.

We are now ready to begin our discussion of the behaviour of grade in relation to exact sequences.

LEMMA 10. *Let $\mathfrak{B}$ be an ideal of R and $0 \to E' \to E$ an exact sequence. If now $\operatorname{Gr}_R\{\mathfrak{B}; E\} > 0$, then $\operatorname{Gr}_R\{\mathfrak{B}; E'\} > 0$.*

This is clear. For if $f \in \mathfrak{B}R[x_1, x_2, \ldots, x_m]$ and is a non-zerodivisor on $E[x_1, x_2, \ldots, x_m]$, then f is also a non-zerodivisor on its submodule $E'[x_1, x_2, \ldots, x_m]$.

LEMMA 11. *Let $\mathfrak{B}$ be an ideal of R and $0 \to E' \to E \to E'' \to 0$ an exact sequence. Then*

$$\operatorname{Gr}_R\{\mathfrak{B}; E\} \geqslant \min[\operatorname{Gr}_R\{\mathfrak{B}; E'\}, \operatorname{Gr}_R\{\mathfrak{B}; E''\}].$$

Proof. Let $k \geqslant 0$ be an integer and suppose that $\operatorname{Gr}_R\{\mathfrak{B}; E'\} \geqslant k$ and $\operatorname{Gr}_R\{\mathfrak{B}; E''\} \geqslant k$. It will suffice to show that $\operatorname{Gr}_R\{\mathfrak{B}; E\} \geqslant k$.

This will be accomplished by induction on k. Note that the case $k = 0$ is trivial.

Now assume that $k \geqslant 1$ and that the desired result has been established for smaller values of the inductive variable. By Theorem 18 Cor., we can find $f \in \mathfrak{B}R[x]$ such that f is not a zero-divisor on either $E'[x]$ or $E''[x]$. Since we have an exact sequence

$$0 \to E'[x] \to E[x] \to E''[x] \to 0$$

we see that, for this part of the proof, we may simplify the notation and suppose that there is an element $\beta \in \mathfrak{B}$ which is a non-zerodivisor on both E' and E''. With this assumption $0 :_{E'} \beta = 0$ and $0 :_{E''} \beta = 0$. Hence, by Lemma 9, $0 :_{E} \beta = 0$ as well. Thus, by Theorem 15,

$$\mathrm{Gr}_R\{\mathfrak{B}; E'/\beta E'\} = \mathrm{Gr}_R\{\mathfrak{B}; E'\} - 1 \geqslant k - 1,$$

$$\mathrm{Gr}_R\{\mathfrak{B}; E''/\beta E''\} = \mathrm{Gr}_R\{\mathfrak{B}; E''\} - 1 \geqslant k - 1,$$

and $\quad \mathrm{Gr}_R\{\mathfrak{B}; E/\beta E\} = \mathrm{Gr}_R\{\mathfrak{B}; E\} - 1.$

Next, because $0 :_{E''} \beta = 0$, it follows from Lemma 9 that the sequence $\quad 0 \to E'/\beta E' \to E/\beta E \to E''/\beta E'' \to 0$

is exact. Consequently the inductive hypothesis shows that

$$\mathrm{Gr}_R\{\mathfrak{B}; E\} - 1 \geqslant k - 1$$

and with this the argument is complete.

LEMMA 12. *Let* $0 \to E' \to E \to E'' \to 0$ *be exact and let* $k \geqslant 0$ *be an integer. If now* $\mathrm{Gr}_R\{\mathfrak{B}; E'\} \geqslant k + 1$ *and* $\mathrm{Gr}_R\{\mathfrak{B}; E\} \geqslant k$, *then* $\mathrm{Gr}_R\{\mathfrak{B}; E''\} \geqslant k$.

Proof. As before we use induction on k and begin by observing that the case $k = 0$ is trivial. We shall therefore assume that $k \geqslant 1$ and that the desired result has been established in the case of smaller values of the inductive variable.

Our first aim will be to show that $\mathrm{Gr}_R\{\mathfrak{B}; E''\} > 0$. By adjoining an indeterminate we can suppose that, to all intents and purposes, there exists $\beta \in \mathfrak{B}$ with the property that β is not a zerodivisor on either E' or E. Thus, in particular, $0 :_{E} \beta = 0$, and therefore, by Lemma 9, we have an exact sequence

$$0 \to (0 :_{E''} \beta) \to E'/\beta E'.$$

Now $\mathrm{Gr}_R\{\mathfrak{B}; E'/\beta E'\} = \mathrm{Gr}_R\{\mathfrak{B}; E'\} - 1 \geqslant k > 0.$

We can therefore choose a finitely generated ideal $\mathfrak{A}$ so that $\beta \in \mathfrak{A} \subseteq \mathfrak{B}$ and $\mathrm{Gr}_R(\mathfrak{A}; E'/\beta E') > 0$. Then $(0:_{E''} \mathfrak{A}) \subseteq (0:_{E''} \beta)$ and therefore $0:_{E''} \mathfrak{A}$ is isomorphic to a submodule of $E'/\beta E'$. Accordingly, by Lemma 10, $\mathrm{Gr}_R\{\mathfrak{A}; 0:_{E''} \mathfrak{A}\} > 0$. But if we put $L = 0:_{E''} \mathfrak{A}$ this means (see Lemma 8) that $0:_L \mathfrak{A} = 0$. However $0:_L \mathfrak{A} = L$. Thus $0:_{E''} \mathfrak{A} = 0$ and therefore $\mathrm{Gr}_R\{\mathfrak{A}; E''\} > 0$. It follows that $\mathrm{Gr}_R\{\mathfrak{B}; E''\} > 0$ as we were aiming to prove.

Now that we know that $\mathrm{Gr}_R\{\mathfrak{B}; E'\} > 0$, $\mathrm{Gr}_R\{\mathfrak{B}; E\} > 0$ and $\mathrm{Gr}_R\{\mathfrak{B}; E''\} > 0$ we can make a fresh start and, by adjoining an indeterminate, place ourselves in the position where there exists $\gamma \in \mathfrak{B}$ with the property that it is not a zerodivisor on any of E', E, E''. Then, by Lemma 9, we have an exact sequence

$$0 \to E'/\gamma E' \to E/\gamma E \; - \; E''/\gamma E'' \to 0.$$

Moreover $\mathrm{Gr}_R\{\mathfrak{B}; E'/\gamma E'\} = \mathrm{Gr}_R\{\mathfrak{B}; E'\} - 1 \geqslant k$

and $\mathrm{Gr}_R\{\mathfrak{B}; E/\gamma E\} = \mathrm{Gr}_R\{\mathfrak{B}; E\} - 1 \geqslant k - 1.$

Accordingly the induction hypothesis shows that

$$\mathrm{Gr}_R\{\mathfrak{B}; E''/\gamma E''\} \geqslant k - 1$$

and therefore

$$\mathrm{Gr}_R\{\mathfrak{B}; E''\} = \mathrm{Gr}_R\{\mathfrak{B}; E''/\gamma E''\} + 1 \geqslant k.$$

This proves the lemma.

LEMMA 13. *Let* $0 \to E' \to E \to E'' \to 0$ *be an exact sequence and* $k \geqslant 1$ *an integer. Suppose that*

$$\mathrm{Gr}_R\{\mathfrak{B}; E\} > \mathrm{Gr}_R\{\mathfrak{B}; E'\} = k.$$

Then $\mathrm{Gr}_R\{\mathfrak{B}; E''\} = k - 1.$

Proof. We use induction on k. First suppose that $k = 1$. By adjoining an indeterminate we may assume that there exists $\beta \in \mathfrak{B}$ with the property that β is not a zerodivisor on E' or E. Lemma 9 then yields an exact sequence

$$0 = (0:_E \beta) \to (0:_{E''} \beta) \to E'/\beta E' \xrightarrow{\phi} E/\beta E.$$

Since $\mathrm{Gr}_R\{\mathfrak{B}; E'\} = 1$, we have $\mathrm{Gr}_R\{\mathfrak{B}; E'/\beta E'\} = 0$. On the other hand $\mathrm{Gr}_R\{\mathfrak{B}; E/\beta E\} > 0$. Let $\mathfrak{A}$ be any finitely generated ideal contained in $\mathfrak{B}$ for which $\text{Gr}_R\{\mathfrak{A}; E/\beta E\} > 0$. Since $\text{Gr}_R(\mathfrak{A}; E'/\beta E') = 0$, there exists $u \in E'/\beta E'$ such that $u \neq 0$ but $\mathfrak{A}u = 0$. It follows that $\mathfrak{A}\phi(u) = 0$ whence $\phi(u) = 0$ because $\text{Gr}_R\{\mathfrak{A}; E/\beta E\} > 0$. Thus there exists $v \in (0:_{E''}\beta)$ with the properties that $v \neq 0$ and $\mathfrak{A}v = 0$. Since $v \in E''$ this implies that $\text{Gr}_R\{\mathfrak{A}; E''\} = 0$. By varying $\mathfrak{A}$ and using Theorem 11, we conclude that $\text{Gr}_R\{\mathfrak{B}; E''\} = 0$. This disposes of the case $k = 1$.

We now assume that $k > 1$ and that the result we wish to prove has been established for smaller values of the inductive variable. First we note that, by Lemma 12, $\text{Gr}_R\{\mathfrak{B}; E''\} \geqslant k-1 > 0$. Thus, by adjoining an indeterminate if necessary, we may suppose that there exists $\gamma \in \mathfrak{B}$ such that γ is not a zerodivisor on any of E', E, E''. In this situation we have an exact sequence

$$0 \to E'/\gamma E' \to E/\gamma E \to E''/\gamma E'' \to 0,$$

where $\qquad \text{Gr}_R\{\mathfrak{B}; E/\gamma E\} > \text{Gr}_R\{\mathfrak{B}; E'/\gamma E'\} = k-1$.

It follows, by induction, that

$$\text{Gr}_R\{\mathfrak{B}; E''/\gamma E''\} = k-2.$$

Since $\text{Gr}_R\{\mathfrak{B}; E''/\gamma E''\} = \text{Gr}_R\{\mathfrak{B}; E''\} - 1$, this completes the proof.

We now reorganize these lemmas in a more convenient form.

Theorem 20 *Let $\mathfrak{B}$ be an ideal of R and $0 \to E' \to E \to E'' \to 0$ an exact sequence. In these circumstances the following statements hold:*

(a) *if* $\text{Gr}_R\{\mathfrak{B}; E'\} > \text{Gr}_R\{\mathfrak{B}; E\}$, *then*

$$\text{Gr}_R\{\mathfrak{B}; E''\} = \text{Gr}_R\{\mathfrak{B}; E\};$$

(b) *if* $\text{Gr}_R\{\mathfrak{B}; E'\} < \text{Gr}_R\{\mathfrak{B}; E\}$, *then*

$$\text{Gr}_R\{\mathfrak{B}; E''\} = \text{Gr}_R\{\mathfrak{B}; E'\} - 1;$$

(c) *if* $\text{Gr}_R\{\mathfrak{B}; E'\} = \text{Gr}_R\{\mathfrak{B}; E\}$, *then*

$$\text{Gr}_R\{\mathfrak{B}; E''\} \geqslant \text{Gr}_R\{\mathfrak{B}; E\} - 1.$$

Remark. It should be noticed that, in some respects, this forms a complement to Chapter 3 Theorem 18. Later† we shall be able to throw some light on why this is so.

† See Chapter 6 Theorem 2.

Proof. (a) Put $\mathrm{Gr}_R\{\mathfrak{B}; E\} = k$. Then, by Lemma 12,

$$\mathrm{Gr}_R\{\mathfrak{B}; E''\} \geqslant k.$$

On the other hand we cannot have $\mathrm{Gr}_R\{\mathfrak{B}; E''\} > k$ for this, by Lemma 11, would imply that $\mathrm{Gr}_R\{\mathfrak{B}; E\} > k$ giving a contradiction.

(b) Put $\mathrm{Gr}_R\{\mathfrak{B}; E'\} = k$. By Lemma 10, we must have $k \geqslant 1$. The desired result therefore follows from Lemma 13.

(c) If $\mathrm{Gr}_R\{\mathfrak{B}; E'\} = \mathrm{Gr}_R\{\mathfrak{B}; E\} = \infty$, then Lemma 12 shows that $\mathrm{Gr}_R\{\mathfrak{B}; E''\} \geqslant k$ for all $k \geqslant 0$. Thus we can confine our attention to the case where

$$\mathrm{Gr}_R\{\mathfrak{B}; E'\} = \mathrm{Gr}_R\{\mathfrak{B}; E\} = m$$

and m is an integer. We may also suppose that $m \geqslant 1$. This ensures that $\mathrm{Gr}_R\{\mathfrak{B}; E'\} \geqslant (m-1)+1$ and $\mathrm{Gr}_R\{\mathfrak{B}; E\} > m-1$. Consequently $\mathrm{Gr}_R\{\mathfrak{B}; E''\} \geqslant m-1$ by Lemma 12. This completes the proof.

EXERCISE 11. *Let E be a finitely generated R-module and $\mathfrak{B}$ an ideal of R such that $E = \mathfrak{B}E$. Show that $\mathrm{Gr}_R\{\mathfrak{B}; E\} = \infty$.*

5.7 A general criterion for exactness

In section (5.7) we consider a complex

$$\mathbb{C}: 0 \to C_n \xrightarrow{d_n} C_{n-1} \xrightarrow{d_{n-1}} C_{n-2} \to \dots \to C_1 \xrightarrow{d_1} C_0 \quad (5.7.1)$$

of R-modules and establish a general criterion for it to be exact. This criterion is due to L. Peskine and C. Szpiro (**39**).

Put $B_k = \mathrm{Im}\, d_{k+1}$ and $Z_k = \mathrm{Ker}\, d_k$. Then $B_k \subseteq Z_k$ and we may set
$$H_k(\mathbb{C}) = Z_k/B_k. \quad (5.7.2)$$

Of course $H_k(\mathbb{C})$ is just the kth *homology module* of $\mathbb{C}$; also $\mathbb{C}$ is exact if and only if $H_k(\mathbb{C}) = 0$ for $k = 1, 2, \dots, n$.

THEOREM 21. *Let $\mathbb{C}$ be the complex (5.7.1), suppose that $n \geqslant 1$, and let $\mathfrak{B}$ be an ideal of R. If now*

(1) $\mathrm{Gr}_R\{\mathfrak{B}; C_k\} \geqslant k$ *for* $1 \leqslant k \leqslant n$,
and

(2) *for each $k (1 \leqslant k \leqslant n)$ either $H_k(\mathbb{C}) = 0$ or $\mathrm{Gr}_R\{\mathfrak{B}; H_k(\mathbb{C})\} = 0$, then $\mathbb{C}$ is exact.*

Proof. Since $H_n(\mathbb{C}) = Z_n \subseteq C_n$, it follows that $H_n(\mathbb{C}) = 0$. (For we know that $\mathrm{Gr}_R\{\mathfrak{B}; C_n\} > 0$ and hence that $\mathrm{Gr}_R\{\mathfrak{B}; Z_n\} > 0$ by Lemma 10. Condition (2) therefore ensures that $H_n(\mathbb{C}) = 0$.) Thus

$$0 \to C_n \xrightarrow{d_n} C_{n-1}$$

is exact; also B_{n-1} is isomorphic to C_n and therefore

$$\mathrm{Gr}_R\{\mathfrak{B}; B_{n-1}\} \geqslant (n-1) + 1.$$

In particular the theorem has been established for the case $n = 1$.

Suppose now that $n > 1$ and that there is an integer s satisfying $0 \leqslant s < n$ for which

(a) $0 \to C_n \xrightarrow{d_n} C_{n-1} \to \dots \to C_{s+1} \xrightarrow{d_{s+1}} C_s$ is exact,

and

(b) $\mathrm{Gr}_R\{\mathfrak{B}; B_k\} \geqslant k+1$ for $s \leqslant k \leqslant n-1$.

For example we already know that (a) and (b) hold when $s = n-1$.

Naturally if $s = 0$, then our goal has been reached. Let us suppose therefore that $1 \leqslant s \leqslant n-1$. Now

$$\mathrm{Gr}_R\{\mathfrak{B}; B_s\} \geqslant s+1 \geqslant 2;$$

furthermore $\mathrm{Gr}_R\{\mathfrak{B}; C_s\} \geqslant s \geqslant 1$ and hence, by Lemma 10, we also have $\mathrm{Gr}_R\{\mathfrak{B}; Z_s\} \geqslant 1$. Consequently from the exact sequence

$$0 \to B_s \to Z_s \to H_s(\mathbb{C}) \to 0$$

and Lemma 12 we conclude that $\mathrm{Gr}_R\{\mathfrak{B}; H_s(\mathbb{C})\} \geqslant 1$ and hence that $H_s(\mathbb{C}) = 0$ by virtue of condition (2). Accordingly

$$0 \to C_n \to C_{n-1} \to \dots \to C_{s+1} \to C_s \to C_{s-1}$$

is exact and we have achieved the desired result if $s = 1$. We assume therefore that $s \geqslant 2$ and consider the exact sequence

$$0 \to B_s \to C_s \to B_{s-1} \to 0.$$

This time Lemma 12 shows that $\mathrm{Gr}_R\{\mathfrak{B}; B_{s-1}\} \geqslant s$. Thus the conditions corresponding to (a) and (b) continue to hold for the extended sequence

$$0 \to C_n \to C_{n-1} \to \dots \to C_s \to C_{s-1}.$$

The theorem now follows.

The final result of this chapter may be regarded as a weakened form of Theorem 21 but it has the advantage of being particularly convenient for applications.

THEOREM 22. *Let $\mathfrak{B}$ be an ideal of R and $\mathbb{C}$ a complex*

$$0 \to C_n \to C_{n-1} \to \ldots \to C_2 \to C_1 \to C_0$$

of R-modules. Assume that for $k = 1, 2, \ldots, n$ both $\mathrm{Gr}_R\{\mathfrak{B}; C_k\} \geqslant k$ and $\mathfrak{B}H_k(\mathbb{C}) = 0$. Then $\mathbb{C}$ is exact.

Proof. If E is an R-module and $\mathfrak{B}E = 0$, then either $E = 0$ or $\mathrm{Gr}_R\{\mathfrak{B}; E\} = 0$. Hence for each value of k either $H_k(\mathbb{C}) = 0$ or $\mathrm{Gr}_R\{\mathfrak{B}; H_k(\mathbb{C})\} = 0$. Theorem 21 therefore yields the desired result.

Solutions to the Exercises on Chapter 5

EXERCISE 1. *Give an example of a commutative ring R, an R-module E, and elements α, β (of R) such that α, β is an R-sequence on E but β, α is not.*

Solution. Let R be any ring containing a zerodivisor α. Then $1, \alpha$ is an R-sequence on R but $\alpha, 1$ is not an R-sequence on R.

A more illuminating example of an R-sequence whose terms cannot be permuted is obtained as follows. Let $R = K[u, v, w]$, where K is a field and u, v, w are indeterminates. Then $u, v(1-u)$, $w(1-u)$ is easily seen to be an R-sequence. However $v(1-u)$, $w(1-u)$ is not an R-sequence and so *a fortiori* the same is true of $v(1-u), w(1-u), u$.

EXERCISE 2. *Let E be an R-module and S a multiplicatively closed subset of R not containing the zero element. Show that if $\alpha_1, \alpha_2, \ldots, \alpha_p$ is an R-sequence on E, then $\alpha_1/1, \alpha_2/1, \ldots, \alpha_p/1$ is an R_S-sequence on E_S.*

Solution. Suppose that $0 \leqslant i < p$ and put

$$K = E/(\alpha_1 E + \alpha_2 E + \ldots + \alpha_i E).$$

Then the R_S-module K_S is isomorphic to

$$E_S \Big/ \Big(\frac{\alpha_1}{1} E_S + \frac{\alpha_2}{1} E_S + \ldots + \frac{\alpha_i}{1} E_S\Big).$$

It will therefore suffice to show that $\alpha_{i+1}/1$ is not a zerodivisor on K_S. Now $0:_K R\alpha_{i+1} = 0$. Consequently, by Chapter 3 Exercise 5,

$$0:_{K_S} R_S \frac{\alpha_{i+1}}{1} = 0$$

as required.

EXERCISE 3. *Let E and E' be R-modules and $m > 0$ an integer. Show that E and E' are isomorphic if and only if $E[x_1, x_2, ..., x_m]$ and $E'[x_1, x_2, ..., x_m]$ are isomorphic as $R[x_1, x_2, ..., x_m]$-modules.*

Solution. We shall assume that

$$E[x_1, x_2, ..., x_m] \quad \text{and} \quad E'[x_1, x_2, ..., x_m]$$

are isomorphic $R[x_1, x_2, ..., x_m]$-modules and deduce that E and E' are isomorphic R-modules. The converse is trivial.

First we note that

$$E[x_1, ..., x_m]/(x_1 E[x_1, ..., x_m] + ... + x_m E[x_1, ..., x_m])$$

and

$$E'[x_1, ..., x_m]/(x_1 E'[x_1, ..., x_m] + ... + x_m E'[x_1, ..., x_m])$$

are isomorphic $R[x_1, x_2, ..., x_m]$-modules. Next we can turn E into an $R[x_1, x_2, ..., x_m]$-module by defining the product of f in $R[x_1, x_2, ..., x_m]$ and e in E to be the result of multiplying e by the constant term of f. On this understanding E and

$$E[x_1, ..., x_m]/(x_1 E[x_1, ..., x_m] + ... + x_m E[x_1, ..., x_m])$$

are isomorphic $R[x_1, x_2, ..., x_m]$-modules. If we also turn E' into an $R[x_1, x_2, ..., x_m]$-module in a similar manner, we now see that E and E' are isomorphic as modules over $R[x_1, x_2, ..., x_m]$. But this clearly implies that they are isomorphic R-modules.

EXERCISE 4. *Show that an idempotent in the polynomial ring $R[x_1, x_2, ..., x_m]$ is necessarily a constant.*

Solution. Suppose that $f^2 = f$, where $f \in R[x_1, x_2, ..., x_m]$, and let η be the constant term of f. Then $\eta^2 = \eta$ and $f = \eta + g$, where g has no constant term. It follows that $(1 - \eta)g = (1 - \eta)f$ is an idempotent without constant term and therefore zero. Thus $g = \eta g$ whence

$$\eta + g = (\eta + g)^2 = \eta^2 + 2\eta g + g^2 = \eta + 2g + g^2$$

and therefore $g^2 = -g$. Accordingly $g^m = (-1)^{m+1}g$ for all $m \geqslant 2$. It follows, because g has no constant term, that $g = 0$ and therefore $f = \eta$.

EXERCISE 5. *Let F be a free R-module and m a positive integer. Show that $F[x; m]$ is a free $R[x; m]$-module and that*

$$\mathrm{rank}_R(F) = \mathrm{rank}_{R[x; m]}(F[x; m]).$$

Solution. Let $\{u_i\}_{i \in I}$ be a base for F over R. Each u_i can be regarded as a constant polynomial in $x_1, x_2, \ldots, x_m$ and as such it belongs to $F[x; m]$. On this understanding $\{u_i\}_{i \in I}$ is also a base for $F[x; m]$ over $R[x; m]$. The relation

$$\mathrm{rank}_R(F) = \mathrm{rank}_{R[x; m]}(F[x; m])$$
follows.

EXERCISE 6. *Let the R-module E have a finite free resolution of finite length. Show that the following statements are equivalent:*
(a) $\mathrm{Char}_R(E) = 0$;
(b) $\mathrm{Ann}_R(E)$ *contains a latent non-zerodivisor.*

Solution. That *(b)* implies *(a)* follows from Chapter 4 Theorem 12.

Assume (a). Then, by Chapter 4 Theorem 8 Cor., $0 :_R \mathfrak{F}(E) = 0$. As $\mathfrak{F}(E)$ is a finitely generated ideal† this shows that it contains a latent non-zerodivisor. However, by Chapter 3 Theorem 5, $\mathfrak{F}(E) \subseteq \mathrm{Ann}_R(E)$. The solution is therefore complete.

EXERCISE 7. *Let $\mathfrak{A}$ be an ideal of R and put $\mathfrak{B} = \mathrm{Rad}\,\mathfrak{A}$. Show that*

$$\mathrm{Rad}\,\{\mathfrak{A}R[x_1, x_2, \ldots, x_m]\} = \mathfrak{B}R[x_1, x_2, \ldots, x_m]$$

for all $m > 0$.

Solution. It is enough to prove the assertion in the case of a single variable x for the general version will follow by induction on the number of variables.

Evidently $\mathfrak{B}R[x] \subseteq \mathrm{Rad}\,\{\mathfrak{A}R[x]\}$. Suppose that

$$f = c_0 + c_1 x + \ldots + c_s x^s$$

---

† Here we use the fact that E is a finitely presented module.

belongs to $\mathrm{Rad}\{\mathfrak{A}R[x]\}$. Then there is a positive integer μ such that $f^{\mu} \in \mathfrak{A}R[x]$. It follows that $c_0^{\mu} \in \mathfrak{A}$ and therefore $c_0 \in \mathfrak{B}$. Accordingly

$$f - c_0 = c_1 x + c_2 x^2 + \ldots + c_s x^s \in \mathrm{Rad}\{\mathfrak{A}R[x]\}$$

and now we see that there is an integer $\nu > 0$ for which

$$(f - c_0)^{\nu} \in \mathfrak{A}R[x].$$

Consequently $c_1^{\nu} \in \mathfrak{A}$ and hence $c_1 \in \mathfrak{B}$. At this stage we have

$$f - c_0 - c_1 x = c_2 x^2 + c_3 x^3 + \ldots + c_s x^s \in \mathrm{Rad}\{\mathfrak{A}R[x]\}.$$

Proceeding in this way we establish that all of $c_0, c_1, \ldots, c_s$ are in $\mathfrak{B}$. Thus $f \in \mathfrak{B}R[x]$ and the desired result follows.

EXERCISE 8. *Let $\mathfrak{B}_1$, $\mathfrak{B}_2$ be ideals of R and let E be an R-module. Show that*

$$\mathrm{Gr}_R\{\mathfrak{B}_1 \cap \mathfrak{B}_2; E\} = \min[\mathrm{Gr}_R\{\mathfrak{B}_1; E\}, \mathrm{Gr}_R\{\mathfrak{B}_2; E\}].$$

Solution. It is trivial that

$$\mathrm{Gr}_R\{\mathfrak{B}_1 \cap \mathfrak{B}_2; E\} \leqslant \min[\mathrm{Gr}_R\{\mathfrak{B}_1; E\}, \mathrm{Gr}_R\{\mathfrak{B}_2; E\}].$$

Suppose now that $k \geqslant 0$ is an integer and that $\mathrm{Gr}_R\{\mathfrak{B}_1; E\} \geqslant k$, $\mathrm{Gr}_R\{\mathfrak{B}_2; E\} \geqslant k$. We shall show, using induction on k, that $\mathrm{Gr}_R\{\mathfrak{B}_1 \cap \mathfrak{B}_2; E\} \geqslant k$. This will complete the solution. Note that when $k = 0$ our contention is obvious.

Assume next that $m \geqslant 1$ and that what we aim to prove has been established when $k = m - 1$. Now suppose that

$$\mathrm{Gr}_R\{\mathfrak{B}_1; E\} \geqslant m \quad \text{and} \quad \mathrm{Gr}_R\{\mathfrak{B}_2; E\} \geqslant m.$$

In these circumstances there exist $f \in \mathfrak{B}_1 R[x]$ and $g \in \mathfrak{B}_2 R[x]$ such that neither f nor g is a zerodivisor on $E[x]$. It follows that fg is in $\mathfrak{B}_1 R[x] \cap \mathfrak{B}_2 R[x]$ and it is not a zerodivisor on $E[x]$. Accordingly, by Theorem 15,

$$\begin{aligned}
\mathrm{Gr}_{R[x]}\{\mathfrak{B}_1 R[x]; E[x]/fgE[x]\} \\
= \mathrm{Gr}_{R[x]}\{\mathfrak{B}_1 R[x]; E[x]\} - 1 \\
= \mathrm{Gr}_R\{\mathfrak{B}_1; E\} - 1 \\
\geqslant m - 1
\end{aligned}$$

and likewise

$$\mathrm{Gr}_{R[x]}\{\mathfrak{B}_2 R[x]; E[x]/fgE[x]\} \geqslant m - 1.$$

Thus, by the induction hypothesis,

$$\text{Gr}_{R[x]}\{\mathfrak{B}_1 R[x] \cap \mathfrak{B}_2 R[x]; E[x]/fgE[x]\} \geqslant m-1$$

and therefore

$$\text{Gr}_{R[x]}\{\mathfrak{B}_1 R[x] \cap \mathfrak{B}_2 R[x]; E[x]\} \geqslant m.$$

But $\mathfrak{B}_1 R[x] \cap \mathfrak{B}_2 R[x] = (\mathfrak{B}_1 \cap \mathfrak{B}_2) R[x]$. Consequently]

$$\text{Gr}_{R[x]}\{(\mathfrak{B}_1 \cap \mathfrak{B}_2) R[x]; E[x]\} \geqslant m$$

which at once yields

$$\text{Gr}_R\{\mathfrak{B}_1 \cap \mathfrak{B}_2; E\} \geqslant m.$$

The solution is now complete.

EXERCISE 9. *Let* $\mathfrak{A} = Ra_0 + Ra_1 + \ldots + Ra_m$ *be an ideal of* R *and let* E *be an* R-*module. Put*

$$f_i = a_0 + a_1 x_i + a_2 x_i^2 + \ldots + a_m x_i^m \quad (i = 1, 2, \ldots).$$

Show that if $s \geqslant 0$, *then the following two statements are equivalent:*
(a) $\text{Gr}_R\{\mathfrak{A}; E\} \geqslant s$;
(b) $f_1, f_2, \ldots, f_s$ *is an* $R[x_1, x_2, \ldots, x_s]$-*sequence on* $E[x_1, x_2, \ldots, x_s]$.

Solution. Obviously (b) implies (a). We shall show that (a) implies (b) by using induction on s. The case $s = 0$ is trivial. When $s = 1$ it suffices to combine Lemma 8 with Theorem 7.

Now suppose that $t \geqslant 1$ and assume that we have shown that (b) follows from (a) when $s = t$. Assume further that

$$\text{Gr}_R\{\mathfrak{A}; E\} \geqslant t+1.$$

If we put $R^* = R[x_1, x_2, \ldots, x_t]$, $E^* = E[x_1, x_2, \ldots, x_t]$ and $\mathfrak{A}^* = \mathfrak{A}R^*$, then

$$\text{Gr}_{R^*}\{\mathfrak{A}^*; E^*\} = \text{Gr}_R\{\mathfrak{A}; E\} \geqslant t+1$$

and $\mathfrak{A}^* = R^*a_0 + R^*a_1 + \ldots + R^*a_m$. Also $f_1, f_2, \ldots, f_t$ belong to $\mathfrak{A}^*$ and, by the induction hypothesis, they form an R^*-sequence on E^*. Accordingly, by Theorem 15,

$$\text{Gr}_{R^*}\{\mathfrak{A}^*; E^*/(f_1 E^* + \ldots + f_t E^*)\} > 0$$

and therefore, by Lemma 8 and Theorem 7, f_{t+1} is a non-zerodivisor on

$$E^*[x_{t+1}]/(f_1 E^*[x_{t+1}] + \ldots + f_t E^*[x_{t+1}]).$$

It follows that $f_1, \ldots, f_t, f_{t+1}$ is an $R^*[x_{t+1}]$-sequence on $E^*[x_{t+1}]$ that is to say it is an $R[x_1, x_2, \ldots, x_{t+1}]$-sequence on

$$E[x_1, x_2, \ldots, x_{t+1}].$$

EXERCISE 10. *Let $\mathfrak{B}$ be an ideal of R, E an R-module, and S a multiplicatively closed subset of R not containing zero. Show that*

$$\mathrm{Gr}_R\{\mathfrak{B}; E\} \leqslant \mathrm{Gr}_{R_S}\{\mathfrak{B}R_S; E_S\}.$$

Solution. Let $k \geqslant 0$ be an integer satisfying $k \leqslant \mathrm{Gr}_R\{\mathfrak{B}; E\}$. Then there is an integer $m > 0$ such that $\mathfrak{B}R[x; m]$ contains an $R[x; m]$-sequence on $E[x; m]$ of length k. Put $R^* = R[x; m]$, $\mathfrak{B}^* = \mathfrak{B}R[x; m]$, and $E^* = E[x; m]$. Then S is a multiplicatively closed subset of R^* and, by Exercise 2, $\mathfrak{B}^*R_S^*$ contains an R_S^*-sequence on E_S^* of length k. Consequently

$$k \leqslant \mathrm{Gr}_{R_S^*}\{\mathfrak{B}^*R_S^*, E_S^*\}.$$

However, as has been noted, we can identify R_S^* with $R_S[x; m]$ and E_S^* with $E_S[x; m]$. Moreover the identification of R_S^* with $R_S[x; m]$ makes $\mathfrak{B}^*R_S^*$ coincide with $\mathfrak{B}_S R_S[x; m]$, where

$$\mathfrak{B}_S = \mathfrak{B}R_S.$$

Thus $k \leqslant \mathrm{Gr}_{R_S[x; m]}\{\mathfrak{B}_S R_S[x; m]; E_S[x; m]\}$

and therefore $k \leqslant \mathrm{Gr}_{R_S}\{\mathfrak{B}_S; E_S\}.$

On letting k tend to $\mathrm{Gr}_R\{\mathfrak{B}; E\}$ from below we obtain the desired result.

EXERCISE 11. *Let E be a finitely generated R-module and $\mathfrak{B}$ an ideal of R such that $E = \mathfrak{B}E$. Show that $\mathrm{Gr}_R\{\mathfrak{B}; E\} = \infty$.*
   Solution. Let $E = Re_1 + Re_2 + \ldots + Re_p$. Then, for each

$$i\,(1 \leqslant i \leqslant p),$$

we have a relation $e_i = b_{i1}e_1 + b_{i2}e_2 + \ldots + b_{ip}e_p$, where $b_{ij} \in \mathfrak{B}$. Hence if δ_{ij} is the Kronecker symbol and D is the determinant of the $p \times p$ matrix $\|\delta_{ij} - b_{ij}\|$, then $De_\mu = 0$ for $\mu = 1, 2, \ldots, p$. Thus $D \in \mathrm{Ann}_R(E)$. But $D = 1 - \beta$ for some $\beta \in \mathfrak{B}$ and therefore $\beta e = e$ for all $e \in E$. It follows that a sequence of the form $\beta, \beta, \ldots, \beta$, whatever its length, is an R-sequence on E. Accordingly $\mathrm{Gr}_R\{\mathfrak{B}; E\} = \infty$.

6. Grade and finite free resolutions

General remarks

We are now ready to continue our study of finite free resolutions of finite length by applying the results of the last chapter. In order to make use of these it is helpful to have results which connect grade with restricted projective dimension. Such a result will be found in Theorem 2 below. This is a classic theorem in our subject. In its original form it dealt with the case where the ring under consideration was a Noetherian quasi-local ring. This form is due to M. Auslander and D. A. Buchsbaum [(1) Theorem 3.7, p. 401]. We shall however require a generalization which dispenses with the Noetherian condition. This extension was discovered by M. Hochster.

Our first application of the Auslander–Buchsbaum–Hochster Theorem will be to MacRae's invariant $\mathfrak{G}(E)$, where E is a module with an elementary resolution of finite length.† It will be shown that $\mathfrak{G}(E)$ is an *integral ideal* of R and not just a fractional ideal. Furthermore the connection between it and the initial Fitting invariant $\mathfrak{F}(E)$ will be made explicit by means of a theory of greatest common divisors. This leads naturally to the study of the connection between finite free resolutions and the factorization of non-zerodivisors into products of irreducible elements.

We shall also return to the consideration of complexes which have the form
$$0 \to F_n \to F_{n-1} \to \ldots \to F_1 \to F_0,$$
where each F_j is a finite free module. Already in Chapter 4 Theorem 8 we have obtained necessary conditions for such a complex to be exact. The theory of grade enables us to go further and give necessary and sufficient conditions for exactness. When

† See section (3.6).

R is Noetherian this theory is due to D. A. Buchsbaum and D. Eisenbud (**13**). However it turns out that the Noetherian condition is superfluous. Once again the extension to non-Noetherian rings is due to M. Hochster.

Throughout this chapter R denotes a commutative ring with a non-zero identity element and we make free use of the notation introduced earlier.

6.1 Finite free resolutions in the quasi-local case

Before coming to grips with the main theorem of this section, we need some preliminary results that are valid in a general (i.e. non-local) setting. We recall that if $\mathfrak{B}$ is an ideal of R, then $\mathrm{Gr}_R\{\mathfrak{B}\}$ is used as an alternative to $\mathrm{Gr}_R\{\mathfrak{B}; R\}$.

LEMMA 1. *Let* $0 \to F \xrightarrow{\phi} G$ *be an exact sequence where* $F \neq 0$, $G \neq 0$ *are finite free R-modules. Further let* $\mathfrak{B}$ *be an ideal of R such that* $\phi(F) \subseteq \mathfrak{B}G$. *Then* $\mathrm{Gr}_R\{\mathfrak{B}\} > 0$.

Proof. There exists a finitely generated ideal $\mathfrak{A}$ such that $\mathfrak{A} \subseteq \mathfrak{B}$ and $\phi(F) \subseteq \mathfrak{A}G$. Evidently it is enough to show that $\mathrm{Gr}_R\{\mathfrak{A}\} > 0$. We shall assume that $\mathrm{Gr}_R\{\mathfrak{A}\} = 0$ and derive a contradiction.

By Chapter 5 Lemma 8, we have $0 :_R \mathfrak{A} \neq 0$. Hence we can choose $r \in R$ so that $r \neq 0$ and $\mathfrak{A}r = 0$. Select any base for F and let x be an element of the base. Then

$$\phi(rx) = r\phi(x) \in r\mathfrak{A}G = 0.$$

Since ϕ is a monomorphism, it follows that $rx = 0$ and hence that $r = 0$. This is the desired contradiction.

THEOREM 1. *Let* $0 \to F \xrightarrow{\phi} G \to E \to 0$ *be an exact sequence where* $F \neq 0$, $G \neq 0$ *are finite free R-modules. Further let* $\mathfrak{B}$ *be an ideal of R such that* $\phi(F) \subseteq \mathfrak{B}G$. *Then* $\mathrm{Gr}_R\{\mathfrak{B}; E\} + 1 = \mathrm{Gr}_R\{B\}$.

Proof. We may suppose that $\mathfrak{B} \neq R$ for otherwise both sides of the equation would be infinite. Lemma 1 shows that $\mathrm{Gr}_R\{\mathfrak{B}\} > 0$ and from Chapter 5 Theorem 18 it follows that

$$\mathrm{Gr}_R\{\mathfrak{B}; F\} = \mathrm{Gr}_R\{\mathfrak{B}\} = \mathrm{Gr}_R\{\mathfrak{B}; G\}.$$

Accordingly, in view of Chapter 5 Theorem 20, we see that

$$\mathrm{Gr}_R\{\mathfrak{B}; E\} \geqslant \mathrm{Gr}_R\{\mathfrak{B}\} - 1.$$

Thus if $\mathrm{Gr}_R\{\mathfrak{B}\}$ is infinite the desired conclusion follows without further argument. We may therefore suppose that $\mathrm{Gr}_R\{\mathfrak{B}\} = k$, where $1 \leqslant k < \infty$. Our previous remarks now show that $\mathrm{Gr}_R\{\mathfrak{B}; E\} \geqslant k - 1$.

Suppose, to begin with, that $k > 1$. Then $\mathrm{Gr}_R\{\mathfrak{B}; F\}$, $\mathrm{Gr}_R\{\mathfrak{B}; G\}$ and $\mathrm{Gr}_R\{\mathfrak{B}; E\}$ are all strictly positive. Let x be an indeterminate and put $R^* = R[x]$, $F^* = F[x]$, $G^* = G[x]$, and $E^* = E[x]$. We then have an exact sequence

$$0 \to F^* \xrightarrow{\ \phi^*\ } G^* \to E^* \to 0,$$

where F^*, G^* are non-zero finite free R^*-modules and we also have $\phi^*(F^*) \subseteq \mathfrak{B}^* G^*$, where $\mathfrak{B}^* = \mathfrak{B}R^*$. By Chapter 5 Theorem 18 Cor., there exists $f \in \mathfrak{B}^*$ with the property that f is not a zero-divisor on any of F^*, G^*, E^*.† Accordingly there results an exact sequence $\quad 0 \to F^*/fF^* \to G^*/fG^* \to E^*/fE^* \to 0 \qquad (6.1.1)$

of R^*-modules.

Put $\bar{R} = R^*/fR^*$, $\bar{F} = F^*/fF^*$, $\bar{G} = G^*/fG^*$ and $\bar{E} = E^*/fE^*$. Then (6.1.1) can be regarded as an exact sequence

$$0 \to \bar{F} \xrightarrow{\ \bar{\phi}\ } \bar{G} \to \bar{E} \to 0$$

of $\bar{R}$-modules, where $\bar{F}$ and $\bar{G}$ are free and have finite bases. Note that $\bar{F}$ and $\bar{G}$ are non-zero because $\mathfrak{B}^* \neq R^*$, and that $\bar{\phi}(\bar{F}) \subseteq \bar{\mathfrak{B}}\bar{G}$, where $\bar{\mathfrak{B}} = \mathfrak{B}^*/fR^*$. Now, by Chapter 5 Theorem 19,

$$\mathrm{Gr}_{\bar{R}}\{\bar{\mathfrak{B}}; \bar{E}\} = \mathrm{Gr}_{R^*}\{\mathfrak{B}^*; E^*/fE^*\}$$
$$= \mathrm{Gr}_{R^*}\{\mathfrak{B}^*; E^*\} - 1$$

and therefore $\quad \mathrm{Gr}_{\bar{R}}\{\bar{\mathfrak{B}}; \bar{E}\} = \mathrm{Gr}_R\{\mathfrak{B}; E\} - 1.$

Similarly it follows that

$$\mathrm{Gr}_{\bar{R}}\{\bar{\mathfrak{B}}\} = \mathrm{Gr}_R\{\mathfrak{B}\} - 1 = k - 1.$$

These observations combine to show that the theorem will follow if we can deal with the case $k = 1$.

† Note that this ensures that f is not a zerodivisor on R^*.

From here on it will be assumed that $\mathrm{Gr}_R\{\mathfrak{B}\} = 1$ and we shall seek to show that $\mathrm{Gr}_R\{\mathfrak{B}; E\} = 0$. In fact we shall suppose that $\mathrm{Gr}_R\{\mathfrak{B}; E\} > 0$ and from this derive a contradiction.

Under the present assumptions we can repeat the construction used above and derive an exact sequence

$$0 \to \overline{F} \overset{\tilde{\phi}}{\longrightarrow} \overline{G},$$

where $\overline{F}$, $\overline{G}$ are non-zero finite free modules with respect to a non-trivial ring $\overline{R}$. Furthermore we obtain $\overline{\phi}(\overline{F}) \subseteq \overline{\mathfrak{B}}\overline{G}$, where $\overline{\mathfrak{B}}$ is an ideal of $\overline{R}$ with the property that

$$\mathrm{Gr}_{\overline{R}}\{\overline{\mathfrak{B}}\} = \mathrm{Gr}_R\{\mathfrak{B}\} - 1 = 0.$$

However this is the contradiction we have been seeking because, by Lemma 1, we must also have $\mathrm{Gr}_{\overline{R}}\{\overline{\mathfrak{B}}\} > 0$.

The next result is the Auslander–Buchsbaum–Hochster theorem.

THEOREM 2. *Let R be a quasi-local ring with maximal ideal M and let $E \neq 0$ be an R-module which admits a finite free resolution of finite length. Then*

$$\mathrm{Gr}_R\{M; E\} + \mathrm{Pd}_R^*(E) = \mathrm{Gr}_R\{M\}.$$

Proof. The hypotheses ensure that $\mathrm{Pd}_R^*(E)$ is finite. For convenience three cases will be distinguished.

Case (1) $\mathrm{Pd}_R^*(E) = 0$. In this case E is a finitely generated projective module and therefore a finite free module because R is a quasi-local ring. Consequently

$$\mathrm{Gr}_R\{M; E\} = \mathrm{Gr}_R\{M\}$$

by virtue of Chapter 5 Theorem 18.

Case (2) $\mathrm{Pd}_R^*(E) = 1$. This time, by using Chapter 4 Exercise 8, we construct an exact sequence

$$0 \to E' \to G \to E \to 0,$$

where G is a non-zero finite free module and E' is a submodule of G contained in MG. Since $\mathrm{Pd}_R^*(E) = 1$, E' is a finitely generated projective module and hence a finite free module because R is quasi-local. Also $E' \neq 0$. (For otherwise E itself would be a finite

free module contradicting the assumption that $\mathrm{Pd}_R^*(E) = 1$.)
We can therefore apply Theorem 1 and so obtain

$$\mathrm{Gr}_R\{M; E\} + 1 = \mathrm{Gr}_R\{M\}$$

which is precisely what we wish to prove.

Case (3) $\mathrm{Pd}_R^*(E) = k$, *where* $1 \leqslant k < \infty$. Here the method is to
use induction on k and we begin by observing that the situation
where $k = 1$ has already been dealt with under Case (2). Suppose
therefore that $k > 1$ and that the result in question has been
proved for non-zero modules with smaller restricted projective
dimension.

Construct an exact sequence

$$0 \to E' \to F \to E \to 0,$$

where F is a finite free module. Then

$$\mathrm{Pd}_R^*(E') = \mathrm{Pd}_R^*(E) - 1 = k - 1$$

and hence, by the induction hypothesis,

$$\mathrm{Gr}_R\{M; E'\} + k - 1 = \mathrm{Gr}_R\{M\}.$$

Let us first examine the case where $\mathrm{Gr}_R\{M\} = \infty$. In these
circumstances both $\mathrm{Gr}_R\{M; E'\} = \infty$ and $\mathrm{Gr}_R\{M; F\} = \infty$.
Consequently, by Chapter 5 Theorem 20, $\mathrm{Gr}_R\{M; E\} = \infty$ and
we have the desired relation. Thus we may assume that

$$\mathrm{Gr}_R\{M\} = m,$$

where $m < \infty$. Accordingly

$$\mathrm{Gr}_R\{M; E'\} = m - k + 1 < m = \mathrm{Gr}_R\{M; F\}.$$

We can therefore apply Theorem 20 of Chapter 5 to the exact
sequence $0 \to E' \to F \to E \to 0$ and thereby obtain

$$\mathrm{Gr}_R\{M; E\} = \mathrm{Gr}_R\{M; E'\} - 1.$$

It follows that

$$\mathrm{Gr}_R\{M; E\} + \mathrm{Pd}_R^*(E) = \mathrm{Gr}_R\{M; E'\} + k - 1 = \mathrm{Gr}_R\{M\}$$

and with this the proof is complete.

EXERCISE 1. *Let E be an R-module and S a multiplicatively* *closed subset of R not containing the zero element. Establish the inequalities* $\text{Pd}_{R_S}(E_S) \leqslant \text{Pd}_R(E)$ *and* $\text{Pd}^*_{R_S}(E_S) \leqslant \text{Pd}^*_R(E)$.

EXERCISE 2. *Let E be an R-module possessing a finite free resolution of finite length. Show that*

$$\text{Pd}^*_R(E) = \sup_M \text{Pd}^*_{R_M}(E_M),$$

where M ranges over all the maximal ideals of R.

We shall now derive a rather general version of an important theorem due to D. Rees.[†] First, however, we need a little extra terminology.

Let E be an R-module and P a prime ideal of R. It is usual to say that P *is associated with the zero submodule of E* or that P *is associated with E* if there exists $e \in E$ such that $P = 0:_R Re$. For a given module E the set of such prime ideals is denoted by $\text{Ass}_R(E)$. In the case of Noetherian rings and modules this concept works very satisfactorily. However in the present circumstances we need to modify it.

DEFINITION. *The prime ideal P will be said to be 'attached to the zero submodule of E' if* $\text{Gr}_{R_P}\{PR_P; E_P\} = 0$.

Naturally the set of prime ideals attached to the zero submodule of E will be denoted by $\text{Att}_R(E)$.

THEOREM 3.[‡] *For an R-module E we have* $\text{Ass}_R(E) \subseteq \text{Att}_R(E)$.

Proof. Let $P \in \text{Ass}_R(E)$ and choose $e \in E$ so that $0:_R Re = P$. Then, by Chapter 2 Theorem 12,

$$0:_{R_P} R_P \frac{e}{1} = PR_P.$$

Accordingly $e/1$ is a non-zero element of E_P and it is annihilated by PR_P. This is more than sufficient to ensure that

$$\text{Gr}_{R_P}\{PR_P; E_P\} = 0$$

and hence that $P \in \text{Att}_R(E)$.

The next group of exercises contains some elementary items of information concerning the new concept.

[†] See [(**40**) Theorem 1.2 Corollary, p. 30].
[‡] It is not hard to show that if E is a Noetherian R-module, then
$$\text{Ass}_R(E) = \text{Att}_R(E).$$

EXERCISE 3. *Let E be an R-module and P a member of* $\mathrm{Att}_R(E)$. *Show that* $\mathrm{Ann}_R(E) \subseteq P$.

EXERCISE 4. *Show that if E is a finitely generated R-module, then every minimal prime ideal of* $\mathrm{Ann}_R(E)$ *belongs to* $\mathrm{Att}_R(E)$.

EXERCISE 5. *Give an example of a ring R and a finitely generated R-module E for which* $\mathrm{Ass}_R(E) \neq \mathrm{Att}_R(E)$.

The next result is Rees's Theorem.

THEOREM 4. *Let $E \neq 0$ be an R-module which possesses a resolution, of finite length, by means of finitely generated projective R-modules. Then*
$$\mathrm{Gr}_R\{\mathrm{Ann}_R(E)\} \leqslant \mathrm{Pd}_R(E).$$
Indeed if $Q \in \mathrm{Att}_R(E)$, then
$$\mathrm{Gr}_R\{Q\} \leqslant \mathrm{Pd}_R(E).$$

Proof. The module E is finitely generated and $\mathrm{Ann}_R(E) \neq R$. It therefore follows, from Exercise 4, that $\mathrm{Att}_R(E)$ is not empty.

Let Q belong to $\mathrm{Att}_R(E)$. By Exercise 3, $\mathrm{Ann}_R(E) \subseteq Q$ and therefore
$$\mathrm{Gr}_R\{\mathrm{Ann}_R(E)\} \leqslant \mathrm{Gr}_R\{Q\} \leqslant \mathrm{Gr}_{R_Q}\{QR_Q\},$$
the second inequality holding by virtue of Chapter 5 Exercise 10. On the other hand, E_Q is non-zero and as an R_Q-module it has a finite *free* resolution of finite length. Accordingly Theorem 2 shows that
$$\mathrm{Gr}_{R_Q}\{QR_Q; E_Q\} + \mathrm{Pd}^*_{R_Q}(E_Q) = \mathrm{Gr}_{R_Q}\{QR_Q\}.$$
But $\mathrm{Gr}_{R_Q}\{QR_Q; E_Q\} = 0$ by the definition of $\mathrm{Att}_R(E)$ and
$$\mathrm{Pd}^*_{R_Q}(E_Q) = \mathrm{Pd}_{R_Q}(E_Q)$$
by Chapter 3 Theorem 15. Thus we see that
$$\mathrm{Gr}_R\{\mathrm{Ann}_R(E)\} \leqslant \mathrm{Gr}_R\{Q\} \leqslant \mathrm{Pd}_{R_Q}(E_Q).$$
The theorem now follows because, by Exercise 1,
$$\mathrm{Pd}_{R_Q}(E_Q) \leqslant \mathrm{Pd}_R(E).$$

The next theorem will be useful when we come to applications of Theorem 2 in the next section.

THEOREM 5. *Let $\mathfrak{A}$ be an ideal generated by an R-sequence of length s ($s \geqslant 0$) and let $\mathfrak{B}$ be any finitely generated ideal for which $\mathfrak{B} \nsubseteq \mathfrak{A}$. If now P is a minimal prime ideal of $\mathfrak{A} :_R \mathfrak{B}$, then $\mathrm{Gr}_{R_P}\{PR_P\} = s$.*

Proof. Let $\mathfrak{A} = R\alpha_1 + R\alpha_2 + \ldots + R\alpha_s$, where $\alpha_1, \alpha_2, \ldots, \alpha_s$ is an R-sequence, and put $I = \mathfrak{A} :_R \mathfrak{B}$. Since $\mathfrak{B} \nsubseteq \mathfrak{A}$, $I \neq R$. By hypothesis, P is a minimal prime ideal of I. Hence, from Chapter 3 Exercise 5, it follows that

$$\mathfrak{A}R_P :_{R_P} \mathfrak{B}R_P = (\mathfrak{A} :_R \mathfrak{B}) R_P = IR_P \subseteq PR_P$$

and therefore $\mathfrak{B}R_P \nsubseteq \mathfrak{A}R_P$. But

$$(IR_P)(\mathfrak{B}R_P) \subseteq \mathfrak{A}R_P$$

and so IR_P annihilates a non-zero element of $R_P/\mathfrak{A}R_P$. This shows that

$$\mathrm{Gr}_{R_P}\{IR_P;\, R_P/\mathfrak{A}R_P\} = 0.$$

However our construction ensures that $\mathrm{Rad}\,(IR_P) = PR_P$. Accordingly, by Chapter 5 Theorem 12,

$$\mathrm{Gr}_{R_P}\{PR_P;\, R_P/\mathfrak{A}R_P\} = 0.$$

Next $\qquad\qquad \mathfrak{A}R_P = R_P \dfrac{\alpha_1}{1} + R_P \dfrac{\alpha_2}{1} + \ldots + R_P \dfrac{\alpha_s}{1}$

and, by Chapter 5 Exercise 2, $\alpha_1/1, \alpha_2/1, \ldots, \alpha_s/1$ is an R_P-sequence. We can therefore use Chapter 5 Theorem 15 to conclude that $\mathrm{Gr}_{R_P}\{PR_P\} = s$. This completes the proof.

6.2   The integral character of MacRae's invariant

Let E be an R-module which admits an elementary resolution of finite length.† For such a module MacRae's invariant was defined in section (3.6). This invariant made its first appearance as a fractional ideal. In fact the definition gave

$$\mathfrak{G}(E) = R\frac{\alpha}{\beta},$$

where $\alpha, \beta \in R$ and are non-zerodivisors. We also know, from Chapter 3 Theorem 27, that when E is an elementary module

<hr>

† See section (3.5).

$\mathfrak{G}(E) = \mathfrak{F}(E)$. It will now be proved that $\mathfrak{G}(E)$ is always an *integral* ideal.

THEOREM 6. *Let the R-module E admit an elementary resolution of finite length. Then $\mathfrak{G}(E)$ is an integral ideal and it is generated by a non-zerodivisor.*

Proof. There exist elements α, β of R, not zerodivisors in the ring, for which

$$\mathfrak{G}(E) = R\frac{\alpha}{\beta}.$$

Clearly it will suffice to show that $R\alpha \subseteq R\beta$. We shall assume that $R\alpha \nsubseteq R\beta$ and derive a contradiction.

Since $R\alpha \nsubseteq R\beta$, the ideal $I = R\beta :_R R\alpha$ is not the whole ring. Let P be one of its minimal prime ideals. Then, by Theorem 5,

$$\mathrm{Gr}_{R_P}\{PR_P\} = 1.$$

Furthermore $\qquad R_P\dfrac{\beta}{1} :_{R_P} R_P\dfrac{\alpha}{1} = IR_P \subseteq PR_P$

and therefore $\qquad R_P\dfrac{\alpha}{1} \nsubseteq R_P\dfrac{\beta}{1}.$

Now, by Chapter 3 Theorem 28, the R_P-module E_P possesses an elementary resolution of finite length and

$$\mathfrak{G}(E_P) = \mathfrak{G}(E)\,R_P = R_P\frac{\alpha/1}{\beta/1}.$$

Accordingly $\mathfrak{G}(E_P)$ is not an integral ideal of R_P. However Theorem 2 shows that

$$\mathrm{Pd}^*_{R_P}(E_P) \leqslant \mathrm{Gr}_{R_P}\{PR_P\} = 1$$

and now we see that E_P is an elementary R_P-module. But this we know implies that $\mathfrak{G}(E_P) = \mathfrak{F}(E_P)$. Since we have already proved that $\mathfrak{G}(E_P)$ is not an integral R_P-ideal, we have arrived at a contradiction and the proof is complete.

The connection between $\mathfrak{G}(E)$ and $\mathfrak{F}(E)$ will now be investigated. We recall that a *principal ideal* of R is one which can be generated by a single element of the ring.

THEOREM 7. *Let E be an R-module which admits an elementary resolution of finite length. Then $\mathfrak{G}(E)$ is a principal ideal generated* *by a non-zerodivisor and* $\mathfrak{F}(E) \subseteq \mathfrak{G}(E)$. *Further if* I *is a principal ideal and* $\mathfrak{F}(E) \subseteq I$, *then* $\mathfrak{G}(E) \subseteq I$.

Proof. That $\mathfrak{G}(E)$ is a principal ideal generated by a non-zerodivisor is precisely Theorem 6. It will now be assumed that $\mathfrak{F}(E) \nsubseteq \mathfrak{G}(E)$ and from this we shall derive a contradiction.

Our assumption ensures that $\mathfrak{G}(E)\!:_R \mathfrak{F}(E) = \mathfrak{A}$ (say) is not the whole ring. Let P be a minimal prime ideal of $\mathfrak{A}$. Since $\mathfrak{F}(E)$ is finitely generated, Theorem 5 shows that $\mathrm{Gr}_{R_P}\{PR_P\} = 1$. Furthermore, by Chapter 3 Theorem 28, the R_P-module E_P admits an elementary resolution of finite length and

$$\mathfrak{G}(E_P)\!:_{R_P} \mathfrak{F}(E_P) = \mathfrak{A}R_P \subseteq PR_P$$

and therefore $\mathfrak{F}(E_P) \neq \mathfrak{G}(E_P)$. On the other hand $\mathrm{Pd}^*_{R_P}(E_P) \leqslant 1$ from Theorem 2. Thus E_P is an elementary R_P-module and therefore $\mathfrak{F}(E_P) = \mathfrak{G}(E_P)$. With this contradiction the first part of the theorem is proved.

Now assume that $\mathfrak{F}(E) \subseteq I$, where I is a principal ideal, but that $\mathfrak{G}(E) \nsubseteq I$. Put $\mathfrak{B} = I\!:_R \mathfrak{G}(E)$ so that $\mathfrak{B} \neq R$ and let Π be a minimal prime ideal of $\mathfrak{B}$. By Chapter 3 Theorem 23, $\mathrm{Ann}_R(E)$ contains a non-zerodivisor and therefore, by Chapter 3 Theorem 5, $\mathfrak{F}(E)$ contains a non-zerodivisor. It follows that I is generated by a non-zerodivisor and, since $\mathfrak{G}(E)$ is singly generated, Theorem 5 implies that $\mathrm{Gr}_{R_\Pi}\{\Pi R_\Pi\} = 1$. From this we conclude, by a now familiar argument, that $\mathfrak{G}(E_\Pi) = \mathfrak{F}(E_\Pi)$. Furthermore

$$IR_\Pi\!:_{R_\Pi} \mathfrak{G}(E_\Pi) = \mathfrak{B}R_\Pi \subseteq \Pi R_\Pi,$$

whence $\mathfrak{G}(E_\Pi) \nsubseteq IR_\Pi$. However

$$\mathfrak{G}(E_\Pi) = \mathfrak{F}(E_\Pi) = \mathfrak{F}(E)R_\Pi \subseteq IR_\Pi.$$

This second contradiction completes the proof.

6.3   Greatest common divisors

Theorem 7 opens the way to the application of finite free resolutions to the study of factorization of non-zerodivisors. We shall make a short excursion into this subject.

Let $a_1, a_2, \ldots, a_s$ be elements of R and consider principal ideals

Rd which contain $Ra_1 + Ra_2 + \ldots + Ra_s$. It may happen that among these principal ideals there is a unique smallest one. When this is the case we shall say that $a_1, a_2, \ldots, a_s$ *have a greatest common divisor*. Assume that we have this situation and that Rd is the smallest principal ideal containing $Ra_1 + Ra_2 + \ldots + Ra_s$. We shall then write

$$d = \text{g.c.d.} \{a_1, a_2, \ldots, a_s\}$$

and we shall refer to d as a *greatest common divisor of $a_1, a_2, \ldots, a_s$*. The connection with the preceding theory arises through

THEOREM 8. *Let $a_1, a_2, \ldots, a_s$ belong to R and suppose that $Ra_1 + Ra_2 + \ldots + Ra_s$ has a finite free resolution of finite length. Then $a_1, a_2, \ldots, a_s$ have a greatest common divisor. Moreover, if $Ra_1 + Ra_2 + \ldots + Ra_s \neq 0$ and $d = \text{g.c.d.} \{a_1, a_2, \ldots, a_s\}$, then d is not a zerodivisor in R.*

Proof. We may suppose that $Ra_1 + Ra_2 + \ldots + Ra_s \neq 0$. Put $E = R/(Ra_1 + Ra_2 + \ldots + Ra_s)$. Then E has a finite free resolution of finite length and, by Chapter 3 Theorem 5,

$$\mathfrak{F}(E) = \text{Ann}_R(E) = Ra_1 + Ra_2 + \ldots + Ra_s.$$

Also $\text{Char}_R(E) = 0$ in view of Chapter 4 Theorem 12.

Assume for the moment that $Ra_1 + Ra_2 + \ldots + Ra_s$ contains a non-zerodivisor. Then E possesses an elementary resolution of finite length. Consequently, by Theorem 7, any generator of $\mathfrak{G}(E)$ will serve as a greatest common divisor of $a_1, a_2, \ldots, a_s$.

Let us now assume merely that $Ra_1 + Ra_2 + \ldots + Ra_s \neq 0$ and let x be an indeterminate. By Chapter 5 Exercise 6, $\text{Ann}_R(E)$ contains a latent non-zerodivisor that is to say

$$a_1 R[x] + a_2 R[x] + \ldots + a_s R[x]$$

contains a non-zerodivisor. It therefore follows (from the first part) that $a_1, a_2, \ldots, a_s$ have a greatest common divisor, say $f = c_0 + c_1 x + c_2 x^2 + \ldots + c_q x^q$, in $R[x]$. Since $a_i \in fR[x]$, we see that $a_i \in Rc_0$. Now suppose that $d \in R$ and

$$Ra_1 + Ra_2 + \ldots + Ra_s \subseteq Rd.$$

It then follows that $fR[x] \subseteq dR[x]$ from which we see that $Rc_0 \subseteq Rd$. Accordingly $c_0 = \text{g.c.d.} \{a_1, a_2, \ldots, a_s\}$.

Finally, still assuming that $Ra_1 + Ra_2 + \ldots + Ra_s \neq 0$, let

$$d = \text{g.c.d.} \{a_1, a_2, \ldots, a_s\}.$$

Then, by Chapter 5 Lemma 8,

$$0:_R d \subseteq 0:_R (Ra_1 + \ldots + Ra_s) = 0$$

because $Ra_1 + Ra_2 + \ldots + Ra_s$ contains a latent non-zerodivisor. This completes the proof.

The following corollary was established in the course of the discussion.

COROLLARY.† *Suppose that the ideal $Ra_1 + Ra_2 + \ldots + Ra$ contains a non-zerodivisor and has a finite free resolution of finite length. Then $R/(Ra_1 + Ra_2 + \ldots + Ra_s)$ possesses an elementary resolution of finite length and any generator of the MacRae invariant of $R/(Ra_1 + Ra_2 + \ldots + Ra_s)$ is a greatest common divisor of $a_1, a_2, \ldots, a_s$.*

We next establish a connection between greatest common divisors and the theory of grade. This will be useful at a later stage.

THEOREM 9. *Let $b_1, b_2, \ldots, b_s$ be elements of R which generate an ideal whose (true) grade is at least two and let α be a non-zerodivisor. Then $\alpha = \text{g.c.d.} \{\alpha b_1, \alpha b_2, \ldots, \alpha b_s\}$.*

Proof. Assume that $\alpha(Rb_1 + Rb_2 + \ldots + Rb_s) \subseteq Rd$, where $d \in R$, then it is sufficient to show that $R\alpha \subseteq Rd$. We shall suppose that $R\alpha \nsubseteq Rd$ and derive a contradiction.

If $r \in R$ and $dr = 0$, then $r(Rb_1 + Rb_2 + \ldots + Rb_s) = 0$. It follows that $r = 0$ because $\text{Gr}_R \{Rb_1 + \ldots + Rb_s\} > 0$. Thus d is a non-zerodivisor and it is contained in $Rb_1 + \ldots + Rb_s + Rd$. On the other hand

$$\alpha(Rb_1 + \ldots + Rb_s + Rd) \subseteq Rd$$

and $\alpha \notin Rd$. This proves that

$$\text{Gr}_R \{Rb_1 + \ldots + Rb_s + Rd; R/Rd\} = 0$$

and therefore, by Chapter 5 Theorem 15,

$$\text{Gr}_R \{Rb_1 + \ldots + Rb_s + Rd\} = 1.$$

† Cf. R. E. MacRae [(**32**) Proposition 5.5, p. 167].

But

$$\mathrm{Gr}_R\{Rb_1 + \ldots + Rb_s + Rd\} \geqslant \mathrm{Gr}_R\{Rb_1 + \ldots + Rb_s\} \geqslant 2.$$

This is the desired contradiction.

EXERCISE 6. *Let α and β be non-zerodivisors in R and suppose that $R\alpha \cap R\beta$ is a principal ideal. Show by means of elementary considerations that α and β have a greatest common divisor.*

EXERCISE 7. *Let α and β be non-zerodivisors in R and let $R\alpha \cap R\beta$ be a principal ideal. Show that $R\alpha + R\beta$ has a finite free resolution of length one.*

In order to combine our observations we introduce the

DEFINITION. *We shall say that R is a 'GCD-ring' if every pair of non-zerodivisors has a greatest common divisor.*

THEOREM 10. *Suppose that whenever α, β are non-zerodivisors in R the ideal $R\alpha + R\beta$ has a finite free resolution of finite length. Then R is a GCD-ring.*

This is an immediate consequence of Theorem 8 and the definition of a GCD-ring.

We shall now investigate the properties of GCD-rings in a certain amount of detail.

EXERCISE 8. *Let R be a GCD-ring and let α, β, γ be non-zerodivisors. If $\delta = \mathrm{g.c.d.}\{\alpha, \beta\}$ establish the following:*
(i) $\delta\gamma = \mathrm{g.c.d.}\{\alpha\gamma, \beta\gamma\}$;
(ii) $1 = \mathrm{g.c.d.}\{\alpha/\delta, \beta/\delta\}$.
Let p be an element of R.

DEFINITION. *We shall say that p is an 'irreducible element' of R provided (i) it is a non-zerodivisor, and (ii) whenever p is expressed as a product $p = uv$, where $u, v \in R$, just one of these is a unit.*

Note that an irreducible element cannot be a unit.

THEOREM 11. *Let R be a GCD-ring and p an irreducible element of R. If now α, β are non-zerodivisors in R and p divides $\alpha\beta$, then either p divides α or it divides β.*

Proof. Assume that p does not divide α. Then $1 = \mathrm{g.c.d.}\{p, \alpha\}$ and therefore, by Exercise 8, $\beta = \mathrm{g.c.d.}\{p\beta, \alpha\beta\}$. But Rp contains $R\alpha\beta$ and $Rp\beta$ and therefore $R\beta \subseteq Rp$. The theorem follows.

We recall that elements α and β, of R, are said to be *associated* if each is a unit times the other. This is an equivalence relation. When α and β are associated in this sense we shall write $\alpha \sim \beta$. Let p and q be irreducible elements of R. Evidently $p \sim q$ if and only if p divides q.

THEOREM 12. *Let R be a* GCD-*ring and suppose that*

$$up_1 p_2 \cdots p_m = vq_1 q_2 \cdots q_n,$$

where u, v are units and p_i, q_j are irreducible. Then $m = n$ and we can renumber the q_j so that $p_i \sim q_i$ for $i = 1, 2, \ldots, m = n$.

Proof. The argument is probably familiar to the reader but we shall include it for the sake of completeness.

Without loss of generality we may suppose that $m \leqslant n$. Now $u_1 p_1 p_2 \cdots p_m = q_1 q_2 \cdots q_n$, where u_1 is a unit. Consequently, by Theorem 11, p_1 divides q_j for some j. By renumbering $q_1, q_2, \ldots, q_n$ we can arrange that p_1 divides q_1. Then $p_1 \sim q_1$, i.e. p_1 is a unit times q_1. This leads to an equation $u_2 p_2 p_3 \cdots p_m = q_2 q_3 \cdots q_n$, where u_2 is a unit. Accordingly p_2 divides one of $q_2, q_3, \ldots, q_n$ and by renumbering these we may suppose that p_2 divides q_2. From this it follows that p_2 is a unit times q_2. We next obtain an equation $u_3 p_3 p_4 \cdots p_m = q_3 q_4 \cdots q_n$, where u_3 is a unit. After m steps the original q_j have been so renumbered that $p_i \sim q_i$ for $i = 1, 2, \ldots, m$. We also see that $m = n$. For otherwise $m + 1 \leqslant n$ and our procedure results in an equation $u_{m+1} = q_{m+1} q_{m+2} \cdots q_n$, where u_{m+1} is a unit. But this would mean that q_{m+1} was a unit contrary to the definition of an irreducible element. The proof is now complete.

We can describe Theorem 12 by saying that *in a* GCD-*ring, whenever a non-zerodivisor can be represented as the product of a unit and a number of irreducible elements, the representation is unique to within multiplication by units.*

Theorems 10 and 12 show that the theory of finite free resolutions has a direct connection with the question of the *uniqueness* of factorizations of non-zerodivisors into products of irreducible elements. The next exercise gives information about conditions which ensure that such factorizations exist.

EXERCISE 9. *Assume that, in R, every increasing sequence of* *principal ideals generated by non-zerodivisors becomes constant. Show that every non-zerodivisor can be expressed as the product of a unit and a finite number of irreducible elements.*

Of course, any commutative Noetherian ring satisfies the hypothesis of Exercise 9.

LEMMA 2. *Let R be a GCD-ring and suppose that α, β are non-zerodivisors. Then among the principal ideals that are* (i) *generated by non-zerodivisors, and* (ii) *contained in $R\alpha \cap R\beta$, there is a unique largest one. Hence in a GCD-ring which is also an integral domain the intersection of two principal ideals is always a principal ideal.*

Proof. Let $\delta = \text{g.c.d.}\{\alpha, \beta\}$ and put $\alpha = \delta a$, $\beta = \delta b$. Then δab is a non-zerodivisor, $R\delta ab \subseteq R\alpha \cap R\beta$ and, by Exercise 8, $1 = \text{g.c.d.}\{a, b\}$.

Suppose now that $R\gamma \subseteq R\alpha \cap R\beta$, where γ is a non-zerodivisor. If $\gamma = \delta c$, then c is a non-zerodivisor and $Rc \subseteq Ra \cap Rb$. Let $c = b\omega$, where $\omega \in R$. By Exercise 8,

$$\omega = \text{g.c.d.}\{a\omega, b\omega\} = \text{g.c.d.}\{a\omega, c\}.$$

Consequently, since $Ra\omega + Rc \subseteq Ra$, $R\omega \subseteq Ra$ and therefore $Rc = Rb\omega \subseteq Rab$. Finally

$$R\gamma = Rc\delta \subseteq Rab\delta.$$

The lemma follows.

THEOREM 13. *Let R be an integral domain. Then the following statements are equivalent:*

(i) *R is a GCD-ring;*

(ii) *the intersection of two principal ideals is always a principal ideal;*

(iii) *every ideal which can be generated by two elements has a finite free resolution of finite length.*

Proof. Lemma 2 and Exercise 6 show that (i) and (ii) are equivalent whereas Theorem 10 proves that (iii) implies (i). Finally (iii) follows from (ii) by virtue of Exercise 7.

The following definition enables us to tie together some loose ends.

DEFINITION. *R will be called a 'UF-ring' if* (i) *every non-zerodivisor can be represented as the product of a unit and a finite* number of irreducible elements, and (ii) *such representations are unique to within multiplication by units.*

EXERCISE 10. *Show that R is a UF-ring if and only if the following two conditions are both satisfied :*

(a) *R is a GCD-ring;*

(b) *every ascending chain $R\alpha_1 \subseteq R\alpha_2 \subseteq R\alpha_3 \subseteq \ldots$ of principal ideals generated by non-zerodivisors terminates.*

6.4 Grade and finite free resolutions

Let $\phi\colon F \to G$ be a homomorphism, where F, G are finite free modules and let $E \neq 0$ be an R-module. Put

$$\mathfrak{A}(\phi, E) = \mathfrak{A}_\nu(\phi) \quad \text{where} \quad \nu = \operatorname{rank}_R(\phi, E), \qquad (6.4.1)$$

and
$$\mathfrak{A}(\phi) = \mathfrak{A}(\phi, R). \qquad (6.4.2)$$

Evidently ϕ is stable relative to E if and only if

$$0:_E \mathfrak{A}(\phi, E) = 0.$$

But $\mathfrak{A}(\phi, E)$ is finitely generated. Consequently ϕ *is stable relative to E if and only if* $\operatorname{Gr}_R\{\mathfrak{A}(\phi, E); E\} > 0$.

EXERCISE 11. *Let R be a quasi-local ring and $\phi\colon F \to G$ a homomorphism, where F, G are finite free R-modules. Show that $\operatorname{Coker}\phi$ is a finite free R-module if and only if $\mathfrak{A}(\phi) = R$.*

EXERCISE 12. *Let $\phi\colon F \to G$ be a stable homomorphism where F, G are finite free R-modules, and let S be a multiplicatively closed subset of R not containing the zero element. Show that*

$$\mathfrak{A}(\phi)\,R_S = \mathfrak{A}(\phi_S).$$

Now suppose that x is an indeterminate. Put $R^* = R[x]$, $F^* = F[x]$, $G^* = G[x]$ and $E^* = E[x]$. Further let $\phi^*\colon F^* \to G^*$ be the R^*-homomorphism induced by ϕ. Then F^*, G^* are finite free R^*-modules,
$$\mathfrak{A}_\nu(\phi^*) = \mathfrak{A}_\nu(\phi)\,R^*, \qquad (6.4.3)$$

and
$$\operatorname{rank}_R(\phi, E) = \operatorname{rank}_{R^*}(\phi^*, E^*). \qquad (6.4.4)$$

It follows therefore that

$$\mathfrak{A}(\phi^*, E^*) = \mathfrak{A}(\phi, E)\,R^* \qquad (6.4.5)$$

and hence that

$$\mathrm{Gr}_{R^*}\{\mathfrak{A}(\phi^*, E^*); E^*\} = \mathrm{Gr}_R\{\mathfrak{A}(\phi, E); E\}. \qquad (6.4.6)$$

We next add a word about the effect of adjoining an indeterminate on the formation of tensor products. To this end let U be an R-module and put $U^* = U[x]$. Then there is an isomorphism

$$U^* \otimes_{R^*} E^* \approx (U \otimes_R E)[x] \qquad (6.4.7)$$

of R^*-modules in which the element $ux^\mu \otimes ex^\nu$ of the left hand side is matched with $(u \otimes e) x^{\mu+\nu}$. (Here, of course, u and e denote elements of U and E respectively.) In particular we may identify $F^* \otimes_{R^*} E^*$ with $(F \otimes_R E)[x]$ and $G^* \otimes_{R^*} E^*$ with $(G \otimes_R E)[x]$. The point to note is that *if these identifications are made, then $\phi^* \otimes E^*$ is just the homomorphism*

$$(F \otimes_R E)[x] \to (G \otimes_R E)[x]$$

induced by $\phi \otimes E$.

In preparation for Lemmas 3 and 4 (below) we observe that in them we shall be concerned with a complex

$$\mathbb{C}\colon 0 \to F_n \xrightarrow{\phi_n} F_{n-1} \xrightarrow{\phi_{n-1}} F_{n-2} \to \dots \to F_1 \xrightarrow{\phi_1} F_0,$$
$$(6.4.8)$$

where $n \geqslant 1$ and each F_j is a finite free R-module. Put

$$M = \mathrm{Coker}\,\phi_1,$$

so that we have an exact sequence

$$F_1 \xrightarrow{\phi_1} F_0 \xrightarrow{\epsilon} M \to 0$$

and note that, because of the right exactness of tensor products,† the sequence

$$F_1 \otimes_R E \xrightarrow{\phi_1 \otimes E} F_0 \otimes_R E \xrightarrow{\epsilon \otimes E} M \otimes_R E \to 0$$

is exact for an arbitrary R-module E.

LEMMA 3. *Let $\mathbb{C}$ be as above and let $E \neq 0$ be an R-module. If now $\mathbb{C} \otimes_R E$ is exact, then*

$$\mathrm{Gr}_R\{\mathfrak{A}(\phi_k, E); E\} > 0$$

for $k = 1, 2, \dots, n$.

† See for example [(35) Theorem 6, p. 44].

Proof. Suppose that $1 \leqslant k \leqslant n$, then, by Chapter 4 Theorem 7, ϕ_k is stable relative to E. The lemma therefore follows by virtue of the remarks made earlier.

Now suppose that $0 \to E' \to E \to E'' \to 0$ is an exact sequence of R-modules and that the element β, of R, is not a zerodivisor on E', E or E''. It then follows (see Chapter 5 Lemma 9) that the induced sequence

$$0 \to E'/\beta E' \to E/\beta E \to E''/\beta E'' \to 0$$

is also exact.

In order to exploit this observation let us assume that

$$0 \to E_n \to E_{n-1} \to \dots \to E_1 \to E_0 \qquad (6.4.9)$$

is an exact sequence and that β is not a zerodivisor on any E_j. By breaking up (6.4.9) into short exact sequences and repeatedly employing the observations of the last paragraph, we find that (6.4.9) gives rise to an exact sequence

$$0 \to E_n/\beta E_n \to E_{n-1}/\beta E_{n-1} \to \dots \to E_1/\beta E_1. \quad (6.4.10)$$

Observe that (6.4.10) has one term fewer than (6.4.9).

LEMMA 4. *Let the complex $\mathbb{C}$ be as before, let E be an R-module, and assume that $\mathbb{C} \otimes_R E$ is exact. Further suppose that the element β, of R, is not a zerodivisor on E. If now $n > 1$ and $\mathbb{C}'$ is the complex*

$$0 \to F_n \xrightarrow{\phi_n} F_{n-1} \to \dots \to F_2 \xrightarrow{\phi_2} F_1,$$

then $\mathbb{C}' \otimes_R (E/\beta E)$ is exact.

Proof. Since $F_j \otimes_R E$ is isomorphic to a direct sum of copies of E, β does not annihilate any of its non-zero elements. Consequently the discussion immediately before the statement of the lemma shows that we have an exact sequence

$$0 \to (F_n \otimes_R E)/\beta(F_n \otimes_R E) \xrightarrow{\psi_n} (F_{n-1} \otimes_R E)/\beta(F_{n-1} \otimes_R E) \to \dots$$

$$\to (F_2 \otimes_R E)/\beta(F_2 \otimes_R E) \xrightarrow{\psi_2} (F_1 \otimes_R E)/\beta(F_1 \otimes_R E),$$

where ψ_j is induced by $\phi_j \otimes E$. Next the exact sequence

$$0 \to E \xrightarrow{\beta} E \to E/\beta E \to 0$$

and the right exactness of tensor products together yield an exact sequence
$$F_j \otimes_R E \to F_j \otimes_R E \to F_j \otimes_R (E/\beta E) \to 0$$
in which the first mapping consists in multiplication by β. Thus there results an isomorphism
$$(F_j \otimes_R E)/\beta(F_j \otimes_R E) \xrightarrow{\sim} F_j \otimes_R (E/\beta E) \qquad (6.4.11)$$
and it is easily verified that the diagram

$$
\begin{array}{ccc}
(F_j \otimes_R E)/\beta(F_j \otimes_R E) & \longrightarrow & F_j \otimes_R (E/\beta E) \\
\Big\downarrow \psi_j & & \Big\downarrow \phi_j \otimes (E/\beta E) \\
(F_{j-1} \otimes_R E)/\beta(F_{j-1} \otimes_R E) & \longrightarrow & F_{j-1} \otimes_R (E/\beta E)
\end{array}
$$

is commutative. (Here the horizontal mappings are as described in (6.4.11).) Accordingly $\mathbb{C}' \otimes_R (E/\beta E)$ is exact as was claimed.

THEOREM 14. *Suppose that*
$$0 \to F_n \xrightarrow{\phi_n} F_{n-1} \xrightarrow{\phi_{n-1}} F_{n-2} \to \ldots \to F_1 \xrightarrow{\phi_1} F_0$$
is a complex of finite free R-modules, where $n \geqslant 1$, and let $E \neq 0$ be an R-module for which
$$0 \to F_n \otimes_R E \to F_{n-1} \otimes_R E \to \ldots \to F_1 \otimes_R E \to F_0 \otimes_R E$$
is exact. Then

   (a) $\mathrm{Gr}_R\{\mathfrak{A}(\phi_k, E); E\} \geqslant k$ *for* $k = 1, 2, \ldots, n$;
   (b) $\mathrm{rank}_R(\phi_n, E) = \mathrm{rank}_R(F_n)$;
   (c) $\mathrm{rank}_R(\phi_{m+1}, E) + \mathrm{rank}_R(\phi_m, E) = \mathrm{rank}_R(F_m)$ *for* $1 \leqslant m < n$.

Remark. It will be noticed that this largely supersedes Theorem 7 of Chapter 4. Indeed in view of that earlier result we need only establish (a).

Proof. As has already been observed we have only to establish (a). This will be accomplished by an argument which uses induction on n. By Lemma 3, we have $\mathrm{Gr}_R\{\mathfrak{A}(\phi_k, E); E\} > 0$ hence there is no problem if $n = 1$. It will therefore be assumed that $n > 1$ and that the theorem has been established for complexes that are shorter than the one under consideration. Observe that, in any case, Lemma 3 ensures that $\mathrm{Gr}_R\{\mathfrak{A}(\phi_1, E); E\} \geqslant 1$.

To begin with we shall assume that for each value of

$$k \ (1 \leqslant k \leqslant n)$$

there is an element $\beta_k \in \mathfrak{A}(\phi_k, E)$ that is not a zerodivisor on E. (This assumption will be removed shortly.) Put $\beta = \beta_1 \beta_2 \ldots \beta_n$. Then $\beta \in \mathfrak{A}(\phi_k, E)$ for all k and β is not a zerodivisor on E. Note that we may suppose that $E/\beta E \neq 0$ for otherwise we should have $\mathrm{Gr}_R\{\mathfrak{A}(\phi_k, E); E\} = \infty$ and there would be nothing to prove.

By Lemma 4, we have an exact sequence

$$0 \to F_n \otimes_R (E/\beta E) \to F_{n-1} \otimes_R (E/\beta E) \to \ldots$$
$$\to F_2 \otimes_R (E/\beta E) \to F_1 \otimes_R (E/\beta E).$$

Consequently $\mathrm{rank}_R(\phi_n, E/\beta E) = \mathrm{rank}_R(F_n)$ and

$$\mathrm{rank}_R(\phi_{m+1}, E/\beta E) + \mathrm{rank}_R(\phi_m, E/\beta E) = \mathrm{rank}_R(F_m)$$

for $2 \leqslant m < n$. Accordingly $\mathrm{rank}_R(\phi_k, E/\beta E) = \mathrm{rank}_R(\phi_k, E)$ and therefore

$$\mathfrak{A}(\phi_k, E/\beta E) = \mathfrak{A}(\phi_k, E)$$

for $2 \leqslant k \leqslant n$. We therefore see that

$$\mathrm{Gr}_R\{\mathfrak{A}(\phi_k, E); E/\beta E\} = \mathrm{Gr}_R\{\mathfrak{A}(\phi_k, E/\beta E); E/\beta E\} \geqslant k-1$$

by virtue of the induction hypothesis. However this implies that

$$\mathrm{Gr}_R\{\mathfrak{A}(\phi_k, E); E\} \geqslant k$$

for $2 \leqslant k \leqslant n$.

To complete the proof we have only to remove the extra assumption that was introduced above. To this end let x be an indeterminate and put $F_k^* = F_k[x]$. Then our complex gives rise to a new complex

$$0 \to F_n^* \xrightarrow{\phi_n^*} F_{n-1}^* \xrightarrow{\phi_{n-1}^*} F_{n-2}^* \to \ldots \to F_1^* \xrightarrow{\phi_1^*} F_0^*$$

composed of finite free R^*-modules, where $R^* = R[x]$. Moreover, in view of the facts established in connection with (6.4.7), this complex becomes exact if we tensor it, over R^*, with E^*. Moreover, we already know that $\mathrm{Gr}_R\{\mathfrak{A}(\phi_k, E); E\} > 0$. Hence $\mathfrak{A}(\phi_k, E)$ contains a latent non-zerodivisor on E and therefore $\mathfrak{A}(\phi_k, E) R^*$ contains a non-zerodivisor on E^*. However, by (6.4.5),

$$\mathfrak{A}(\phi_k, E) R^* = \mathfrak{A}(\phi_k^*, E^*). \tag{6.4.12}$$

Consequently each of the ideals $\mathfrak{A}(\phi_k^*, E^*)$ contains a non-zerodivisor on the R^*-module E^* and therefore we are in the special situation envisaged above. It therefore follows that

$$\mathrm{Gr}_{R^*}\{\mathfrak{A}(\phi_k^*, E^*); E^*\} \geqslant k$$

and hence, in view of (6.4.12), that

$$\mathrm{Gr}_R\{\mathfrak{A}(\phi_k, E); E\} \geqslant k.$$

This completes the proof.

In our next theorem we establish the converse of Theorem 14 in the special case obtained by taking $E = R$, and in Appendix B the converse is established in full generality but using methods not developed in the main text.

EXERCISE 13. *Let* $\phi: F \to G$ *and* $\phi': F' \to G$ *be homomorphisms, where* F, F', G *are finite free R-modules, and let*

$$\phi(F) = \phi'(F').$$

Show that $\mathfrak{A}_\nu(\phi) = \mathfrak{A}_\nu(\phi')$ *for all* $\nu \geqslant 0$, $\mathrm{rank}_R(\phi) = \mathrm{rank}_R(\phi')$ *and* $\mathfrak{A}(\phi) = \mathfrak{A}(\phi')$.

THEOREM 15. *Suppose that*

$$0 \to F_n \xrightarrow{\phi_n} F_{n-1} \xrightarrow{\phi_{n-1}} F_{n-2} \to \dots \to F_1 \xrightarrow{\phi_1} F_0,$$

where $n \geqslant 1$, *is a complex of finite free R-modules. Then in order that the complex be exact it is necessary and sufficient that the following three conditions be satisfied:*

(a) $\mathrm{Gr}_R\{\mathfrak{A}(\phi_k)\} \geqslant k$ *for* $k = 1, 2, \dots, n$;
(b) $\mathrm{rank}_R(\phi_n) = \mathrm{rank}_R(F_n)$;
(c) $\mathrm{rank}_R(\phi_{m+1}) + \mathrm{rank}_R(\phi_m) = \mathrm{rank}_R(F_m)$ *for* $1 \leqslant m < n$.

Proof. If the complex is exact, then (a), (b) and (c) follow from Theorem 14 on taking $E = R$.

Suppose then that (a), (b), (c) are all satisfied and let S be a multiplicatively closed subset of R not containing the zero element. Put $R^* = R_S$, $F_k^* = (F_k)_S$ and $\phi_k^* = (\phi_k)_S$. Condition (a) ensures that ϕ_k is stable. Consequently $\phi_k^*: F_k^* \to F_{k-1}^*$ is also stable and it has the same rank as ϕ_k. Thus

$$\mathrm{rank}_{R^*}(\phi_k^*) = \mathrm{rank}_R(\phi_k)$$

and, by Exercise 12,
$$\mathfrak{A}(\phi_k^*) = \mathfrak{A}(\phi_k)\, R^*.$$

But we also have $\mathrm{rank}_{R^*}(F_m^*) = \mathrm{rank}_R(F_m)$ and, by Chapter 5 Exercise 10,
$$\mathrm{Gr}_{R^*}\{\mathfrak{A}(\phi_k)\, R^*\} \geqslant \mathrm{Gr}_R\{\mathfrak{A}(\phi_k)\}.$$

It follows that on forming fractions with respect to S the counterparts of conditions (a), (b), (c) continue to hold.

We recall that we wish to show the three conditions together imply the exactness of our complex. Assume the implication does not hold and among all the counter-examples select one for which n is minimal. (In considering counter-examples we allow not only the complex but also R itself to vary.)

Suppose now that
$$0 \to F_n \xrightarrow{\phi_n} F_{n-1} \xrightarrow{\phi_{n-1}} F_{n-2} \to \ldots \to F_1 \xrightarrow{\phi_1} F_0$$

is the chosen counter-example. Then (a), (b), (c) all hold but the complex is not exact. It therefore follows, from Chapter 4 Theorem 4, that $n \geqslant 2$. Put $B = \mathrm{Im}\, \phi_2$ and $C = \mathrm{Ker}\, \phi_1$ so that $B \subseteq C$. The minimality of n ensures that
$$0 \to F_n \xrightarrow{\phi_n} F_{n-1} \xrightarrow{\phi_{n-1}} F_{n-2} \to \ldots \to F_2 \xrightarrow{\phi_2} F_1 \quad (6.4.13)$$

is exact and also that $C/B \neq 0$. Let $K \neq 0$ be a finitely generated submodule of C/B and put $\mathfrak{B} = \mathrm{Ann}_R(K)$. Then $\mathfrak{B} \neq R$, $\mathfrak{B}$ has a minimal prime ideal P, and we have
$$\mathrm{Ann}_{R_P}(K_P) = \mathfrak{B}R_P \subseteq \mathrm{Rad}\,(\mathfrak{B}R_P) = PR_P.$$

Thus $K_P \neq 0$ and if I is a finitely generated ideal of R_P that is contained in PR_P, then $I^m K_P = 0$ for some positive integer m. It follows that I itself annihilates a non-zero element of K_P. But $K \subseteq C/B \subseteq F_1/B$. Consequently I annihilates a non-zero element of $(F_1/B)_P$ and therefore we may conclude that
$$\mathrm{Gr}_{R_P}\{PR_P;\, (F_1/B)_P\} = 0.$$

Thus $(F_1/\mathrm{Im}\, \phi_2)_P \neq 0$ and
$$\mathrm{Gr}_{R_P}\{PR_P;\, (F_1/\mathrm{Im}\, \phi_2)_P\} = 0.$$

At this point we localize at P and then drop P as a suffix. In the situation so realized

(1) conditions (a), (b), (c) still hold;

(2) the complex

$$0 \to F_n \xrightarrow{\phi_n} F_{n-1} \xrightarrow{\phi_{n-1}} F_{n-2} \to \dots \to F_1 \xrightarrow{\phi_1} F_0$$

is not exact and n still has its minimal value;

(3) R is a quasi-local ring with maximal ideal P say;

(4) $F_1/\operatorname{Im}\phi_2 \neq 0$ and $\operatorname{Gr}_R\{P; F_1/\operatorname{Im}\phi_2\} = 0$.

Now the minimality of n ensures that the sequence

$$0 \to F_n \xrightarrow{\phi_n} F_{n-1} \to \dots \to F_2 \xrightarrow{\phi_2} F_1 \to F_1/\operatorname{Im}\phi_2 \to 0$$

is exact and hence that $\operatorname{Pd}_R^*(F_1/\operatorname{Im}\phi_2) \leqslant n-1$. It therefore follows, from (4) and Theorem 2, that $\operatorname{Gr}_R\{P\} \leqslant n-1$. But, by (a), $\operatorname{Gr}_R\{\mathfrak{A}(\phi_n)\} \geqslant n$. This shows that $\mathfrak{A}(\phi_n) = R$ and therefore, by Exercise 11, $\operatorname{Coker}\phi_n$ *is a finite free module.*

Next $\phi_{n-1}\colon F_{n-1} \to F_{n-2}$ induces a homomorphism

$$\psi_{n-1}\colon \operatorname{Coker}\phi_n \to F_{n-2}$$

and thus we arrive at a new complex

$$0 \to \operatorname{Coker}\phi_n \xrightarrow{\psi_{n-1}} F_{n-2} \xrightarrow{\phi_{n-2}} F_{n-3} \to \dots \to F_1 \xrightarrow{\phi_1} F_0 \tag{6.4.14}$$

of finite free modules. By construction, this is not exact. Moreover $\operatorname{Im}\phi_{n-1} = \operatorname{Im}\psi_{n-1}$ and therefore, by Exercise 13,

$$\operatorname{rank}_R(\psi_{n-1}) = \operatorname{rank}_R(\phi_{n-1}) \quad \text{and} \quad \mathfrak{A}(\phi_{n-1}) = \mathfrak{A}(\psi_{n-1}).$$

In particular we have

$$\operatorname{Gr}_R\{\mathfrak{A}(\psi_{n-1})\} \geqslant n-1.$$

Next the exact sequence

$$0 \to F_n \xrightarrow{\phi_n} F_{n-1} \to \operatorname{Coker}\phi_n \to 0$$

shows that

$$\begin{aligned}\operatorname{rank}_R(\operatorname{Coker}\phi_n) &= \operatorname{rank}_R(F_{n-1}) - \operatorname{rank}_R(F_n)\\&= \operatorname{rank}_R(F_{n-1}) - \operatorname{rank}_R(\phi_n)\\&= \operatorname{rank}_R(\phi_{n-1})\\&= \operatorname{rank}_R(\psi_{n-1}).\end{aligned}$$

Here, of course, we have made use of condition (*c*). Furthermore

$$\text{rank}_R(\psi_{n-1}) + \text{rank}_R(\phi_{n-2}) = \text{rank}_R(\phi_{n-1}) + \text{rank}_R(\phi_{n-2})$$
$$= \text{rank}_R(F_{n-2}).$$

Thus (6.4.14) is a non-exact complex of finite free modules and it satisfies conditions analogous to (*a*), (*b*), (*c*). This however contradicts the minimal property of the integer *n* and now the proof is complete.

We conclude this chapter by establishing some additional facts concerning the ideals $\mathfrak{A}(\phi_k)$. For these investigations, which are more general in character than the preceding ones, a few preparatory remarks will be helpful.

Let *H* be a finite free module. Then, by Chapter 3 Exercise 1, the smallest non-zero Fitting invariant of *H* is *R*. It therefore follows (see Chapter 3 Exercise 3) that if *M* is a finitely generated *R*-module, then the smallest non-zero Fitting invariant of *M* is the same as that of $M \oplus H$.

Now suppose that we have an exact sequence

$$F \xrightarrow{\phi} G \to E \to 0, \tag{6.4.15}$$

where *F*, *G* are finite free *R*-modules. It follows at once from (4.2.15) that $\mathfrak{A}(\phi)$ is the smallest non-zero Fitting invariant of *E*. Since we know that *E* and $E \oplus H$ have the same smallest non-zero Fitting invariant, it follows that if

$$F \xrightarrow{\phi*} G \oplus H \to E \oplus H \to 0$$

is the exact sequence derived from (6.4.15) in the obvious manner, then $\mathfrak{A}(\phi*) = \mathfrak{A}(\phi)$.

THEOREM 16. *Suppose that*

$$F_n \xrightarrow{\phi_n} F_{n-1} \xrightarrow{\phi_{n-1}} F_{n-2} \to \ldots \to F_1 \xrightarrow{\phi_1} F_0 \to E \to 0$$

and

$$F'_n \xrightarrow{\psi_n} F'_{n-1} \xrightarrow{\psi_{n-1}} F'_{n-2} \to \ldots \to F'_1 \xrightarrow{\psi_1} F'_0 \to E \to 0$$

are exact sequences, where each F_j and each F'_j is a finite free module. Then $\mathfrak{A}(\phi_k) = \mathfrak{A}(\psi_k)$ for $k = 1, 2, \ldots, n$.

Remark. It is worth-while observing that we do not assume

that ϕ_n and ψ_n are monomorphisms. A similar observation applies to certain of the other results that follow.†

Proof. First we note that $\mathfrak{A}(\phi_1) = \mathfrak{A}(\psi_1)$ because each is the smallest non-zero Fitting invariant of E. In particular this proves the theorem for the case $n = 1$. We shall now assume that $n > 1$ and that the theorem has been proved in all cases where the sequences involved have shorter lengths.

Put $E_1 = \mathrm{Ker}\,(F_0 \to E)$ and $E_1' = \mathrm{Ker}\,(F_0' \to E)$. By Chapter 2 Theorem 3 we have an isomorphism $E_1 \oplus F_0' \approx E_1' \oplus F_0$ and therefore these modules may be identified. Next we construct in an obvious manner exact sequences

$$F_n \xrightarrow{\phi_n} F_{n-1} \to \dots \to F_3 \xrightarrow{\phi_3} F_2 \xrightarrow{\phi_2^*} F_1 \oplus F_0' \to E_1 \oplus F_0' \to 0$$

and

$$F_n' \xrightarrow{\psi_n} F_{n-1}' \to \dots \to F_3' \xrightarrow{\psi_3} F_2' \xrightarrow{\psi_2^*} F_1' \oplus F_0 \to E_1' \oplus F_0 \to 0.$$

Induction now shows that $\mathfrak{A}(\phi_k) = \mathfrak{A}(\psi_k)$ for $k = 3, \dots, n$ and also that $\mathfrak{A}(\phi_2^*) = \mathfrak{A}(\psi_2^*)$. However, by the discussion immediately preceding the statement of the theorem, we have $\mathfrak{A}(\phi_2^*) = \mathfrak{A}(\phi_2)$ and $\mathfrak{A}(\psi_2^*) = \mathfrak{A}(\psi_2)$. Accordingly the theorem is established.

THEOREM 17. *Suppose that*

$$F_n \xrightarrow{\phi_n} F_{n-1} \xrightarrow{\phi_{n-1}} F_{n-2} \to \dots \to F_1 \xrightarrow{\phi_1} F_0$$

is an exact sequence, that each F_j is a finite free R-module, and that $\phi_1, \phi_2, \dots, \phi_n$ are all stable. Then

$$\mathrm{Rad}\,(\mathfrak{A}(\phi_1)) \subseteq \mathrm{Rad}\,(\mathfrak{A}(\phi_2)) \subseteq \dots \subseteq \mathrm{Rad}\,(\mathfrak{A}(\phi_n)).$$

Proof. Assume that $1 \leqslant k < n$ and suppose that it has already been proved that

$$\mathrm{Rad}\,(\mathfrak{A}(\phi_1)) \subseteq \mathrm{Rad}\,(\mathfrak{A}(\phi_2)) \subseteq \dots \subseteq \mathrm{Rad}\,(\mathfrak{A}(\phi_k)).$$

Assume further that P is a prime ideal such that $\mathfrak{A}(\phi_{k+1}) \subseteq P$. Since the radical of an ideal is the intersection of all the prime ideals that contain it, the theorem will follow provided that we can show that $\mathfrak{A}(\phi_k) \subseteq P$. We shall in fact suppose that $\mathfrak{A}(\phi_k) \nsubseteq P$ and derive a contradiction.

To this end we localize at P and then drop P as a suffix in order to simplify the notation. We then find ourselves with an exact sequence

$$F_n \xrightarrow{\phi_n} F_{n-1} \xrightarrow{\phi_{n-1}} F_{n-2} \to \ldots \to F_{k+1} \xrightarrow{\phi_{k+1}} F_k \xrightarrow{\phi_k} F_{k-1}$$

of finite free modules over a quasi-local ring. Moreover, Exercise 12 shows that $\mathfrak{A}(\phi_k) = R$, whereas $\mathfrak{A}(\phi_{k+1}) \neq R$. Next, by Exercise 11, $\operatorname{Coker} \phi_k$ is a finite free module whence

$$\operatorname{Im} \phi_k \approx \operatorname{Coker} \phi_{k+1}$$

is also free. However, since $\mathfrak{A}(\phi_{k+1}) \neq R$, Exercise 11 shows that $\operatorname{Coker} \phi_{k+1}$ cannot be free and with this contradiction the theorem is proved.

THEOREM 18. *Assume that*

$$F_n \xrightarrow{\phi_n} F_{n-1} \xrightarrow{\phi_{n-1}} F_{n-2} \to \ldots \to F_1 \xrightarrow{\phi_1} F_0 \qquad (6.4.16)$$

is an exact sequence of finite free R-modules. If now (i) $\phi_1, \phi_2, \ldots, \phi_n$ *are all stable,* (ii) $\operatorname{rank}_R(\phi_1) = \operatorname{rank}_R(F_0)$, *and*

$$\text{(iii)} \quad 1 \leqslant k \leqslant \min[n, \operatorname{Gr}_R(\mathfrak{A}(\phi_1))],$$

then

$$\operatorname{Rad}(\mathfrak{A}(\phi_1)) = \operatorname{Rad}(\mathfrak{A}(\phi_2)) = \ldots = \operatorname{Rad}(\mathfrak{A}(\phi_k)).$$

Proof. In view of Theorem 17 we need only establish that $\operatorname{Rad}(\mathfrak{A}(\phi_k)) \subseteq \operatorname{Rad}(\mathfrak{A}(\phi_1))$. Let P be a prime ideal having the property that $\mathfrak{A}(\phi_1) \subseteq P$, then it will suffice to show that $\mathfrak{A}(\phi_k) \subseteq P$. In what follows we suppose that $\mathfrak{A}(\phi_k) \nsubseteq P$ and arrive at a contradiction.

As in the proof of the last theorem, we localize at P and then omit P as a suffix. As a result (6.4.16) remains exact; furthermore the conditions

$$\operatorname{rank}_R(\phi_1) = \operatorname{rank}_R(F_0) \quad \text{and} \quad 1 \leqslant k \leqslant \min[n, \operatorname{Gr}_R(\mathfrak{A}(\phi_1))]$$

continue to hold. (Here we use Exercise 12 and Chapter 5 Exercise 10, as well as the fact that the rank of a stable homomorphism is unchanged by localization.) We also have $\mathfrak{A}(\phi_k) = R$ and $\mathfrak{A}(\phi_1) \neq R$.

Consider the exact sequence

$$0 \to \operatorname{Coker} \phi_k \to F_{k-2} \xrightarrow{\phi_{k-2}} F_{k-3} \to \ldots \to F_1 \xrightarrow{\phi_1} F_0.$$

Since $\mathfrak{A}(\phi_k) = R$, Exercise 11 shows that $\operatorname{Coker} \phi_k$ is a finite free module and, by the same exercise, $\operatorname{Coker} \phi_1$ cannot be projective. Thus $\operatorname{Coker} \phi_1 \neq 0$ and

$$\operatorname{Pd}_R^*(\operatorname{Coker} \phi_1) \leqslant k-1 < \infty.$$

Accordingly, by Theorem 4,

$$\operatorname{Gr}_R\{\operatorname{Ann}_R(\operatorname{Coker} \phi_1)\} \leqslant k-1.$$

Next, because $\operatorname{rank}_R(\phi_1) = \operatorname{rank}_R(F_0)$, we have†

$$\mathfrak{F}(\operatorname{Coker} \phi_1) = \mathfrak{A}(\phi_1)$$

and hence

$$\operatorname{Rad}(\mathfrak{A}(\phi_1)) = \operatorname{Rad}(\mathfrak{F}(\operatorname{Coker} \phi_1)) = \operatorname{Rad}(\operatorname{Ann}_R(\operatorname{Coker} \phi_1)).$$
$$(6.4.17)$$

Accordingly, by Chapter 5 Theorem 12,

$$\operatorname{Gr}_R\{\mathfrak{A}(\phi_1)\} = \operatorname{Gr}_R\{\operatorname{Ann}_R(\operatorname{Coker} \phi_1)\} \leqslant k-1$$

contrary to our original hypothesis. The theorem is therefore proved.

THEOREM 19. *Suppose that*

$$0 \to F_n \xrightarrow{\phi_n} F_{n-1} \xrightarrow{\phi_{n-1}} F_{n-2} \to \ldots \to F_1 \xrightarrow{\phi_1} F_0 \to M \to 0$$

is a finite free resolution of the R-module M and suppose also that $\operatorname{Char}_R(M) = 0$. *Then*

$$\operatorname{Rad}(\mathfrak{A}(\phi_k)) = \operatorname{Rad}(\operatorname{Ann}_R(M))$$

provided that $1 \leqslant k \leqslant \min[n, \operatorname{Gr}_R(\operatorname{Ann}_R(M))]$.

Proof. Theorem 8 of Chapter 4 shows that $\phi_1, \phi_2, \ldots, \phi_n$ are all stable, and $\operatorname{rank}_R(\phi_1) = \operatorname{rank}_R(F_0)$ by the same result. Next $\operatorname{Coker} \phi_1 = M$ and therefore as in (6.4.17)

$$\operatorname{Rad}(\mathfrak{A}(\phi_1)) = \operatorname{Rad}(\operatorname{Ann}_R(M)).$$

† See (4.2.15).

Theorem 12 of Chapter 5 now shows that

$$\mathrm{Gr}_R\{\mathfrak{A}(\phi_1)\} = \mathrm{Gr}_R\{\mathrm{Ann}_R(M)\}.$$

Consequently the desired result follows without further consideration from Theorem 18.

Solutions to the Exercises on Chapter 6

EXERCISE 1. *Let E be an R-module and S a multiplicatively closed subset of R not containing the zero element. Establish the inequalities* $\mathrm{Pd}_{R_S}(E_S) \leqslant \mathrm{Pd}_R(E)$ *and* $\mathrm{Pd}_{R_S}^*(E_S) \leqslant \mathrm{Pd}_R^*(E)$.

Solution. In dealing with the first inequality we may assume that $\mathrm{Pd}_R(E) = k$, where $0 \leqslant k < \infty$. Let

$$0 \to P_k \to P_{k-1} \to \ldots \to P_1 \to P_0 \to E \to 0$$

be a projective resolution of E of length k. On forming fractions using S, we obtain a projective resolution of the R_S-module E_S of the same length. Accordingly $\mathrm{Pd}_{R_S}(E_S) \leqslant k$.

The second inequality may be established in a similar manner using resolutions which involve only supplementable projective modules. (Note that if K is a supplementable projective R-module, then K_S is a supplementable projective R_S-module.)

EXERCISE 2. *Let E be an R-module possessing a finite free resolution of finite length. Show that*

$$\mathrm{Pd}_R^*(E) = \sup_M \mathrm{Pd}_{R_M}^*(E_M),$$

where M ranges over all the maximal ideals of R.

Solution. Throughout the solution M will denote a typical maximal ideal of R. By Exercise 1,

$$\mathrm{Pd}_{R_M}^*(E_M) \leqslant \mathrm{Pd}_R^*(E)$$

hence we need only show that

$$\mathrm{Pd}_R^*(E) \leqslant \sup_M \mathrm{Pd}_{R_M}^*(E_M).$$

Now if $E_M = 0$ for all M then, by Chapter 4 Exercise 10, we have $E = 0$. Consequently we may assume that

$$\sup_M \mathrm{Pd}_{R_M}^*(E_M) = k,$$

where $0 \leqslant k < \infty$.

E has a finite free resolution of finite length. Accordingly we can construct an exact sequence

$$0 \to N \to F_{k-1} \to F_{k-2} \to \ldots \to F_1 \to F_0 \to E \to 0,$$

where each F_j is a finite free R-module. Next from

$$\mathrm{Pd}^*_{R_M}(E_M) \leqslant k$$

it follows that N_M is a projective R_M-module. But $\mathrm{Pd}^*_R(E) < \infty$ and therefore $\mathrm{Pd}^*_R(N) < \infty$ as well. In particular N is finitely presented. We can therefore use Chapter 2 Theorem 14 to deduce that N is a projective R-module. Finally, by Chapter 3 Theorem 15,
$$\mathrm{Pd}^*_R(E) = \mathrm{Pd}_R(E) \leqslant k$$

and with this the solution is complete.

EXERCISE 3. *Let E be an R-module and P a member of* $\mathrm{Att}_R(E)$. *Show that* $\mathrm{Ann}_R(E) \subseteq P$.

Solution. Assume that $\mathrm{Ann}_R(E) \nsubseteq P$. Then $E_P = 0$ and hence $\mathrm{Gr}_{R_P}\{PR_P; E_P\} = \infty$. This however is contrary to the definition of $\mathrm{Att}_R(E)$.

EXERCISE 4. *Show that if E is a finitely generated R-module, then every minimal prime ideal of* $\mathrm{Ann}_R(E)$ *belongs to* $\mathrm{Att}_R(E)$.

Solution. Let P be a minimal prime ideal of $\mathrm{Ann}_R(E)$. Since E is finitely generated, PR_P is a minimal prime ideal of $\mathrm{Ann}_{R_P}(E_P)$ and therefore $E_P \neq 0$; furthermore

$$PR_P = \mathrm{Rad}\{\mathrm{Ann}_{R_P}(E_P)\}.$$

Accordingly if $I \subseteq PR_P$ is a finitely generated ideal of R_P, then $I^m E_P = 0$ for some $m > 0$. It follows that I annihilates a non-zero element of E_P. Thus $\mathrm{Gr}_{R_P}\{I; E_P\} = 0$ and now

$$\mathrm{Gr}_{R_P}\{PR_P; E_P\} = 0$$

by Chapter 5 Theorem 11. But this is what we have to prove.

EXERCISE 5. *Give an example of a ring R and a finitely generated R-module E for which* $\mathrm{Ass}_R(E) \neq \mathrm{Att}_R(E)$.

Solution. Let K be a field and $x_1, x_2, x_3, \ldots$ an infinite sequence of indeterminates. Put $R = K[x_1, x_2, x_3, \ldots]$. Further let P be the prime ideal generated by the x_j and Q the ideal generated by the squares of the indeterminates. If now $E = R/Q$, then E is singly generated and $P \in \mathrm{Att}_R(E)$ by Exercise 4. On the other hand, no non-zero element of E is annihilated by P and therefore $P \notin \mathrm{Ass}_R(E)$.

EXERCISE 6. *Let α and β be non-zerodivisors in R and suppose that $R\alpha \cap R\beta$ is a principal ideal. Show by means of elementary considerations that α and β have a greatest common divisor.*

Solution. Let $R\alpha \cap R\beta = R\gamma$. Then $\alpha\beta = \gamma\delta$, where $\delta \in R$. Note that $R\alpha + R\beta \subseteq R\delta$ and that neither γ nor δ can be a zerodivisor.

Now assume that ω belongs to R and that $R\alpha + R\beta \subseteq R\omega$. We can then write $\alpha = \omega x$, $\beta = \omega y$, $\gamma = \omega z$, where x, y, z are in R. Since $R\alpha \cap R\beta = R\gamma$ and ω is not a zerodivisor, it follows that $Rx \cap Ry = Rz$. Again $\omega^2 xy = \omega z\delta$ whence $\omega(xy) = z\delta$; furthermore $xy = zu$ for some $u \in R$. Thus $\omega z u = z\delta$ which yields $\omega u = \delta$ because z is not a zerodivisor. Accordingly $R\delta \subseteq R\omega$ and we have shown that $\delta = \mathrm{g.c.d.}\{\alpha, \beta\}$.

EXERCISE 7. *Let α and β be non-zerodivisors in R and let $R\alpha \cap R\beta$ be a principal ideal. Show that $R\alpha + R\beta$ has a finite free resolution of length one.*

Solution. Let $R\alpha \cap R\beta = R\gamma$. If now $\gamma = \alpha a = \beta b$, where $a, b \in R$, then we have an exact sequence

$$0 \to R \xrightarrow{\ \lambda\ } R \oplus R \xrightarrow{\ \mu\ } R\alpha + R\beta \to 0$$

in which $\lambda(r) = (ra, rb)$ and $\mu(r_1, r_2) = r_1\alpha - r_2\beta$.

EXERCISE 8. *Let R be a GCD-ring and let α, β, γ be non-zerodivisors. If $\delta = \mathrm{g.c.d.}\{\alpha, \beta\}$ establish the following:*
  (i) $\delta\gamma = \mathrm{g.c.d.}\{\alpha\gamma, \beta\gamma\}$;
  (ii) $1 = \mathrm{g.c.d.}\{\alpha/\delta, \beta/\delta\}$.

Solution. Obviously $R\alpha\gamma + R\beta\gamma \subseteq R\delta\gamma$. Let $\omega = \mathrm{g.c.d.}\{\alpha\gamma, \beta\gamma\}$. Then $R\omega \subseteq R\delta\gamma$ so we can write $\omega = r\delta\gamma$, where $r \in R$, and now $R\alpha + R\beta \subseteq Rr\delta$. Since $\delta = \mathrm{g.c.d.}\{\alpha, \beta\}$, it follows that $Rr\delta = R\delta$ and this shows that r is a unit. Accordingly $\delta\gamma = \mathrm{g.c.d.}\{\alpha\gamma, \beta\gamma\}$ and (i) is established.

Now assume that $u = \mathrm{g.c.d.}\{\alpha/\delta, \beta/\delta\}$. By (i) δu is a greatest common divisor of α and β and therefore $R\delta u = R\delta$. Thus u is a unit and $1 = $ g.c.d. $\{\alpha/\delta, \beta/\delta\}$ as required.

EXERCISE 9. *Assume that, in R, every increasing sequence of principal ideals generated by non-zerodivisors becomes constant. Show that every non-zerodivisor can be expressed as the product of a unit and a finite number of irreducible elements.*

Solution. Suppose that there is a non-zerodivisor α_1 which cannot be expressed in the manner described. Then α_1 cannot be a unit nor can it be irreducible. Hence $\alpha_1 = \alpha_2'\alpha_2''$, where α_2', α_2'' are non-units and also non-zerodivisors. Now $R\alpha_1 \subseteq R\alpha_2'$ and indeed the inclusion is strict, that is $R\alpha_1 \subset R\alpha_2'$. (For otherwise α_2'' would be a unit and this is not the case.) Likewise $R\alpha_1 \subset R\alpha_2''$. Further at least one of α_2' and α_2'' cannot be expressed in the desired form for otherwise their product, namely α_1, could be so expressed. Thus to sum up, there exists a non-zerodivisor α_2 such that $R\alpha_1 \subset R\alpha_2$ and α_2 is not the product of a unit and a finite number of irreducible elements.

We can now repeat the argument using α_2 in place of α_1. In this way we generate an infinite strictly increasing sequence

$$R\alpha_1 \subset R\alpha_2 \subset R\alpha_3 \subset \dots$$

of ideals, where each α_j is a non-zerodivisor. This, however, is contrary to our original hypothesis.

EXERCISE 10. *Show that R is a UF-ring if and only if the following two conditions are both satisfied:*
(a) R is a GCD-ring;
(b) every ascending chain $R\alpha_1 \subseteq R\alpha_2 \subseteq R\alpha_3 \subseteq \dots$ of principal ideals generated by non-zerodivisors terminates.

Solution. First suppose that (a) and (b) both hold. Then Exercise 9 shows that any non-zerodivisor can be represented as a unit times the product of a finite number of irreducible elements; furthermore such a representation is virtually unique in view of Theorem 12. Accordingly when (a) and (b) hold R is a UF-ring.

Next suppose that R is a UF-ring. If α is a non-zerodivisor and α is expressed in the form $\alpha = up_1p_2 \dots p_k$, where u is a unit and each p_j is irreducible, then the number k of irreducible factors depends only on α. In what follows this number is denoted by $n(\alpha)$. Thus $n(\alpha) = 0$ if and only if α is a unit. Also if α' is a second non-zerodivisor, then $n(\alpha\alpha') = n(\alpha) + n(\alpha')$.

Assume now that α, β are non-zerodivisors. *We claim that α and β have a greatest common divisor.* To establish this claim we use induction on $\mu = \min\,[n(\alpha), n(\beta)]$. Obviously if $\mu = 0$, then $1 = \mathrm{g.c.d.}\,\{\alpha, \beta\}$. Suppose therefore that $\mu > 0$ and that the existence of a greatest common divisor has been established for smaller values of the inductive variable. If no irreducible element divides both α and β, then of course $1 = \mathrm{g.c.d.}\,\{\alpha, \beta\}$. Let us assume therefore that there is an irreducible element p which divides α and β. Then $\min\,[n(\alpha/p), n(\beta/p)]$ is smaller than μ and so there exists a non-zerodivisor γ such that $\gamma = \mathrm{g.c.d.}\,\{\alpha/p, \beta/p\}$. Obviously $R\alpha + R\beta \subseteq Rp\gamma$.

Now suppose that $R\alpha + R\beta \subseteq R\omega$, where $\omega \in R$. If p divides ω, then $R(\alpha/p) + R(\beta/p) \subseteq R(\omega/p)$ whence $R\gamma \subseteq R(\omega/p)$ and therefore $Rp\gamma \subseteq R\omega$. Consider the case where p does not divide ω. We can find $a \in R$ so that $\alpha = a\omega$. Evidently p divides a and hence $R(\alpha/p) \subseteq R\omega$. Similarly $R(\beta/p) \subseteq R\omega$. Accordingly $R\gamma \subseteq R\omega$ so again we may conclude that $Rp\gamma \subseteq R\omega$. This proves that $p\gamma = \mathrm{g.c.d.}\,\{\alpha, \beta\}$ and we have established that R is a GCD-ring.

Finally let
$$R\alpha_1 \subseteq R\alpha_2 \subseteq R\alpha_3 \subseteq \ldots$$

be an increasing sequence of principal ideals where the α_j are non-zerodivisors. Then
$$n(\alpha_1) \geqslant n(\alpha_2) \geqslant n(\alpha_3) \geqslant \ldots$$

and there is an integer s such that $n(\alpha_t) = n(\alpha_s)$ for all $t \geqslant s$. It follows that when $t \geqslant s$ the element α_t is a unit times α_s and therefore $R\alpha_t = R\alpha_s$. Accordingly the sequence terminates. The solution is now complete.

EXERCISE 11. *Let R be a quasi-local ring and $\phi\colon F \to G$ a homomorphism, where F, G are finite free R-modules. Show that $\mathrm{Coker}\,\phi$ is a finite free R-module if and only if $\mathfrak{A}(\phi) = R$.*

Solution. The condition $\mathfrak{A}(\phi) = R$ means that, for each $\nu \geqslant 0$, $\mathfrak{A}_\nu(\phi)$ is either the zero ideal or it is the whole ring. Since R is quasi-local, this is equivalent to saying that each of the ideals

$\mathfrak{A}_\nu(\phi)$ is generated by an idempotent. On the other hand the hypotheses ensure that Coker ϕ is a finite free module if and only if it is projective. The desired result therefore follows from Chapter 4 Theorem 18.

EXERCISE 12. *Let* $\phi\colon F \to G$ *be a stable homomorphism, where* F, G *are finite free* R-*modules, and let* S *be a multiplicatively closed subset of* R *not containing the zero element. Show that*

$$\mathfrak{A}(\phi)\, R_S = \mathfrak{A}(\phi_S).$$

Solution. Put $\nu = \operatorname{rank}_R(\phi)$. Since ϕ is stable we also have $\nu = \operatorname{rank}_{R_S}(\phi_S)$. Finally

$$\mathfrak{A}(\phi_S) = \mathfrak{A}_\nu(\phi_S) = \mathfrak{A}_\nu(\phi)\, R_S = \mathfrak{A}(\phi)\, R_S$$

by virtue of the counterpart of (1.7.4).

EXERCISE 13. *Let* $\phi\colon F \to G$ *and* $\phi'\colon F' \to G$ *be homomorphisms, where* F, F', G *are finite free* R-*modules, and let*

$$\phi(F) = \phi'(F').$$

Show that $\mathfrak{A}_\nu(\phi) = \mathfrak{A}_\nu(\phi')$ *for all* $\nu \geqslant 0$, $\operatorname{rank}_R(\phi) = \operatorname{rank}_R(\phi')$ *and* $\mathfrak{A}(\phi) = \mathfrak{A}(\phi')$.

Solution. Put $q = \operatorname{rank}_R(G)$. Then, by (4.2.15),

$$\mathfrak{F}_\mu(\operatorname{Coker}\phi) = \begin{cases} \mathfrak{A}_{q-\mu}(\phi) & \text{for} \quad 0 \leqslant \mu \leqslant q, \\ R & \text{for} \quad \mu > q, \end{cases}$$

and a similar statement holds concerning the Fitting invariants of Coker ϕ'. But Coker $\phi = $ Coker ϕ'. Consequently

$$\mathfrak{A}_\nu(\phi) = \mathfrak{A}_\nu(\phi')$$

for all $\nu \geqslant 0$. The remaining assertions now follow.

7.   *The multiplicative structure*

General remarks

Suppose that we have a finite free resolution

$$0 \to F_n \xrightarrow{\phi_n} F_{n-1} \xrightarrow{\phi_{n-1}} F_{n-2} \to \ldots \to F_1 \xrightarrow{\phi_1} F_0 \to E \to 0$$

of the R-module E. It is a consequence of the key result (Theorem 4) of this chapter that there exist ideals $\mathfrak{B}_0, \mathfrak{B}_1, \ldots, \mathfrak{B}_n$ which, among their other properties, satisfy $\mathfrak{B}_k \mathfrak{B}_{k-1} = \mathfrak{A}(\phi_k)$ for $k = 1, 2, \ldots, n$ and $\mathrm{Gr}_R\{\mathfrak{B}_h\} \geqslant h+1$ for $h = 0, 1, \ldots, n$. These ideals, here called *factorization ideals*, increase our insight into the structure of finite free resolutions. For example, if

$$\mathrm{Char}_R(E) = 0,$$

then $\mathfrak{B}_0$ turns out to be a principal ideal, and if E possesses an elementary resolution of finite length, then $\mathfrak{B}_0$ coincides with the MacRae invariant† of the module. We use this fact to re-define the invariant so as to make it applicable to all modules for which $\mathrm{Char}_R(E) = 0$. This provides a more general and more aesthetic theory.

The discovery of the factorization ideals is due to D. A. Buchsbaum and D. Eisenbud (**14**). A highly special case is implicit in the work of Lindsay Burch (**15**) concerned with the description of ideals that have a finite free resolution of length one. The study of such ideals has an interesting history which goes back to D. Hilbert (**23**). For completeness an up-to-date version of their theory has been included. As in the other investigations, this avoids all use of Noetherian conditions.

† See section (3.6).

7.1 Grade and proportionality

Let E be an R-module and $a_1, a_2, ..., a_n$ respectively $e_1, e_2, ..., e_n$ elements of R respectively E. Let us say that $e_1, e_2, ..., e_n$ are *proportional* to $a_1, a_2, ..., a_n$ if there exists $e \in E$ such that $a_i e = e_i$ for $i = 1, 2, ..., n$. A necessary condition for this state of affairs is that $a_i e_j = a_j e_i$ for all i and j or, equivalently, that all the 2×2 minors of the matrix

$$\left\| \begin{matrix} a_1 & a_2 & ... & a_n \\ e_1 & e_2 & ... & e_n \end{matrix} \right\|$$

be zero. (However it should be noted that here we are using the term *matrix* in an extended sense.)

The next result gives sufficient conditions for proportionality.

THEOREM 1. *Let E be an R-module and suppose that all the 2×2 minors of the matrix*

$$\left\| \begin{matrix} a_1 & a_2 & ... & a_n \\ e_1 & e_2 & ... & e_n \end{matrix} \right\|$$

are zero, where $a_i \in R$ and $e_j \in E$. If now

$$\mathrm{Gr}_R \{ Ra_1 + Ra_2 + ... + Ra_n; E \} \geqslant 2,$$

then there exists a unique $e \in E$ such that $e_i = a_i e$ for $i = 1, 2, ..., n$.

Proof. First suppose that $Ra_1 + Ra_2 + ... + Ra_n$ contains elements a, a' which form an R-sequence on E of length two. Then we have relations

$$a = r_1 a_1 + r_2 a_2 + ... + r_n a_n \quad \text{and} \quad a' = r_1' a_1 + r_2' a_2 + ... + r_n' a_n,$$

where r_i and r_i' belong to R. Put $\eta = r_1 e_1 + r_2 e_2 + ... + r_n e_n$ and $\eta' = r_1' e_1 + r_2' e_2 + ... + r_n' e_n$. Evidently $a\eta' = a'\eta$ and therefore η belongs to $aE :_E a' = aE$, say $\eta = ae$. Next $ae_i = a_i \eta = a_i ae$ and therefore $e_i = a_i e$. This holds for $i = 1, 2, ..., n$. Since the equation $\eta = ae$ determines e uniquely, there can be at most one element of E for which $e_i = a_i e$ for $1 \leqslant i \leqslant n$.

We now remove the condition introduced at the beginning of the last paragraph. Let x, y be indeterminates. Then, by Chapter 5 Exercise 9, the ideal $a_1 R[x, y] + a_2 R[x, y] + ... + a_n R[x, y]$

contains an $R[x, y]$-sequence on $E[x, y]$ of length two. Hence by our earlier considerations there is a unique $\phi \in E[x, y]$ such that $e_i = a_i \phi$ for $i = 1, 2, ..., n$. Because ϕ is unique, these relations imply that ϕ is a constant. The theorem follows.

Note that the condition $\mathrm{Gr}_R \{Ra_1 + Ra_2 + ... + Ra_n; E\} \geqslant 2$ in the theorem cannot be replaced by

$$\mathrm{Gr}_R \{Ra_1 + Ra_2 + ... + Ra_n; E\} \geqslant 1.$$

For example, take R to be the ring of integers. Then the determinant of the matrix

$$\left\| \begin{array}{cc} 2 & 4 \\ 3 & 6 \end{array} \right\|$$

is zero and each row generates an ideal of grade one. However neither row can be obtained from the other by multiplying by an integer.

The next two lemmas provide the means whereby Theorem 1 can be applied to the study of finite free resolutions. In them A denotes a $p \times q$ R-matrix and B a $q \times t$ R-matrix, where p, q, t are positive integers. Also μ, ν designate non-negative integers and J, K, M, N variable members of $S_\mu^p, S_\mu^q, S_\nu^q, S_\nu^t$ respectively, the notation being that introduced in section (1.3). Thus M is a subsequence of $\{1, 2, ..., q\}$. As on previous occasions we shall use M' to describe the residual sequence obtained from $\{1, 2, ..., q\}$ by striking out the members of M. Accordingly M followed by M' gives a permutation of the integers $1, 2, ..., q$. The signature of this permutation will be denoted by $\mathrm{sgn}\,(M, M')$.

LEMMA 1. *Suppose that* $\mu + \nu = q$ *and assume that*

$$U^{-1} A V^{-1} = \left\| \begin{array}{c|c} 0 & I_\mu \\ \hline 0 & 0 \end{array} \right\| \quad and \quad V B W^{-1} = \left\| \begin{array}{c|c} 0 & I_\nu \\ \hline 0 & 0 \end{array} \right\|,$$

where U, V, W *are unimodular matrices and* I_μ, I_ν *identity matrices of orders* μ, ν *respectively. Then there exist elements* u_J, v_K, w_M, z_N *in* R *such that* (i) $w_M = \mathrm{sgn}\,(M, M')\,v_{M'}$ *for all* M, *and* (ii)

$$A_{JK}^{(\mu)} = u_J v_K, \quad B_{MN}^{(\nu)} = w_M z_N$$

for all J, K, M, N.

Proof. Since

$$A = U \left\| \begin{array}{c|c} 0 & I_\mu \\ \hline 0 & 0 \end{array} \right\| V \quad and \quad B = \Lambda \left\| \begin{array}{c|c} 0 & I_\nu \\ \hline 0 & 0 \end{array} \right\| W,$$

where $\Lambda = V^{-1}$, it follows that

$$A^{(\mu)}_{JK} = U^{(\mu)}_{J\{1,2,\ldots,\mu\}} V^{(\mu)}_{\{\nu+1,\ldots,q\}K}$$

and

$$B^{(\nu)}_{MN} = \Lambda^{(\nu)}_{M\{1,2,\ldots,\nu\}} W^{(\nu)}_{\{t-\nu+1,\ldots,t\}N}.$$

Again, by Chapter 1 Theorem 9,

$$\det(V)\,\Lambda^{(\nu)}_{M\{1,2,\ldots,\nu\}} = \operatorname{sgn}(M,M')\,V^{(\mu)}_{\{\nu+1,\ldots,q\}M'}.$$

Hence if we put

$$u_J = U^{(\mu)}_{J\{1,2,\ldots,\mu\}}, \quad v_K = V^{(\mu)}_{\{\nu+1,\ldots,q\}K},$$

$$w_M = \det(V)\,\Lambda^{(\nu)}_{M\{1,2,\ldots,\nu\}} = \operatorname{sgn}(M,M')\,v_{M'},$$

and

$$z_N = \det(V^{-1})\,W^{(\nu)}_{\{t-\nu+1,\ldots,t\}N},$$

we obtain the required result.

At this point we require some additional notation. Let us suppose, as in Lemma 1, that $\mu + \nu = q$. Further let $J \in S^p_\mu$, M_1, $M_2 \in S^q_\nu$ and $N \in S^t_\nu$. Now put

$$D_{JM_1M_2N} = \begin{vmatrix} \operatorname{sgn}(M_1,M_1')\,A^{(\mu)}_{JM_1'} & \operatorname{sgn}(M_2,M_2')\,A^{(\mu)}_{JM_2'} \\ B^{(\nu)}_{M_1N} & B^{(\nu)}_{M_2N} \end{vmatrix},$$
$$(7.1.1)$$

where M_1', M_2' denote the complements of M_1, M_2 in $\{1,2,\ldots,q\}$.

LEMMA 2. *Let the situation be as above. Assume that $AB = 0$, $\mu + \nu = q$, and $\mu = \operatorname{rank}_R(A)$. Then some power of $\mathfrak{A}_\mu(A)\,\mathfrak{A}_\nu(B)$ annihilates $D_{JM_1M_2N}$.*

Proof. Let Δ be a $\mu \times \mu$ minor of A and Δ' a $\nu \times \nu$ minor of B. Since $\mathfrak{A}_\mu(A)\,\mathfrak{A}_\nu(B)$ is generated by a finite number of products such as $\Delta\Delta'$, it will suffice to show that some power of $\Delta\Delta'$ annihilates $D_{JM_1M_2N}$. From here on we shall assume the contrary and look for a contradiction.

By considering the multiplicatively closed subset of R that consists of the powers of $\Delta\Delta'$, we see that there exists a prime ideal P which contains the annihilator of $D_{JM_1M_2N}$ but which does not contain $\Delta\Delta'$. Observe that if we localize at P, then A becomes a matrix with a $\mu \times \mu$ minor which is a unit. Since on localization the rank of a matrix does not increase,† the rank of A remains equal to μ.

† See (3.2.11).

Let us localize at P and then, to simplify the notation, drop P as suffix. We still have $AB = 0$ and $\mu = \operatorname{rank}_R(A)$. In addition the ring R is now quasi-local, $\mathfrak{A}_\mu(A) = R = \mathfrak{A}_\nu(B)$, and

$$D_{JM_1M_2N} \neq 0.$$

From these data we shall derive a contradiction.

In the new situation the matrix A is stable and it is clear that $\operatorname{rank}_R(B) \geqslant \nu$. Consequently, by Chapter 4 Theorem 2,

$$\operatorname{rank}_R(B) = \nu.$$

It therefore follows, from Chapter 1 Theorem 12, that there exist unimodular matrices V_0, W such that

$$V_0 B W = \left\| \begin{array}{c|c} 0 & I_\nu \\ \hline 0 & 0 \end{array} \right\|.$$

However $AB = 0$. Accordingly

$$A V_0^{-1} = \left\| \begin{array}{c|c} 0 & A' \end{array} \right\|,$$

where A' is a $p \times \mu$ matrix of rank μ with the property that $\mathfrak{A}_\mu(A') = \mathfrak{A}_\mu(A) = R$. Similar considerations now show that there exist unimodular matrices U, V' such that

$$U A' V' = \left\| \begin{array}{c} I_\mu \\ \hline 0 \end{array} \right\|.$$

Put

$$V_1 = \left\| \begin{array}{c|c} I_\nu & 0 \\ \hline 0 & V' \end{array} \right\|.$$

Then V_1 is a unimodular matrix,

$$U A V_0^{-1} V_1 = \left\| \begin{array}{c|c} 0 & I_\mu \\ \hline 0 & 0 \end{array} \right\|,$$

and

$$V_1^{-1} V_0 B W = \left\| \begin{array}{c|c} 0 & I_\nu \\ \hline 0 & 0 \end{array} \right\|.$$

But here we have a situation in which Lemma 1 is applicable. Hence, with the notation of that result, there exist elements u_J, v_K, w_M, z_N in R such that

$$\operatorname{sgn}(M, M') A_{JM'}^{(\mu)} = u_J w_M,$$

and

$$B_{MN}^{(\nu)} = w_M z_N$$

for all $J \in S_\mu^p$, $M \in S_\nu^q$ and $N \in S_\nu^t$. Thus

$$D_{JM_1M_2N} = \begin{vmatrix} u_J w_{M_1} & u_J w_{M_2} \\ w_{M_1} z_N & w_{M_2} z_N \end{vmatrix} = 0$$

as required.

THEOREM 2. *Let A be a $p \times q$ matrix and B a $q \times t$ matrix, where p, q, t are positive integers, and let $E \neq 0$ be an R-module. Suppose that A and B are stable relative to E and that $AB = 0$ exactly on E. Put $\mu = \mathrm{rank}_R (A, E)$, $\nu = \mathrm{rank}_R (B, E)$ so that, by Chapter 4 Theorem 3, $\mu + \nu = q$. Then, with the notation of (7.1.1), the element $D_{JM_1M_2N}$ belongs to $\mathrm{Ann}_R (E)$ for all choices of J, M_1, M_2, and N.*

Proof. Put $\bar{R} = R/\mathrm{Ann}_R (E)$ and let $\bar{A}$ respectively $\bar{B}$ be the matrices obtained by applying the canonical homomorphism $R \to \bar{R}$ to the elements of A respectively B. Clearly $\bar{R}$ is a nontrivial ring and $\bar{A}\bar{B} = 0$ exactly on E. Also $\mathfrak{A}_\lambda(\bar{A}) = \mathfrak{A}_\lambda(A)\,\bar{R}$ for all λ. It follows that $\bar{A}$ is stable relative to E and that

$$\mu = \mathrm{rank}_R (A, E) = \mathrm{rank}_{\bar{R}} (\bar{A}, E).$$

But $\mathrm{Ann}_{\bar{R}} (E) = 0$ and so we see that $\mu = \mathrm{rank}_{\bar{R}} (\bar{A})$. Similar considerations show that $\bar{B}$ is stable relative to E. Note that if $\bar{D}_{JM_1 M_2 N}$ is defined by applying (7.1.1) to the matrices $\bar{A}$ and $\bar{B}$, then it can also be obtained by acting on $D_{JM_1 M_2 N}$ with the homomorphism $R \to \bar{R}$.

It follows from Lemma 2 that some power of $\mathfrak{A}_\mu(\bar{A})\,\mathfrak{A}_\nu(\bar{B})$ annihilates $\bar{D}_{JM_1 M_2 N}$ and therefore such a power annihilates $\bar{D}_{JM_1M_2N}E$. But $\bar{A}$ is stable relative to E and $\mathrm{rank}_{\bar{R}} (\bar{A}, E) = \mu$. Accordingly $\mathfrak{A}_\mu(\bar{A})$ does not annihilate any non-zero element of E. This shows that there is a positive integer m such that

$$(\mathfrak{A}_\nu(\bar{B}))^m\,\bar{D}_{JM_1M_2N}E = 0.$$

However $\bar{B}$ is also stable relative to E and $\mathrm{rank}_{\bar{R}} (\bar{B}, E) = \nu$. We may therefore conclude that $\bar{D}_{JM_1M_2N}E = 0$, i.e. that $D_{JM_1M_2N}$ belongs to $\mathrm{Ann}_R (E)$.

EXERCISE 1. *Let $E \neq 0$ be an R-module and A a $p \times q$ matrix which is stable relative to E. Show that if $\mathrm{Ann}_R (E) = 0$, then A is stable relative to R and $\mathrm{rank}_R (A) = \mathrm{rank}_R (A, E)$.*

EXERCISE 2. *Let $A = \|a_{jk}\|$ be a $p \times q$ matrix and $U = \|u_j\|$ a $p \times 1$ matrix both with entries in R. Suppose that*

$$\mathrm{Gr}_R\{Ru_1 + Ru_2 + \ldots + Ru_p\} \geqslant 2,$$

and
$$\begin{vmatrix} u_{j_1} & u_{j_2} \\ a_{j_1 k} & a_{j_2 k} \end{vmatrix} = 0$$

for all choices of j_1, j_2, k. Show that there is a unique $1 \times q$ matrix V such that $A = UV$.

EXERCISE 3. *Let $\alpha \in R$ and suppose that $\mathrm{Gr}_R\{R\alpha\} \geqslant 2$. Show that α is a unit.*

A more general result than Exercise 3, which also has applications, is provided by

EXERCISE 4. *Let $\mathfrak{A}$ be an invertible integral ideal of R and suppose that $\mathrm{Gr}_R\{\mathfrak{A}\} \geqslant 2$. Show that $\mathfrak{A} = R$.*

7.2 Factorization of matrices

Let $C = \|c_{jk}\|$ be an $r \times s$ R-matrix. By a *complete factorization* of C we shall mean a pair consisting of an $r \times 1$ matrix $U = \|u_j\|$ and a $1 \times s$ matrix $V = \|v_k\|$ such that $C = UV$ and therefore $c_{jk} = u_j v_k$ for all j and k. Two such factorizations of C, say $C = UV$ and $C = U^*V^*$, will be called *equivalent* if there is a unit ϵ of the ring R such that $U^* = \epsilon U$ and $V^* = \epsilon^{-1}V$. Clearly this is an equivalence relation on the (possibly empty) set of all complete factorizations of C.

Suppose for the moment that C is a $1 \times s$ matrix, i.e. a row matrix of length s, and let ϵ be a unit of R. Then $C = \|\epsilon\| (\epsilon^{-1}C)$ is certainly a complete factorization of C. Such a complete factorization will be called a *canonical complete factorization* of C. Obviously any two canonical complete factorizations of a row matrix are equivalent.

After these preliminaries let A be a $p \times q$ matrix and B a $q \times t$ matrix both with entries in R. Further suppose that

$$0 \leqslant \mu \leqslant \min(p, q), \quad 0 \leqslant \nu \leqslant \min(q, t) \quad \text{and} \quad \mu + \nu = q.$$

Then $A^{(\mu)}$ is an $S_\mu^p \times S_\mu^q$ matrix, $B^{(\nu)}$ is an $S_\nu^q \times S_\nu^t$ matrix, and the number of columns in $A^{(\mu)}$ is equal to the number of rows in $B^{(\nu)}$.

Now assume that we are given column matrices $U = \|u_J\|$, $W = \|w_M\|$ and row matrices $V = \|v_K\|$, $Z = \|z_N\|$ such that

$$A^{(\mu)} = UV \tag{7.2.1}$$

and $$B^{(\nu)} = WZ, \tag{7.2.2}$$

i.e. suppose that complete factorizations of $A^{(\mu)}$ and $B^{(\nu)}$ are prescribed. These will be said to be *complementary* if there is a unit η, in R, for which

$$w_M = \eta \operatorname{sgn}(M, M') v_{M'} \tag{7.2.3}$$

for all $M \in S_\nu^q$. (As usual, M' is obtained from $1, 2, \ldots, q$ by deleting the terms of M.)

Next let $\Omega, \Omega^*, \Omega^{**}$ be unimodular matrices of orders p, q, t respectively and put

$$\bar{A} = \Omega A \Omega^{*-1} \tag{7.2.4}$$

$$\bar{B} = \Omega^* B \Omega^{**-1} \tag{7.2.5}$$

that is $\bar{A} = \Omega A \Lambda^*$ and $\bar{B} = \Omega^* B \Lambda^{**}$, where we have set $\Lambda^* = \Omega^{*-1}$ and $\Lambda^{**} = \Omega^{**-1}$. Then

$$\bar{A}^{(\mu)} = \Omega^{(\mu)} A^{(\mu)} \Lambda^{*(\mu)},$$

$$\bar{B}^{(\nu)} = \Omega^{*(\nu)} B^{(\nu)} \Lambda^{**(\nu)}.$$

Consequently if (7.2.1) and (7.2.2) are complementary complete factorizations of $A^{(\mu)}$ and $B^{(\nu)}$, then

$$\bar{A}^{(\mu)} = \bar{U}\,\bar{V} \tag{7.2.6}$$

and $$\bar{B}^{(\nu)} = \bar{W}\bar{Z}, \tag{7.2.7}$$

where $\bar{U} = \Omega^{(\mu)} U$, $\bar{V} = V \Lambda^{*(\mu)}$, $\bar{W} = \Omega^{*(\nu)} W$ and $\bar{Z} = Z \Lambda^{**(\nu)}$, are complete factorizations of $\bar{A}^{(\mu)}$ and $\bar{B}^{(\nu)}$. These are also complementary. For if $M \in S_\nu^q$, then

$$\bar{w}_M = \sum_{P \in S_\nu^q} \Omega_{MP}^{*(\nu)} w_P$$

$$= \eta \sum_{P \in S_\nu^q} \Omega_{MP}^{*(\nu)} \operatorname{sgn}(P, P') v_{P'}$$

for some unit η. But $\Omega^{*(\mu)}$ is the inverse of $\Lambda^{*(\mu)}$ and therefore $V = \bar{V}\Omega^{*(\mu)}$. Thus

$$\bar{w}_M = \eta \sum_{Q \in S_\nu^q} \bar{v}_{Q'} \sum_{P \in S_\nu^q} \operatorname{sgn}(P, P') \Omega_{MP}^{*(\nu)} \Omega_{Q'P'}^{*(\mu)}.$$

Now, by Chapter 1 Theorem 8,

$$\sum_{P \in S_\nu^q} \operatorname{sgn}(P, P') \, \Omega_{MP}^{*(\nu)} \, \Omega_{Q'P'}^{*(\mu)} = \det(\Omega^*) \operatorname{sgn}(M, M') \, \delta_{MQ}$$

so the formula for $\overline{w}_M$ reduces to

$$\overline{w}_M = \eta \det(\Omega^*) \operatorname{sgn}(M, M') \, \overline{v}_{M'}.$$

Thus to sum up: *if* (7.2.1) *and* (7.2.2) *are complementary complete factorizations of* $A^{(\mu)}$ *and* $B^{(\nu)}$, *then* (7.2.6) *and* (7.2.7) *are complementary complete factorizations of* $\overline{A}^{(\mu)}$ *and* $\overline{B}^{(\nu)}$.

THEOREM 3. *Let* $F \xrightarrow{\phi} G \xrightarrow{\psi} H$ *be an exact sequence, where* F, G, H *are finite free R-modules and* ϕ, ψ *are stable homomorphisms. Suppose that* $\operatorname{Gr}_R\{\mathfrak{A}(\phi)\} \geqslant 2$ *and put* $\mu = \operatorname{rank}_R(\phi)$, $\nu = \operatorname{rank}_R(\psi)$ *and* $q = \operatorname{rank}_R(G)$ *so that, by Chapter 4 Theorem 6, $\mu + \nu = q$. Choose a base for each of F, G, H and let the matrix of ϕ respectively ψ be A respectively B. Finally assume that a complete factorization of $A^{(\mu)}$ is given. Then $B^{(\nu)}$ possesses a complete factorization complementary to that of $A^{(\mu)}$ and this is unique to within equivalence.*

Remark. In this theorem we have to allow for the possibility that either A or B is *degenerate* i.e. has an empty set of rows or columns. In this connection we recall our convention† for non-degenerate matrices which gives $A^{(0)} = \|1_R\|$. This must now be extended to include degenerate situations as well.

Proof. We first dispose of the degenerate cases. If $\mu = 0$, then $A^{(\mu)}$ is just $\|1_R\|$. Accordingly a complete factorization of $A^{(\mu)}$ is essentially a representation of 1_R as the product of two units. Now $B^{(\nu)}$ is a row matrix. The complete factorizations of $B^{(\nu)}$ complementary to the given factorization of $A^{(\mu)}$ are none other than the canonical ones.

Next suppose that $\nu = 0$. Then $A^{(\mu)}$ is a column matrix and the given complete factorization of $A^{(\mu)}$ represents it as a product of a column matrix and a 1×1 matrix $\|\beta\|$. Moreover the entries in $A^{(\mu)}$ generate $\mathfrak{A}(\phi)$ and this shows that $\operatorname{Gr}_R\{R\beta\} \geqslant 2$ whence β is a unit by Exercise 3. Since on this occasion $B^{(\nu)}$ is a 1×1 matrix, the claims of the theorem are obvious in this special case.

From here on we suppose that both $\mu > 0$ and $\nu > 0$. Put $p = \mathrm{rank}_R(F)$, $t = \mathrm{rank}_R(H)$. Then p, q, t are all positive, A is a $p \times q$ matrix and B a $q \times t$ matrix. These matrices are stable and $AB = 0$ exactly. Hence Theorem 2 (with $E = R$) yields $D_{JM_1M_2N} = 0$ for all choices of J, M_1, M_2, N, where the notation is the same as before. Let

$$A^{(\mu)}_{JM'} = u_J v_{M'} \quad (J \in S^p_\mu, M \in S^q_\nu)$$

be the given complete factorization of $A^{(\mu)}$. Then the equation $D_{JM_1M_2N} = 0$ can be written as

$$u_J \begin{vmatrix} \mathrm{sgn}\,(M_1, M_1')\,v_{M_1'} & \mathrm{sgn}\,(M_2, M_2')\,v_{M_2'} \\ B^{(\nu)}_{M_1N} & B^{(\nu)}_{M_2N} \end{vmatrix} = 0.$$

But the ideal generated by the u_J contains $\mathfrak{A}(\phi)$ and therefore it is of positive grade. It follows that

$$\begin{vmatrix} \mathrm{sgn}\,(M_1, M_1')\,v_{M_1'} & \mathrm{sgn}\,(M_2, M_2')\,v_{M_2'} \\ B^{(\nu)}_{M_1N} & B^{(\nu)}_{M_2N} \end{vmatrix} = 0$$

for all choices of M_1, M_2, N. Let us fix N and observe that the elements $\mathrm{sgn}\,(M, M')\,v_{M'}$ $(M \in S^q_\nu)$ generate an ideal which, since it too contains $\mathfrak{A}(\phi)$, has grade at least equal to two. Hence, by Theorem 1, there is a *unique* $z_N \in R$ such that

$$B^{(\nu)}_{MN} = \mathrm{sgn}\,(M, M')\,v_{M'}z_N$$

for all M. This clearly implies the assertions of the theorem.

We are now ready to renew our attack on the theory of finite free resolutions.

7.3 Factorization ideals

Consider a finite free resolution

$$0 \to F_n \xrightarrow{\phi_n} F_{n-1} \xrightarrow{\phi_{n-1}} F_{n-2} \to \ldots \to F_1 \xrightarrow{\phi_1} F_0 \to E \to 0 \quad (7.3.1)$$

of an R-module E. Put

$$\mathrm{rank}_R(F_k) = q_k \quad (0 \leqslant k \leqslant n), \qquad (7.3.2)$$

$$\mathrm{rank}_R(\phi_h) = \mu_h \quad (1 \leqslant h \leqslant n) \qquad (7.3.3)$$

and before proceeding note that, by Chapter 4 Theorem 8,

$$\mu_n = q_n \tag{7.3.4}$$

and
$$\mu_{h+1} + \mu_h = q_h \quad (1 \leqslant h < n). \tag{7.3.5}$$

Let us select a base for each of the free modules and for the moment keep these bases fixed. Then $\phi_h \colon F_h \to F_{h-1}$ will be represented by a matrix A_h which arises from the chosen bases for F_h and F_{h-1}. Observe that even if some of the free modules are null, each of the exterior powers $A_h^{(\mu_h)}$ has at least one row and one column. Also, by (7.3.4), $A_n^{(\mu_n)}$ is a *row matrix*.

THEOREM 4. *Suppose that*

$$0 \to F_n \xrightarrow{\phi_n} F_{n-1} \xrightarrow{\phi_{n-1}} F_{n-2} \to \dots \to F_1 \xrightarrow{\phi_1} F_0 \to E \to 0,$$

where $n \geqslant 1$, is a finite free resolution of the R-module E and let a base be chosen for each of the free modules in the resolution. Then there exist complete factorizations

$$A_h^{(\mu_h)} = U_h V_h \quad (h = 1, 2, \dots, n),$$

where the factorization of the row matrix $A_n^{(\mu_n)}$ is canonical and, for $1 \leqslant h < n$, the factorizations of $A_{h+1}^{(\mu_{h+1})}$ and $A_h^{(\mu_h)}$ are complementary. Moreover the complete factorizations are uniquely determined by these conditions to within equivalence.

Proof. For the moment let us leave aside the question of uniqueness. Certainly $A_n^{(\mu_n)}$ can be given a canonical factorization. Now suppose that we have obtained complete factorizations for $A_n^{(\mu_n)}$, $A_{n-1}^{(\mu_{n-1})}, \dots, A_{h+1}^{(\mu_{h+1})}$ and that these have the desired properties. Consider the exact sequence

$$F_{h+1} \xrightarrow{\phi_{h+1}} F_h \xrightarrow{\phi_h} F_{h-1}.$$

We know, from Chapter 4 Theorem 8, that ϕ_{h+1} and ϕ_h are stable and Chapter 6 Theorem 15 shows that $\mathrm{Gr}_R\{\mathfrak{A}(\phi_{h+1})\} \geqslant 2$. Hence, by Theorem 3, there is a complete factorization of $A_h^{(\mu_h)}$ that is complementary to the one already obtained for $A_{h+1}^{(\mu_{h+1})}$.

This establishes the existence of a system of complete factorizations of the desired kind. However canonical factorizations of row matrices are unique to within equivalence and the other complete factorizations are also unique in this sense by virtue of Theorem 3. Thus the theorem is proved.

We shall now extract some information which is particularly useful for applications. Let the assumptions be as in Theorem 4 and let the notation be the same as that employed in the theorem and its proof. The ideal $\mathfrak{A}_1(U_h)$ respectively $\mathfrak{A}_1(V_h)$ is generated by the entries in U_h respectively V_h. Since U_h and V_h are uniquely determined to within multiplication by units of R, there is no ambiguity about these ideals once bases for $F_0, F_1, \ldots, F_n$ have been chosen. Again $A_h^{(\mu_h)} = U_h V_h$ and $\mathfrak{A}(\phi_h)$ is the ideal generated by the entries in $A_h^{(\mu_h)}$. Consequently

$$\mathfrak{A}(\phi_h) = \mathfrak{A}_1(U_h)\,\mathfrak{A}_1(V_h).$$

Note that because U_n is a 1×1 matrix whose sole entry is a unit, we also have
$$\mathfrak{A}_1(U_n) = R.$$

Next we observe that as successive factorizations are complementary, we have
$$\mathfrak{A}_1(V_{h+1}) = \mathfrak{A}_1(U_h).$$

Put
$$\mathfrak{B}_k = \mathfrak{A}_1(U_k) = \mathfrak{A}_1(V_{k+1}) \tag{7.3.6}$$

it being understood that $\mathfrak{B}_n = \mathfrak{A}_1(U_n)$ and $\mathfrak{B}_0 = \mathfrak{A}_1(V_1)$. Then

$$\mathfrak{A}(\phi_h) = \mathfrak{B}_h \mathfrak{B}_{h-1} \tag{7.3.7}$$

for $h = 1, 2, \ldots, n$, and
$$\mathfrak{B}_n = R. \tag{7.3.8}$$

Finally U_k is a column matrix with $\binom{q_k}{\mu_k}$ entries and V_k a row matrix with $\binom{q_{k-1}}{\mu_k}$ entries. Accordingly $\mathfrak{B}_k$ *can be generated by* $\binom{q_k}{\mu_k} = \binom{q_k}{\mu_{k+1}}$ *elements.*

Let us now examine what happens to the ideals $\mathfrak{B}_0, \mathfrak{B}_1, \ldots, \mathfrak{B}_n$ when we change the bases of $F_0, F_1, \ldots, F_n$. The homomorphism ϕ_k will then be represented by a new matrix $\bar{A}_k$. These new matrices also factorize completely in the manner described in Theorem 4 and yield ideals $\bar{\mathfrak{B}}_0, \bar{\mathfrak{B}}_1, \ldots, \bar{\mathfrak{B}}_n$ that are the counterparts of $\mathfrak{B}_0, \mathfrak{B}_1, \ldots, \mathfrak{B}_n$. Next associated with the change of bases there will be certain unimodular matrices $\Omega_0, \Omega_1, \ldots, \Omega_n$ and these will relate the matrices $\bar{A}_h$ to the original A_h by means of formulae

$$\bar{A}_h = \Omega_h A_h \Omega_{h-1}^{-1} \quad (h = 1, 2, \ldots, n).$$

Put $\Lambda_j = \Omega_j^{-1}$. Then from (7.2.6) and (7.2.7) it follows that

$$\bar{A}_h^{(\mu_h)} = \bar{U}_h \bar{V}_h,$$

where $\bar{U}_h = \Omega_h^{(\mu_h)} U_h$ and $\bar{V}_h = V_h \Lambda_{h-1}^{(\mu_h)}$. Furthermore the discussion in the paragraph following (7.2.3) shows that we have here complementary complete factorizations of $\bar{A}_h^{(\mu_h)}$ and $\bar{A}_{h-1}^{(\mu_{h-1})}$ and, since it is clear that $\bar{A}_n^{(\mu_n)} = \bar{U}_n \bar{V}_n$ is a canonical complete factorization, we may conclude that $\bar{\mathfrak{B}}_k = \mathfrak{A}_1(\bar{U}_k) = \mathfrak{A}_1(\bar{V}_{k+1})$. However, the exterior powers of a unimodular matrix are unimodular,† and so it follows that

$$\bar{\mathfrak{B}}_k = \mathfrak{B}_k \quad (k = 0, 1, \ldots, n).$$

Thus $\mathfrak{B}_0, \mathfrak{B}_1, \ldots, \mathfrak{B}_n$ do not depend on which bases are chosen for $F_0, F_1, \ldots, F_n$ and therefore we are justified in making the

DEFINITION. *Let a finite free resolution*

$$0 \to F_n \to F_{n-1} \to \ldots \to F_1 \to F_0 \to E \to 0$$

be given. Then the ideals $\mathfrak{B}_0, \mathfrak{B}_1, \ldots, \mathfrak{B}_n$ *(constructed as above) will be called the 'factorization ideals' of the resolution.*

Sometimes we shall simply say that $\mathfrak{B}_0, \mathfrak{B}_1, \ldots, \mathfrak{B}_n$ is a *system of factorization ideals for E* without specifying the resolution which produces them. The theorem which follows summarizes the principal properties of factorization ideals that have been established so far.

THEOREM 5. *Suppose that*

$$0 \to F_n \xrightarrow{\phi_n} F_{n-1} \xrightarrow{\phi_{n-1}} F_{n-2} \to \ldots \to F_1 \xrightarrow{\phi_1} F_0 \to E \to 0,$$

where $n \geqslant 1$, *is a finite free resolution of the R-module E. Put* $\mathrm{rank}_R(F_k) = q_k$, $\mathrm{rank}_R(\phi_h) = \mu_h$, *and let* $\mathfrak{B}_0, \mathfrak{B}_1, \ldots, \mathfrak{B}_n$ *be the factorization ideals associated with the resolution. Then* $\mathfrak{B}_n = R$,

$$\mathfrak{B}_h \mathfrak{B}_{h-1} = \mathfrak{A}(\phi_h)$$

† See Chapter 1 Theorem 9.

for $h = 1, 2, ..., n$, *and* $\mathfrak{B}_k$ *can be generated by* $\begin{pmatrix} q_k \\ \mu_k \end{pmatrix} = \begin{pmatrix} q_k \\ \mu_{k+1} \end{pmatrix}$ *elements. In particular* $\mathfrak{B}_0$ *can be generated by* $\begin{pmatrix} q_0 \\ \mu_1 \end{pmatrix}$ *elements.*

COROLLARY 1. *Let the assumptions and notation be as in Theorem 5. Then* $\mathrm{Gr}_R\{\mathfrak{B}_h\} \geqslant h+1$ *for* $h = 0, 1, ..., n$.

Proof. We may suppose that $0 \leqslant h < n$. On this understanding

$$\mathfrak{A}(\phi_{h+1}) = \mathfrak{B}_{h+1}\mathfrak{B}_h \subseteq \mathfrak{B}_h$$

and, by Chapter 6 Theorem 15, $\mathrm{Gr}_R\{\mathfrak{A}(\phi_{h+1})\} \geqslant h+1$.

COROLLARY 2. *Let the assumptions and notation again be as in the theorem and put*

$$I = \mathfrak{A}(\phi_1)\,\mathfrak{A}(\phi_3)\,\mathfrak{A}(\phi_5)\,...,$$
$$J = \mathfrak{A}(\phi_2)\,\mathfrak{A}(\phi_4)\,\mathfrak{A}(\phi_6)\,....$$

Then $I = \mathfrak{B}_0 J$.
This is obvious.

AN EXAMPLE. Theorem 4 is concerned with a module having a *terminating* resolution involving finite free modules. It is worth noting that the result can fail completely in the case of a module having a non-terminating resolution, even when the ranks of the free modules concerned are bounded. We shall digress in order to illustrate this point by means of an example.†

Let K be a field and X, Y indeterminates. For each

$$\Phi(X, Y) \in K[X, Y]$$

there exist unique $\Phi_0(X)$ and $\Phi_1(X)$ in $K[X]$ such that

$$\Phi(X, Y) \equiv \Phi_0(X) + \Phi_1(X)\,Y \quad (\mathrm{mod}\,(Y^2 - X^3)).$$

It follows readily that $Y^2 - X^3$ is irreducible. We also have

$$K[X] \cap (Y^2 - X^3) = (0) = K[Y] \cap (Y^2 - X^3).$$

Next $(X, Y)/(X, Y)^2$ is a *two-dimensional* K-space and (X, Y) is a maximal ideal. From this we conclude that if

$$\Phi(X, Y) \in K[X, Y]$$

† See (**14**).

and $(X, Y) \subseteq (Y^2 - X^3, \Phi(X, Y))$,

then $(Y^2 - X^3, \Phi(X, Y)) = K[X, Y]$.

Now put $R = K[X, Y]/(Y^2 - X^3)$ and let x respectively y be the natural image of X respectively Y in R. Then R is an integral domain, x and y are non-zero elements with $y^2 = x^3$, and each $\phi \in R$ has a unique representation in the form $\phi = \phi_0 + \phi_1 y$ with $\phi_0, \phi_1 \in K[x]$. Further $Rx + Ry$ is a maximal ideal and if

$$Rx + Ry \subseteq R\phi,$$

then ϕ is a unit.

Suppose that $x\phi + y\psi = 0$ and write

$$\phi = \phi_0 + \phi_1 y, \quad \psi = \psi_0 + \psi_1 y,$$

where $\phi_i, \psi_i \in K[x]$. Then

$$x(\phi_0 + x^2\psi_1) + y(x\phi_1 + \psi_0) = 0$$

whence $\phi_0 = -x^2\psi_1$ and $\psi_0 = -x\phi_1$. It follows that the general solution of $x\phi + y\psi = 0$ is given by

$$\phi = -x^2\lambda + y\mu,$$
$$\psi = -x\mu + y\lambda,$$

where λ, μ denote arbitrary elements of R.

Define the homomorphism $f: R^2 \to R^2$ by

$$f(1, 0) = (y, -x),$$
$$f(0, 1) = (x^2, -y).$$

Then $f^2 = 0$. Now suppose that $f(\phi, \psi) = 0$. This implies that $x\phi + y\psi = 0$ and therefore $\phi = -x^2\lambda + y\mu$, $\psi = -x\mu + y\lambda$ for suitable $\lambda, \mu \in R$. Thus $(\phi, \psi) = f(\mu, -\lambda) \in \mathrm{Im} f$ and hence

$$R^2 \xrightarrow{\ f\ } R^2 \xrightarrow{\ f\ } R^2$$

is exact. Accordingly the semi-infinite sequence

$$\ldots \to R^2 \xrightarrow{\ f\ } R^2 \xrightarrow{\ f\ } R^2 \xrightarrow{\ f\ } R^2$$

is exact.

Next $\mathrm{rank}_R(f) = 1$ and

$$A = \left\| \begin{array}{cc} y & -x \\ x^2 & -y \end{array} \right\|$$

is the matrix of f, so $A^{(1)} = A$. *We claim that this matrix does not factor completely.* For assume that

$$\left\| \begin{array}{cc} y & -x \\ x^2 & -y \end{array} \right\| = \left\| \begin{array}{cc} u\xi & u\eta \\ v\xi & v\eta \end{array} \right\|,$$

where $u, v, \xi, \eta \in R$. Then

$$Rx + Ry = Ru\xi + Ru\eta \subseteq Ru$$

and likewise $Rx + Ry \subseteq R\eta$. Hence, by our earlier remarks, u and η are units and therefore $u\eta = -x$ is also a unit. This is the contradiction we have been seeking.

We now return to the main discussion. The two theorems which follow describe the effect on factorization ideals of the operations of forming fractions and adjoining indeterminates.

THEOREM 6. *Suppose that*

$$0 \to F_n \to F_{n-1} \to \dots \to F_1 \to F_0 \to E \to 0$$

is a finite free resolution of E and let $\mathfrak{B}_0, \mathfrak{B}_1, \dots, \mathfrak{B}_n$ be the corresponding factorization ideals. If now S is a multiplicatively closed subset of R not containing the zero element, then the factorization ideals associated with the resolution

$$0 \to (F_n)_S \to (F_{n-1})_S \to \dots \to (F_1)_S \to (F_0)_S \to E_S \to 0$$

are $\mathfrak{B}_0 R_S, \mathfrak{B}_1 R_S, \dots, \mathfrak{B}_n R_S$.

Proof. Let us choose bases for the F_j and use the same notation as in Theorem 4. The chosen bases give rise to bases of the free R_S-modules $(F_j)_S$. Now $(A_k)_S$ is the matrix of $(\phi_k)_S$ with respect to the acquired bases of $(F_k)_S$ and $(F_{k-1})_S$. Furthermore, because ϕ_k is stable, we have

$$\operatorname{rank}_{R_S}((\phi_k)_S) = \operatorname{rank}_R(\phi_k) = \mu_k.$$

Thus the complete factorizations

$$A_k^{(\mu_k)} = U_k V_k$$

carry over to similar complete factorizations associated with

$$0 \to (F_n)_S \to (F_{n-1})_S \to \dots \to (F_1)_S \to (F_0)_S \to E_S \to 0.$$

Accordingly the kth factorization ideal of this resolution is generated by the entries in either $(U_k)_S$ or $(V_{k+1})_S$. The theorem follows.

Rather similar considerations provide a proof of the next theorem, but the argument is simpler. We leave the reader to check the details.

THEOREM 7. *Assume that the finite free resolution*

$$0 \to F_n \to F_{n-1} \to \ldots \to F_1 \to F_0 \to E \to 0$$

of E has $\mathfrak{B}_0, \mathfrak{B}_1, \ldots, \mathfrak{B}_n$ *as its factorization ideals. If now $x_1, x_2, \ldots, x_m$ are indeterminates and* $\mathfrak{B}_0^*, \mathfrak{B}_1^*, \ldots, \mathfrak{B}_n^*$ *are the factorization ideals belonging to the resolution obtained by adjoining the indeterminates to $F_0, F_1, \ldots, F_n$ and E, then* $\mathfrak{B}_k^* = \mathfrak{B}_k R[x_1, x_2, \ldots, x_m]$ *for all k.*

The next result is more unexpected.

THEOREM 8. *Suppose that*

$$0 \to F_n \xrightarrow{\phi_n} F_{n-1} \xrightarrow{\phi_{n-1}} F_{n-2} \to \ldots \to F_1 \xrightarrow{\phi_1} F_0 \to E \to 0$$

is a finite free resolution of the R-module E and let $\mathfrak{B}_0, \mathfrak{B}_1, \ldots, \mathfrak{B}_n$ *be the associated factorization ideals. If now $1 < k \leqslant n$, then* $\mathrm{Rad}\,(\mathfrak{B}_{k-1}) = \mathrm{Rad}\,(\mathfrak{A}(\phi_k))$.

Remark. It should be noted that the value $k = 1$ is excluded.

Proof. Since $\mathfrak{A}(\phi_k) = \mathfrak{B}_k \mathfrak{B}_{k-1} \subseteq \mathfrak{B}_{k-1}$, it suffices to show that $\mathrm{Rad}\,(\mathfrak{B}_{k-1}) \subseteq \mathrm{Rad}\,(\mathfrak{A}(\phi_k))$. Let P be a minimal prime ideal of $\mathfrak{A}(\phi_k)$. In view of Chapter 4 Theorem 10 it is enough to prove that $\mathfrak{B}_{k-1} \subseteq P$. We shall in fact assume that $\mathfrak{B}_{k-1} \nsubseteq P$ and derive a contradiction.

To this end we localize at P and then drop P as a suffix. R is now a quasi-local ring with maximal ideal P, and, by Theorem 6, $\mathfrak{B}_{k-1} = R$. Furthermore, since the original ϕ_k was stable (Chapter 4 Theorem 8), it follows from Chapter 6 Exercise 12 that, after localization, P is a minimal prime ideal of $\mathfrak{A}(\phi_k)$ and therefore $\mathrm{Rad}\,(\mathfrak{A}(\phi_k)) = P$. In particular $\mathfrak{A}(\phi_k) \neq R$ and hence (Chapter 6 Exercise 11) Coker ϕ_k is *not* projective. Note that this implies that $\phi_k \neq 0$, $F_k \neq 0$ and $F_{k-1} \neq 0$.

We now select a base for each of $F_0, F_1, \ldots, F_n$. To simplify the notation we let A denote the matrix of ϕ_k and put $p = \mathrm{rank}_R (F_k)$,

$q = \operatorname{rank}_R(F_{k-1})$, $\mu = \operatorname{rank}_R(\phi_k)$. Then p, q, μ are all positive and A is a $p \times q$ matrix.

Next there is a column matrix $U = \|u_J\|$ and a row matrix $V = \|v_K\|$ such that (i) $A^{(\mu)}_{JK} = u_J v_K$ for all $J \in S^p_\mu$ and $K \in S^q_\mu$; (ii) the elements u_J generate $\mathfrak{B}_k = \mathfrak{B}_k \mathfrak{B}_{k-1} = \mathfrak{A}(\phi_k)$; (iii) the elements v_K generate $\mathfrak{B}_{k-1} = R$. Since R is quasi-local, (iii) implies that there exists $\overline{K} \in S^q_\mu$ such that $v_{\overline{K}}$ is a unit.

Let $\omega_1, \omega_2, \ldots, \omega_q$ be the chosen base for F_{k-1} and F'_{k-1} the free direct summand generated by the elements ω_l, where $l \in \overline{K}$. Further let $\pi \colon F_{k-1} \to F'_{k-1}$ be the natural projection and put $\psi = \pi \phi_k$ so that $\psi \colon F_k \to F'_{k-1}$. If now B is the matrix of ψ with respect to the obvious bases, then B is a $p \times \mu$ matrix; in fact B is obtained from A by selecting the columns corresponding to the members of $\overline{K}$. Accordingly

$$\operatorname{rank}_R(\psi) = \operatorname{rank}_R(B) \leqslant \operatorname{rank}_R(A) = \operatorname{rank}_R(\phi_k) = \mu,$$

and
$$B^{(\mu)}_{J\{1, \ldots, \mu\}} = A^{(\mu)}_{J\overline{K}} = u_J v_{\overline{K}}.$$

Thus $\mathfrak{A}_\mu(\psi) = \mathfrak{A}(\phi_k)$ and this is non-zero. It follows that

$$\operatorname{rank}_R(\psi) = \mu = \operatorname{rank}_R(\phi_k)$$

and that $\mathfrak{A}(\psi) = \mathfrak{A}(\phi_k)$. From these facts and Chapter 6 Theorem 15, we can now conclude that the complex

$$0 \to F_n \xrightarrow{\phi_n} F_{n-1} \to \ldots \to F_{k+1} \xrightarrow{\phi_{k+1}} F_k \xrightarrow{\psi} F'_{k-1} \to \operatorname{Coker} \psi \to 0$$

is exact.

Consider the exact sequences

$$0 \to \operatorname{Coker} \phi_{k+1} \to F'_{k-1} \to \operatorname{Coker} \psi \to 0$$

and
$$0 \to \operatorname{Coker} \phi_{k+1} \to F_{k-1} \to \operatorname{Coker} \phi_k \to 0.$$

Neither $\operatorname{Coker} \psi$ nor $\operatorname{Coker} \phi_k$ is projective because $\mathfrak{A}(\psi) = \mathfrak{A}(\phi_k)$ is not R. (Here we use Chapter 6 Exercise 11.) Accordingly

$$\operatorname{Pd}^*_R(\operatorname{Coker} \psi) = \operatorname{Pd}^*_R(\operatorname{Coker} \phi_{k+1}) + 1 = \operatorname{Pd}^*_R(\operatorname{Coker} \phi_k).$$

Again the exact sequence

$$0 \to \operatorname{Coker} \phi_k \to F_{k-2} \to \ldots \to F_1 \to F_0 \to E \to 0$$

shows that $\quad \operatorname{Pd}^*_R(\operatorname{Coker} \phi_k) = \operatorname{Pd}^*_R(E) - k + 1;$

moreover $k > 1$ and $E \neq 0$ since otherwise $\operatorname{Coker} \phi_k$ would be projective. These observations taken in conjunction with Chapter 6 Theorem 2 combine to show that

$$\operatorname{Pd}_R^*(\operatorname{Coker} \psi) < \operatorname{Pd}_R^*(E) \leqslant \operatorname{Gr}_R\{P\}$$

and now we find, because $\operatorname{Coker} \psi \neq 0$, that

$$\operatorname{Gr}_R\{\operatorname{Ann}_R(\operatorname{Coker} \psi)\} < \operatorname{Gr}_R\{P\}$$

this time by using Chapter 6 Theorem 4. But $\operatorname{Ann}_R(\operatorname{Coker} \psi)$ and $\mathfrak{F}(\operatorname{Coker} \psi)$ have the same radical and

$$\mathfrak{F}(\operatorname{Coker} \psi) = \mathfrak{A}(\psi) = \mathfrak{A}(\phi_k)$$

because $\operatorname{rank}_R(\psi) = \mu = \operatorname{rank}_R(F'_{k-1}).$

This shows that $\operatorname{Ann}_R(\operatorname{Coker} \psi)$ has P as its radical and therefore, by Chapter 5 Theorem 12, we have $\operatorname{Gr}_R\{P\} < \operatorname{Gr}_R\{P\}$. Here, of course, we have the required contradiction.

A particularly interesting situation arises when $\mathfrak{B}_0$ is a principal ideal. This is shown by

THEOREM 9. *Let E possess a finite free resolution of finite length and let $k = \operatorname{Char}_R(E)$. Further let $\mathfrak{B}_0, \mathfrak{B}_1, ..., \mathfrak{B}_n$ be a system of factorization ideals for E. Then $\mathfrak{F}_k(E) = \mathfrak{B}_0 \mathfrak{B}_1 \subseteq \mathfrak{B}_0$. Moreover, if $\mathfrak{B}_0$ is a principal ideal, then it is the (unique) smallest principal ideal containing $\mathfrak{F}_k(E)$.*

Proof. Suppose that

$$0 \to F_n \to F_{n-1} \to ... \to F_1 \xrightarrow{\phi_1} F_0 \to E \to 0$$

is a resolution of E giving rise to $\mathfrak{B}_0, \mathfrak{B}_1, ..., \mathfrak{B}_n$. Put $q = \operatorname{rank}_R(F_0)$. Then, by Chapter 4 Theorem 8, $\operatorname{rank}_R(\phi_1) = q - k$ and hence $\mathfrak{A}(\phi_1) = \mathfrak{A}_{q-k}(\phi_1)$. Next, by (4.2.15),

$$\mathfrak{F}_k(E) = \mathfrak{A}_{q-k}(\phi_1) = \mathfrak{A}(\phi_1) = \mathfrak{B}_0 \mathfrak{B}_1 \subseteq \mathfrak{B}_0.$$

Thus the first assertion is proved.

Now assume that $\mathfrak{B}_0$ is the principal ideal $R\alpha$ and let

$$c_1, c_2, ..., c_s$$

generate $\mathfrak{B}_1$. By Theorem 5 Cor. 1, $\operatorname{Gr}_R\{\mathfrak{B}_0\} \geqslant 1$, so α is a non-zerodivisor, and $\operatorname{Gr}_R\{Rc_1 + Rc_2 + ... + Rc_s\} \geqslant 2$.

Thus from Chapter 6 Theorem 9 it follows that

$$\alpha = \text{g.c.d.} \{\alpha c_1, \alpha c_2, ..., \alpha c_s\}.$$

Since the ideal generated by $\alpha c_1, \alpha c_2, ..., \alpha c_s$ is $\mathfrak{B}_0 \mathfrak{B}_1 = \mathfrak{F}_k(E)$ this completes the proof.

7.4 Extension of the theory of MacRae's invariant

Theorem 9 contains a strong hint that there should be a connection between the factorization ideal $\mathfrak{B}_0$ and MacRae's invariant $\mathfrak{G}(E)$ as defined in section (3.6). This is indeed the case as we shall now show. We recall that, so far, $\mathfrak{G}(E)$ has been defined only for modules E which possess elementary resolutions of finite length. Such modules form a subclass of those having Euler characteristic zero.

THEOREM 10. *Let the module E have a finite free resolution of finite length and let $\mathfrak{B}_0, \mathfrak{B}_1, ..., \mathfrak{B}_n$ be a system of factorization ideals for E. If now $\text{Char}_R(E) = 0$, then $\mathfrak{B}_0$ is a principal ideal and it is the (unique) smallest principal ideal containing the initial Fitting invariant $\mathfrak{F}(E)$. Further, if E possesses an elementary resolution of finite length, then $\mathfrak{B}_0 = \mathfrak{G}(E)$.*

Proof. Suppose that

$$0 \to F_n \to F_{n-1} \to ... \to F_1 \xrightarrow{\phi_1} F_0 \to E \to 0$$

is a finite free resolution giving rise to the ideals $\mathfrak{B}_0, \mathfrak{B}_1, ..., \mathfrak{B}_n$. By Chapter 4 Theorem 8 we have $\text{rank}_R(F_0) = \text{rank}_R(\phi_1)$ and now Theorem 5 shows that $\mathfrak{B}_0$ is generated by a single element. This proves the first assertion and the second follows from Theorem 9. An appeal to Chapter 6 Theorem 7 completes the proof.

Theorem 10 enables us to extend the definition of MacRae's invariant. Once again E is assumed to have a finite free resolution of finite length.

NEW DEFINITION. *If $\text{Char}_R(E) = 0$, then the MacRae invariant of E is (from here on) defined to be the smallest principal ideal containing $\mathfrak{F}(E)$. This invariant will be denoted by $\mathfrak{G}(E)$. It is necessarily generated by a non-zerodivisor.*

Note that Theorem 10 ensures that $\mathfrak{G}(E)$ is well-defined and that the new definition agrees with the old whenever the two overlap. Again we know from Chapter 4 Theorem 8 Cor., that $0:_R \mathfrak{F}(E) = 0$. This implies that any generator of $\mathfrak{G}(E)$ must be a non-zerodivisor.

To prevent confusion we proceed to verify that results previously established for $\mathfrak{G}(E)$ continue to hold for the enlarged class of modules for which it is now defined. This is certainly the case for Theorem 7 of Chapter 6. We now turn our attention to Theorem 28 of Chapter 3.

THEOREM 11. *Suppose that* $\text{Char}_R (E) = 0$ *and let* S *be a multiplicatively closed subset of* R *not containing zero. Then* $\text{Char}_{R_S} (E_S) = 0$ *and* $\mathfrak{G}(E_S) = \mathfrak{G}(E) R_S$.

Since (with the usual notation) $\mathfrak{B}_0 = \mathfrak{G}(E)$, this follows from Theorem 6.

The only remaining result which requires extension is Theorem 26 of Chapter 3. We deal with this in

THEOREM 12. *Let* $0 \to E' \to E \to E'' \to 0$ *be an exact sequence of* R-*modules each having Euler characteristic zero. Then*

$$\mathfrak{G}(E) = \mathfrak{G}(E') \mathfrak{G}(E'').$$

Proof. Let x be an indeterminate. Then the sequence

$$0 \to E' \to E \to E'' \to 0$$

gives rise to an exact sequence

$$0 \to E'[x] \to E[x] \to E''[x] \to 0$$

of $R[x]$-modules. Obviously $\mathfrak{G}(E[x])$ is defined and in fact Theorem 7 shows that $\mathfrak{G}(E[x]) = \mathfrak{G}(E) R[x]$. Naturally similar remarks apply to $\mathfrak{G}(E'[x])$ and $\mathfrak{G}(E''[x])$. Next

$$\text{Ann}_{R[x]} (E[x]) = \text{Ann}_R (E) R[x]$$

and this contains a non-zerodivisor because, by Chapter 5 Exercise 6, $\text{Ann}_R (E)$ contains a latent non-zerodivisor. Accordingly $E[x]$ possesses an elementary resolution of finite length† and the same is true of $E'[x]$ and $E''[x]$. We may therefore apply Theorem 26 of Chapter 3. This shows that

$$\mathfrak{G}(E)\, R[x] = (\mathfrak{G}(E')\, R[x])\, (\mathfrak{G}(E'')\, R[x]) = (\mathfrak{G}(E')\, \mathfrak{G}(E''))\, R[x]$$

and therefore $\mathfrak{G}(E) = \mathfrak{G}(E')\, \mathfrak{G}(E'')$ as required.

7.5   The Hilbert–Burch theorem

In this section we shall consider *ideals* which have finite free resolutions. Our previous results at once yield some general information.

THEOREM 13. *Let $\mathfrak{A} \neq 0$ be an ideal with a finite free resolution of finite length. Then $\mathfrak{A}$ can be represented in the form $\mathfrak{A} = \alpha\mathfrak{B}$, where $\alpha \in R$ and is a non-zerodivisor and $\mathfrak{B}$ is an ideal of grade at least two. In this representation $\mathfrak{B}$ is unique and α is unique to within multiplication by a unit.*

Proof. Construct a finite free resolution of $R/\mathfrak{A}$ and let $\mathfrak{B}_0$, $\mathfrak{B}_1$, etc. be the factorization ideals associated with the resolution. Since $\mathrm{Char}_R (R/\mathfrak{A}) = 0$ and $\mathfrak{F}(R/\mathfrak{A}) = \mathfrak{A}$, Theorems 9 and 10 show that
$$\mathfrak{A} = \mathfrak{B}_0\mathfrak{B}_1 = \mathfrak{G}(R/\mathfrak{A})\,\mathfrak{B}_1.$$
But $\mathfrak{G}(R/\mathfrak{A})$ is generated by a non-zerodivisor and $\mathrm{Gr}_R\{\mathfrak{B}_1\} \geqslant 2$ by Theorem 5 Cor. 1. Thus $\mathfrak{A}$ can certainly be represented in the manner described.

Next assume that $\mathfrak{A} = \alpha\mathfrak{B}$, where α and $\mathfrak{B}$ are as stated. Then $\mathfrak{B}$ is finitely generated (it is isomorphic to $\mathfrak{A}$) and Theorem 9 of Chapter 6 shows that α is a greatest common divisor of any finite set of generators of $\mathfrak{A}$. This establishes the essential uniqueness of α and that of $\mathfrak{B}$ is now clear.

From here on we shall be taking a look at the ideals $\mathfrak{A}$ for which $\mathrm{Pd}_R^*(\mathfrak{A}) < \infty$. In fact little is known about the subject but what is known is full of interest. Naturally one expects the situation to become more complicated as $\mathrm{Pd}_R^*(\mathfrak{A})$ becomes larger. We therefore begin with the rather trivial

THEOREM 14. *Let $\mathfrak{A} \neq 0$ be an ideal of R. Then $\mathrm{Pd}_R^*(\mathfrak{A}) = 0$ if and only if $\mathfrak{A}$ is a principal ideal generated by a non-zerodivisor.*

† See Chapter 3 Theorem 23.

Proof. Assume that $\mathrm{Pd}_R^*(\mathfrak{A}) = 0$. Then $R/\mathfrak{A}$ has a finite free resolution of length one and its Euler characteristic is zero because its annihilator is not zero. Consequently $R/\mathfrak{A}$ is an elementary module. Accordingly, by Chapter 3 Theorem 27,

$$\mathfrak{A} = \mathfrak{F}(R/\mathfrak{A}) = \mathfrak{G}(R/\mathfrak{A})$$

and therefore $\mathfrak{A}$ is generated by a non-zerodivisor. This establishes part of the theorem and what remains is obvious.

EXERCISE 5. *Let $\mathfrak{A}$ be an invertible integral ideal. Show that if $\mathfrak{A}$ has a finite free resolution of finite length, then $\mathfrak{A}$ is a principal ideal generated by a non-zerodivisor.*

We now turn our attention to the case where $\mathrm{Pd}_R^*(\mathfrak{A}) \leqslant 1$. A special instance of this was considered by D. Hilbert [(**23**) §4, p. 516] and more recently interest in these matters has been revived by Lindsay Burch (**15**) who treated the subject in a much more modern and abstract context.†. The next result is the Hilbert–Burch Theorem referred to at the beginning of this section.

THEOREM 15. *Let $\mathfrak{A} \neq 0$ be an ideal of R having a finite free resolution of length one. Then there exists an $m \times (m+1)$ matrix A and a non-zerodivisor α such that* (i) $\mathfrak{A} = \alpha \mathfrak{A}_m(A)$ *and* (ii) $\mathrm{Gr}_R\{\mathfrak{A}_m(A)\} \geqslant 2$.

Remarks. In this theorem m denotes a positive integer. Note that $\mathfrak{A}_m(A)$ is the ideal generated by the set of $m \times m$ minors of A. It therefore follows from (i) that $\mathfrak{A}$ itself can be generated by the $m \times m$ minors of an $m \times (m+1)$ matrix. However the fuller statement given here will enable us later to provide a converse.

Proof. Since $\mathrm{Pd}_R^*(\mathfrak{A}) \leqslant 1$, there exists an exact sequence

$$0 \to G \xrightarrow{\phi} F \xrightarrow{\psi} R \to R/\mathfrak{A} \to 0,$$

where F and G are non-zero finite free R-modules. Let $\mathfrak{B}_0, \mathfrak{B}_1, \mathfrak{B}_2$ be the associated factorization ideals. Now $\mathrm{Char}_R(R/\mathfrak{A}) = 0$. Hence if $\mathrm{rank}_R(G) = m$, then $\mathrm{rank}_R(F) = m+1$. Choose a base for F and a base for G. Then the matrix of ϕ with respect to these bases will be an $m \times (m+1)$ matrix A and, since

$$\mathrm{rank}_R(\phi) = \mathrm{rank}_R(G) = m,$$

† See also W. Gröbner [(**22**) §155, p. 204].

we have $\mathfrak{A}(\phi) = \mathfrak{A}_m(A)$. Now, by Theorem 5, $\mathfrak{B}_2 = R$ and $\mathfrak{B}_2\mathfrak{B}_1 = \mathfrak{A}(\phi)$. Thus $\mathfrak{B}_1 = \mathfrak{A}_m(A)$ and now we see from Theorem 5 Cor. 1 that $\mathrm{Gr}_R\{\mathfrak{A}_m(A)\} \geqslant 2$. Next $\mathrm{rank}_R(\psi) = 1$ and therefore $\mathfrak{A}(\psi) = \mathfrak{A}$. Also $\mathfrak{B}_0 = \mathfrak{G}(R/\mathfrak{A})$ and therefore $\mathfrak{B}_0 = R\alpha$ for some non-zerodivisor α. Finally

$$\mathfrak{A} = \mathfrak{A}(\psi) = \mathfrak{B}_0\mathfrak{B}_1 = \alpha\mathfrak{A}_m(A)$$

and with this the proof is complete.

We have already remarked that Theorem 15 possesses a converse. We shall now prepare to demonstrate this. Until further notice $A = \|a_{ij}\|$ will denote an $m \times (m+1)$ matrix with entries in R, where m is a positive integer. Let F_2 be a free R-module with a base $u_1, u_2, \ldots, u_m$, F_1 a free R-module with a base $v_1, v_2, \ldots, v_{m+1}$, and consider the sequence

$$0 \to F_2 \xrightarrow{d_2} F_1 \xrightarrow{d_1} R \xrightarrow{\epsilon} R/\mathfrak{A}_m(A) \to 0, \qquad (7.5.1)$$

where

$$d_2(u_i) = \sum_{j=1}^{m+1} a_{ij}v_j, \qquad (7.5.2)$$

$$d_1(v_j) = (-1)^j \det(A_j), \qquad (7.5.3)$$

and ϵ is the natural homomorphism. Here A_j denotes the matrix obtained from A by deleting the jth column. Evidently (7.5.1) is a complex† and $\mathrm{Im}\, d_1 = \mathrm{Ker}\, \epsilon$.

EXERCISE 6. *With the notation of (7.5.1) show that d_2 is a monomorphism if and only if $\mathrm{Gr}_R\{\mathfrak{A}_m(A)\} \geqslant 1$.*

LEMMA 3. *With the above notation $\mathfrak{A}_m(A)$ annihilates all the homology modules of the complex (7.5.1).*

Proof. It is enough to show that $\mathfrak{A}_m(A)$ annihilates $\mathrm{Ker}\, d_2$ and $(\mathrm{Ker}\, d_1)/(\mathrm{Im}\, d_2)$.

First suppose that $\alpha_1 u_1 + \alpha_2 u_2 + \ldots + \alpha_m u_m$ belongs to $\mathrm{Ker}\, d_2$. Then

$$\sum_{i=1}^m \alpha_i a_{ij} = 0 \quad (j = 1, 2, \ldots, m+1)$$

whence, by considering the equations corresponding to

$$j = 1, \ldots, k-1, k+1, \ldots, m+1,$$

† This complex belongs to a range of complexes that are discussed in Appendix C.

we see that $\det{(A_k)}\alpha_i = 0$. As this holds for all i and k the assertion that $\mathfrak{A}_m(A)$ annihilates $\mathrm{Ker}\,d_2$ is proved.

Next assume that an integer μ satisfying $1 \leqslant \mu \leqslant m+1$ has been given and define a homomorphism $\sigma_0 : R \to F_1$ by the requirement that

$$\sigma_0(1_R) = (-1)^\mu v_\mu \qquad (7.5.4)$$

and note that

$$\sigma_0 d_1(v_j) = (-1)^{\mu+j} \det{(A_j)}\, v_\mu \qquad (7.5.5)$$

for all values of j. We also need a certain homomorphism $\sigma_1 : F_1 \to F_2$. To this end we introduce elements

$$c_{ij} \quad (1 \leqslant i \leqslant m,\ 1 \leqslant j \leqslant m+1)$$

given by

$$c_{ij} = \begin{cases} \text{the cofactor of } a_{ij} \text{ in } A_\mu \text{ when } j \neq \mu, \\ 0 \quad \text{when} \quad j = \mu. \end{cases}$$

The homomorphism σ_1 is now defined by

$$\sigma_1(v_j) = \sum_{i=1}^m c_{ij} u_i. \qquad (7.5.6)$$

Note that $\sigma_1(v_\mu) = 0$.

The point of introducing σ_0 and σ_1 is that we now have

$$(d_2\sigma_1 + \sigma_0 d_1)\,(v_j) = \det{(A_\mu)}\, v_j \qquad (7.5.7)$$

for all j. To see this we observe that we may suppose that $j \neq \mu$. On this understanding

$$d_2\sigma_1(v_j) = \sum_{k=1}^{m+1} \left(\sum_{i=1}^m a_{ik} c_{ij} \right) v_k$$

$$= \sum_{k\neq\mu} \delta_{kj} \det{(A_\mu)}\, v_k + \left(\sum_{i=1}^m a_{i\mu} c_{ij} \right) v_\mu$$

$$= \det{(A_\mu)}\, v_j + \det{(A_\mu^*)}\, v_\mu.$$

Here A^* is the matrix obtained from A by first interchanging the μth and jth columns and then deleting the μth column of the resulting matrix. It follows that

$$\det{(A_\mu^*)} = (-1)^{\mu+j+1} \det{(A_j)}$$

and therefore

$$d_2\sigma_1(v_j) = \det{(A_\mu)}\, v_j - (-1)^{\mu+j} \det{(A_j)}\, v_\mu.$$

In view of (7.5.5) this establishes (7.5.7). Note that we at once obtain

$$(d_2 \sigma_1 + \sigma_0 d_1)(\xi) = \det(A_\mu)\,\xi, \qquad (7.5.8)$$

where ξ is an arbitrary element of F_1.

We are now ready to complete the proof of the lemma. Suppose that $\xi \in \operatorname{Ker} d_1$. Then, by (7.5.8), $\det(A_\mu)\,\xi \in \operatorname{Im} d_2$ and as this holds for all μ we see that $\mathfrak{A}_m(A)\,\xi \subseteq \operatorname{Im} d_2$. Thus $\mathfrak{A}_m(A)$ annihilates $(\operatorname{Ker} d_1)/(\operatorname{Im} d_2)$ and with this the lemma is proved.

THEOREM 16. *Let A be an $m \times (m+1)$ matrix with entries in R, where m is a positive integer. If now $\operatorname{Gr}_R\{\mathfrak{A}_m(A)\} \geqslant 2$ then the sequence* (7.5.1) *is exact.*

Proof. For any non-zero free module F we have

$$\operatorname{Gr}_R\{\mathfrak{A}_m(A);\, F\} = \operatorname{Gr}_R\{\mathfrak{A}_m(A)\} \geqslant 2.$$

Hence, in view of Lemma 3, we may conclude from Chapter 5 Theorem 22 that

$$0 \to F_2 \xrightarrow{\ d_2\ } F_1 \xrightarrow{\ d_1\ } R$$

is exact. The theorem follows.

We can now establish the converse of Theorem 15 namely

THEOREM 17. *Let A be an $m \times (m+1)$ matrix with entries in R. Suppose that $\operatorname{Gr}_R\{\mathfrak{A}_m(A)\} \geqslant 2$, and let α be a non-zerodivisor in R. Then the ideal $\alpha\mathfrak{A}_m(A)$ possesses a finite free resolution of length one.*

For $\mathfrak{A}_m(A)$ and $\alpha\mathfrak{A}_m(A)$ are isomorphic and, by Theorem 16, $\operatorname{Pd}_R^*(\mathfrak{A}_m(A)) \leqslant 1$.

Theorems 15 and 17 provide a complete description of ideals $\mathfrak{A}$ for which $\operatorname{Pd}_R^*(\mathfrak{A}) \leqslant 1$ and the next step would appear to be the investigation of those which satisfy $\operatorname{Pd}_R^*(\mathfrak{A}) \leqslant 2$. This however is relatively unexplored territory though a start has been made. The reader who wishes to follow up this line of enquiry should consult D. A. Buchsbaum and D. Eisenbud (**14**).

EXERCISE 7. *Let A be an $m \times (m+1)$ matrix with entries in R. Show that if $\mathfrak{A}_m(A) \neq R$, then $\operatorname{Gr}_R\{\mathfrak{A}_m(A)\} \leqslant 2$.*

Solutions to the Exercises on Chapter 7

EXERCISE 1. *Let $E \neq 0$ be an R-module and A a $p \times q$ matrix which is stable relative to E. Show that if $\mathrm{Ann}_R(E) = 0$, then A is stable relative to R and $\mathrm{rank}_R(A) = \mathrm{rank}_R(A, E)$.*

Solution. Put $\mu = \mathrm{red.rank}_R(A)$ and $\nu = \mathrm{rank}_R(A, E)$. Then there exists an element $r \neq 0$ of R such that $\mathfrak{A}_{\mu+1}(A)r = 0$. Now $rE \neq 0$ and therefore there exists $e \in E$ such that $re \neq 0$. Thus $re \neq 0$ and $\mathfrak{A}_{\mu+1}(A)re = 0$. Accordingly

$$\mathrm{rank}_R(A, E) = \mathrm{red.rank}_R(A, E) \leqslant \mu = \mathrm{red.rank}_R(A).$$

Next $\mathfrak{A}_{\nu+1}(A)E = 0$ and therefore $\mathfrak{A}_{\nu+1}(A) = 0$. This time we conclude that

$$\mathrm{red.rank}_R(A) \leqslant \mathrm{rank}_R(A) \leqslant \nu = \mathrm{rank}_R(A, E).$$

Thus $\mathrm{red.rank}_R(A) = \mathrm{rank}_R(A) = \mathrm{rank}_R(A, E)$

as required.

EXERCISE 2. *Let $A = \|a_{jk}\|$ be a $p \times q$ matrix and $U = \|u_j\|$ a $p \times 1$ matrix both with entries in R. Suppose that*

$$\mathrm{Gr}_R\{Ru_1 + Ru_2 + \ldots + Ru_p\} \geqslant 2,$$

and
$$\begin{vmatrix} u_{j_1} & u_{j_2} \\ a_{j_1 k} & a_{j_2 k} \end{vmatrix} = 0$$

for all choices of j_1, j_2, k. Show that there exists a unique $1 \times q$ matrix V such that $A = UV$.

Solution. Theorem 1 applied to the matrix

$$\left\| \begin{matrix} u_1 & u_2 & \ldots & u_p \\ a_{1k} & a_{2k} & \ldots & a_{pk} \end{matrix} \right\|$$

shows that there is a unique $v_k \in R$ such that $a_{jk} = u_i v_k$ for all i. If now V is the row matrix formed by the v_k, then $A = UV$. The uniqueness of V follows from that of the v_k.

EXERCISE 3. *Let $\alpha \in R$ and suppose that $\mathrm{Gr}_R\{R\alpha\} \geqslant 2$. Show that α is a unit.*

Solution. The element α must be a unit for otherwise $R\alpha \neq R$ and therefore $\mathrm{Gr}_R\{R\alpha\} \leqslant 1$ by Chapter 5 Theorem 13. Thus we have a contradiction.

EXERCISE 4. *Let $\mathfrak{A}$ be an invertible integral ideal of R and suppose that $\mathrm{Gr}_R\{\mathfrak{A}\} \geqslant 2$. Show that $\mathfrak{A} = R$.*

Solution. Put $\mathfrak{B} = \mathfrak{A}^{-1}$. By Chapter 3 Exercise 10, $\mathfrak{A}$ contains a non-zerodivisor α say. Let $\xi \in \mathfrak{B}$. Then $\xi = c/d$, where $c, d \in R$ and d is not a zerodivisor. Furthermore $c\mathfrak{A} \subseteq Rd$ and therefore $\alpha c \mathfrak{A} \subseteq R\alpha d$. We claim that $\alpha c \in R\alpha d$. For if not then $\mathfrak{A}$ annihilates a non-zero element of $R/R\alpha d$ and therefore $\mathrm{Gr}_R\{\mathfrak{A}; R/R\alpha d\} = 0$. Hence, by Chapter 5 Theorem 15, $\mathrm{Gr}_R\{\mathfrak{A}\} = 1$ and we have a contradiction. This establishes our claim and shows that $\xi \in R$. Thus $\mathfrak{A}^{-1} \subseteq R$ and hence $R \subseteq \mathfrak{A}$. Since $\mathfrak{A}$ is an integral ideal we now have $\mathfrak{A} = R$ as required.

EXERCISE 5. *Let $\mathfrak{A}$ be an invertible integral ideal. Show that if $\mathfrak{A}$ has a finite free resolution of finite length, then $\mathfrak{A}$ is a principal ideal generated by a non-zerodivisor.*

Solution. Chapter 3 Exercise 10 shows that $\mathfrak{A}$ is projective as an R-module. Since $\mathrm{Pd}_R^*(\mathfrak{A}) < \infty$ it follows, from Chapter 3 Theorem 14, that $\mathfrak{A}$ is a supplementable projective module and therefore $\mathrm{Pd}_R^*(\mathfrak{A}) = 0$. The desired conclusion now follows from Theorem 14.

EXERCISE 6. *With the notation of (7.5.1) show that d_2 is a monomorphism if and only if $\mathrm{Gr}_R\{\mathfrak{A}_m(A)\} \geqslant 1$.*

Solution. Let $\alpha_1, \alpha_2, \ldots, \alpha_m$ be elements of R. Then

$$d_2(\alpha_1 u_1 + \alpha_2 u_2 + \ldots + \alpha_m u_m) = \sum_j \left(\sum_i \alpha_i a_{ij} \right) v_j,$$

hence $\mathrm{Ker}\, d_2 \neq 0$ if and only if the equations

$$x_1 a_{1j} + x_2 a_{2j} + \ldots + x_m a_{mj} = 0 \quad (j = 1, 2, \ldots, m+1)$$

have a non-trivial solution in R. Now, by Chapter 3 Theorem 6, this is the case if and only if $0 :_R \mathfrak{A}_m(A) \neq 0$, that is if and only if $\mathrm{Gr}_R\{\mathfrak{A}_m(A)\} = 0$. This solves the exercise.

EXERCISE 7. *Let A be an $m \times (m+1)$ matrix with entries in R. Show that if $\mathfrak{A}_m(A) \neq R$, then $\mathrm{Gr}_R\{\mathfrak{A}_m(A)\} \leqslant 2$.*

Solution. Let us assume that $\mathrm{Gr}_R\{\mathfrak{A}_m(A)\} > 2$. Then

$$R/\mathfrak{A}_m(A) \neq 0$$

and, by Theorem 16, $\mathrm{Pd}_R^*(R/\mathfrak{A}_m(A)) \leqslant 2$. However Theorem 4 of Chapter 6 now shows that $\mathrm{Gr}_R\{\mathfrak{A}_m(A)\} \leqslant 2$ and this is the contradiction we are seeking.

Appendix A:
A non-free supplementable projective module†

General remarks

Let R be a commutative ring with a non-zero identity element. We recall that, according to the definition given in section (3.3), an R-module P is called a *supplementable projective module* if there exist finite free modules F, G such that $P \oplus G$ and F are isomorphic. It is a trivial observation that any finite free module is a supplementable projective module, but it is not obvious that others do exist. The purpose of this appendix is to show that supplementable projective modules can form a genuinely larger class than that formed by finite free modules. We shall in fact show that this is the case when R is either (i) the ring of continuous functions on a sphere, or (ii)

$$\mathbb{Z}[X_1, X_2, X_3]/(X_1^2 + X_2^2 + X_3^2 - 1),$$

where $\mathbb{Z}$ denotes the integers and X_1, X_2, X_3 are indeterminates.

A. 1 A non-free supplementable projective module

The following simple lemma will provide the basis for our example.

LEMMA 1. *Let a_1, a_2, a_3 belong to R and satisfy*

$$Ra_1 + Ra_2 + Ra_3 = R,$$

and assume that there is no invertible 3×3 R-matrix having (a_1, a_2, a_3) as its first row. Then R possesses a non-free supplementable projective module.

† Supplementable projective modules are often called *stably free* modules. However the title of the appendix will indicate why the nomenclature has been changed.

Proof. Let F be a free R-module with a base $\omega_1, \omega_2, \omega_3$ of three elements, and let $\phi: F \to F$ be the homomorphism in which

$$\phi(\omega_1) = a_1\omega_1 + a_2\omega_2 + a_3\omega_3 = \xi \quad \text{(say)}$$

and $\phi(\omega_2) = \phi(\omega_3) = 0$. Then $\mathfrak{A}_0(\phi) = \mathfrak{A}_1(\phi) = R$, and $\mathfrak{A}_\nu(\phi) = 0$ when $\nu > 1$. It therefore follows, from Chapter 4 Theorem 18, that Coker ϕ is a projective module. Thus $\phi(F)$ is a direct summand of F and therefore $F = \phi(F) \oplus P$ for some R-module P. Next $\phi(F)$ is a free R-module having ξ as a base. Accordingly P is a supplementable projective module. From here on we assume that P is free and derive a contradiction.

Since P is free, the relation $F = R\xi \oplus P$ shows that

$$\text{rank}_R(P) = 2.$$

Let $\eta = b_1\omega_1 + b_2\omega_2 + b_3\omega_3$ and $\zeta = c_1\omega_1 + c_2\omega_2 + c_3\omega_3$ be a base for P. Then ξ, η, ζ form a base for F and therefore, by Chapter 2 Lemma 4,

$$\left\| \begin{array}{ccc} a_1 & a_2 & a_3 \\ b_1 & b_2 & b_3 \\ c_1 & c_2 & c_3 \end{array} \right\|$$

is a unimodular matrix. But this is contrary to our original assumption concerning a_1, a_2, a_3 and so the lemma is proved. Observe that we have shown that $F/R\xi$ is a non-free supplementable projective module.

THEOREM 1. *Let R be the ring of continuous real-valued functions on the sphere $x_1^2 + x_2^2 + x_3^2 = 1$. Then R possesses a non-free supplementable projective module.*

Proof. Each of the coordinate functions induces a continuous function on the sphere. These functions will be denoted by x_1, x_2, x_3. Thus, in this sense, $x_i \in R$. Note that $Rx_1 + Rx_2 + Rx_3 = R$ because $x_1^2 + x_2^2 + x_3^2$ is the identity element of R.

Consider the matrix

$$\left\| \begin{array}{ccc} x_1 & x_2 & x_3 \\ f_1 & f_2 & f_3 \\ g_1 & g_2 & g_3 \end{array} \right\|$$

where $f_i, g_i \in R$. In view of Lemma 1 it will suffice to show that this matrix cannot be invertible.

Assume the contrary. Then the determinant of the matrix vanishes nowhere on the sphere and hence the first two rows are never proportional. Put

$$\xi_i = f_i - (x_1 f_1 + x_2 f_2 + x_3 f_3)\, x_i$$

for $i = 1, 2, 3$. Then the ξ_i provide the components of a non-vanishing field of tangent vectors on the sphere thereby contradicting a well known theorem of topology.†

In the next theorem $\mathbb{Z}$ denotes the ring of integers.

THEOREM 2. *Let* X_1, X_2, X_3 *be indeterminates and put*

$$T = \mathbb{Z}[X_1, X_2, X_3]/(X_1^2 + X_2^2 + X_3^2 - 1).$$

Then T *possesses a non-free supplementable projective module.*

Proof. Let $\overline{X}_i$ denote the natural image of X_i in T and let R be the ring in Theorem 1. Then $T\overline{X}_1 + T\overline{X}_2 + T\overline{X}_3 = T$ and there exists a ring-homomorphism $T \to R$ in which $\overline{X}_i \mapsto x_i$. Further there can be no invertible 3×3 T-matrix with $(\overline{X}_1, \overline{X}_2, \overline{X}_3)$ as its first row; for otherwise by applying the homomorphism $T \to R$ we could obtain an invertible 3×3 R-matrix with (x_1, x_2, x_3) as its first row and this we know is impossible. Theorem 2 now follows from Lemma 1.

† See for example [(**24**) Theorem 5.8.6, p. 219].

Appendix B:
A further result on exact sequences

General remarks

Suppose that we have a complex

$$\mathbb{C}: 0 \to F_n \xrightarrow{\phi_n} F_{n-1} \xrightarrow{\phi_{n-1}} F_{n-2} \to \ldots \to F_1 \xrightarrow{\phi_1} F_0$$

of finite free R-modules, and let $E \neq 0$ be a further R-module. If $\mathbb{C} \otimes_R E$ is exact, then we know (Chapter 6 Theorem 14) that the following conditions hold:

(a) $\mathrm{Gr}_R \{\mathfrak{A}(\phi_k, E); E\} \geqslant k$ for $1 \leqslant k \leqslant n$;
(b) $\mathrm{rank}_R (\phi_n, E) = \mathrm{rank}_R (F_n)$;
(c) $\mathrm{rank}_R (\phi_{m+1}, E) + \mathrm{rank}_R (\phi_m, E)$
$$= \mathrm{rank}_R (F_m) \text{ for } 1 \leqslant m < n.$$

Furthermore, by Chapter 6 Theorem 15, the converse of this result is valid in the special case where $E = R$.

It is the aim of this appendix to show that, in fact, the converse holds without any special restriction being placed on the non-zero R-module E. When R is a Noetherian ring and E is finitely generated, the result that $\mathbb{C} \otimes_R E$ is exact if and only if conditions $(a), (b), (c)$ hold is one of the original theorems of D. A. Buchsbaum and D. Eisenbud (13). The general result has been dealt with here (rather than in the main text) because the argument uses the theory of dimension for commutative rings and this is not needed elsewhere. For some background reading in dimension theory the reader may like to consult (37).

As usual R always denotes a commutative ring with a non-zero identity element.

B. 1   Considerations involving a change of ring

It will be advantageous, in dealing with the main result of this appendix, to transform the problem with which we start into one where R has a number of additional properties.

To this end suppose that we are given a ring-homomorphism

$$\chi \colon R \to \bar{R}, \tag{B.1.1}$$

where $\bar{R}$ is a non-trivial commutative ring. (χ need not be surjective.) The homomorphism enables us to regard $\bar{R}$ as an R-module; hence if U is an R-module then we can form the $\bar{R}$-module $\bar{U} = U \otimes_R \bar{R}$. (Here if $u \in U$ and $\alpha, \beta \in \bar{R}$, then $\alpha(u \otimes \beta) = u \otimes \alpha\beta$.) Note that if $f \colon U \to V$ is a homomorphism of R-modules, $\bar{V} = V \otimes_R \bar{R}$ and $\bar{f} = f \otimes \bar{R}$, then $\bar{f} \colon \bar{U} \to \bar{V}$ is a homomorphism of $\bar{R}$-modules.

Now suppose that F, G are finite free R-modules, that $x_1, x_2, \ldots, x_p$ is a base of F and $y_1, y_2, \ldots, y_q$ a base of G. Then $\bar{F} = F \otimes_R \bar{R}$ and $\bar{G} = G \otimes_R \bar{R}$ are finite free $\bar{R}$-modules with bases $x_1 \otimes 1_{\bar{R}}, x_2 \otimes 1_{\bar{R}}, \ldots, x_p \otimes 1_{\bar{R}}$ and $y_1 \otimes 1_{\bar{R}}, y_2 \otimes 1_{\bar{R}}, \ldots, y_q \otimes 1_{\bar{R}}$ respectively. In particular

$$\operatorname{rank}_R(F) = \operatorname{rank}_{\bar{R}}(\bar{F}). \tag{B.1.2}$$

Next assume that $\phi \colon F \to G$ is an R-homomorphism and that $A = \|a_{jk}\|$ is the matrix of ϕ with respect to the given bases of F and G. Put

$$\bar{\phi} = \phi \otimes \bar{R}. \tag{B.1.3}$$

Then the matrix of $\bar{\phi}$ with respect to the induced bases of $\bar{F}$ and $\bar{G}$ is $\|\chi(a_{jk})\|$. It follows that

$$\mathfrak{A}_\nu(\bar{\phi}) = \mathfrak{A}_\nu(\phi)\, \bar{R} \tag{B.1.4}$$

for all $\nu \geqslant 0$.

Let $E \neq 0$ be an $\bar{R}$-module. Then, of course, E can be regarded as an R-module and, in view of (B.1.4),

$$\mathfrak{A}_\nu(\phi)\, E = \mathfrak{A}_\nu(\bar{\phi})\, E.$$

Hence $\operatorname{rank}_{\bar{R}}(\bar{\phi}, E) = \operatorname{rank}_R(\phi, E)$ (B.1.5)

and therefore $\mathfrak{A}(\bar{\phi}, E) = \mathfrak{A}(\phi, E)\, \bar{R}.$ (B.1.6)

Accordingly, by Chapter 5 Theorem 19,

$$\operatorname{Gr}_R \{\mathfrak{A}(\phi, E); E\} = \operatorname{Gr}_{\bar{R}} \{\mathfrak{A}(\bar{\phi}, E); E\}. \tag{B.1.7}$$

After these initial observations, let us turn our attention to a complex

$$\mathbb{C}: 0 \to F_n \xrightarrow{\phi_n} F_{n-1} \xrightarrow{\phi_{n-1}} F_{n-2} \to \dots \to F_1 \xrightarrow{\phi_1} F_0$$
(B.1.8)

of finite free R-modules. Put $\overline{\mathbb{C}} = \mathbb{C} \otimes_R \overline{R}$ so that, with a self-explanatory notation,

$$\mathbb{C}: 0 \to \overline{F}_n \xrightarrow{\overline{\phi}_n} \overline{F}_{n-1} \xrightarrow{\overline{\phi}_{n-1}} \overline{F}_{n-2} \to \dots \to \overline{F}_1 \xrightarrow{\overline{\phi}_1} \overline{F}_0$$
(B.1.9)

is a complex of finite free $\overline{R}$-modules.

LEMMA 1. *Let $\mathbb{C}$ and $\overline{\mathbb{C}}$ be as above and let $E \neq 0$ be an $\overline{R}$-module. If now conditions*

  (a) $\mathrm{Gr}_R\{\mathfrak{A}(\phi_k, E); E\} \geqslant k$ *for* $k = 1, 2, \dots, n$;
  (b) $\mathrm{rank}_R(\phi_n, E) = \mathrm{rank}_R(F_n)$;
  (c) $\mathrm{rank}_R(\phi_{m+1}, E) + \mathrm{rank}_R(\phi_m, E) = \mathrm{rank}_R(F_m)$ *for*
$$1 \leqslant m < n;$$

all hold, then so do conditions

  ($\overline{a}$) $\mathrm{Gr}_{\overline{R}}(\mathfrak{A}(\overline{\phi}_k, E); E\} \geqslant k$ *for* $k = 1, 2, \dots, n$;
  ($\overline{b}$) $\mathrm{rank}_{\overline{R}}(\overline{\phi}_n, E) = \mathrm{rank}_{\overline{R}}(\overline{F}_n)$;
  ($\overline{c}$) $\mathrm{rank}_{\overline{R}}(\overline{\phi}_{m+1}, E) + \mathrm{rank}_{\overline{R}}(\overline{\phi}_m, E) = \mathrm{rank}_{\overline{R}}(\overline{F}_m)$ *for*
$$1 \leqslant m < n;$$

and vice versa.

This is an immediate consequence of our previous remarks.

As before let U be an R-module and E an $\overline{R}$-module. Then $(U \otimes_R \overline{R}) \otimes_{\overline{R}} E$ is an $\overline{R}$-module and $U \otimes_R E$ is an R-module, so we can regard both of them as R-modules. On this understanding we have an isomorphism $(U \otimes_R \overline{R}) \otimes_{\overline{R}} E \approx U \otimes_R E$, that is an isomorphism $\qquad \overline{U} \otimes_{\overline{R}} E \approx U \otimes_R E,$ (B.1.10)

in which $(u \otimes \overline{r}) \otimes e$ is matched with $u \otimes \overline{r}e$. An easy verification shows that if $f: U \to V$ is a homomorphism of R-modules, then the isomorphisms

$$\overline{U} \otimes_{\overline{R}} E \approx U \otimes_R E \quad \text{and} \quad \overline{V} \otimes_{\overline{R}} E \approx V \otimes_R E$$

are compatible (in an obvious sense) with the homomorphisms

$$\overline{f} \otimes E: U \otimes_{\overline{R}} E \to \overline{V} \otimes_{\overline{R}} E$$

and $\qquad\qquad f \otimes E: U \otimes_R E \to V \otimes_R E.$

The next lemma follows at once from these observations.

LEMMA 2. *Let* $\mathbb{C}$ *and* $\overline{\mathbb{C}}$ *be as before and let* E *be an* $\overline{R}$-*module. Then* $\mathbb{C} \otimes_R E$ *is exact at* $F_k \otimes_R E$ *if and only if* $\overline{\mathbb{C}} \otimes_{\overline{R}} E$ *is exact at* $\overline{F}_k \otimes_{\overline{R}} E$. *In particular, in order that* $\mathbb{C} \otimes_R E$ *be exact it is necessary and sufficient that* $\overline{\mathbb{C}} \otimes_{\overline{R}} E$ *be exact.*

We next examine the effect of forming fractions. Let S be a multiplicatively closed subset of R not containing the zero element, and let

$$\mathbb{Q}: 0 \to Q_m \xrightarrow{\psi_m} Q_{m-1} \xrightarrow{\psi_{m-1}} Q_{m-2} \to \ldots \to Q_1 \xrightarrow{\psi_1} Q_0 \tag{B.1.11}$$

be a complex of R-modules. We put

$$H_k(\mathbb{Q}) = \operatorname{Ker} \psi_k / \operatorname{Im} \psi_{k+1} \tag{B.1.12}$$

and use $\mathbb{Q}_S$ to denote the complex of R_S-modules obtained from (B.1.11) by forming fractions. (In the case where S is the complement of a prime ideal P, we also use $\mathbb{Q}_P$ as an alternative to $\mathbb{Q}_S$.)

LEMMA 3. *Suppose that* $1 \leqslant k \leqslant m$. *Then the* R_S-*modules* $H_k(\mathbb{Q}_S)$ *and* $[H_k(\mathbb{Q})]_S$ *are isomorphic.*

This is a consequence of the fact that fraction formation preserves exact sequences. See [(**35**) Theorem 1, p. 93].

LEMMA 4. *Suppose that* $1 \leqslant k \leqslant m$. *Then* $H_k(\mathbb{Q}) = 0$ *if and only if* $H_k(\mathbb{Q}_M) = 0$ *for every maximal ideal* M.

Proof. By Lemma 3, we need only show that $H_k(\mathbb{Q}) = 0$ when and only when $[H_k(\mathbb{Q})]_M = 0$ for every maximal ideal M. However this follows from Chapter 4 Exercise 10.

LEMMA 5. *Let* R, $\mathbb{C}$ *and* $E \neq 0$ *be as in Lemma* 1 *and suppose that conditions* (*a*), (*b*), (*c*) *of that lemma are satisfied. Then* $\phi_1, \phi_2, \ldots, \phi_n$ *are all stable relative to* E. *Further if* S *is a multiplicatively closed subset of* R *and* $E_S \neq 0$, *then* R_S, $\mathbb{C}_S$ *and* E_S *satisfy conditions analogous to* (*a*), (*b*), (*c*).

Proof. Suppose that $1 \leqslant k \leqslant n$. By (*a*) we have

$$\operatorname{Gr}_R \{\mathfrak{A}(\phi_k, E); E\} > 0$$

and this, as was remarked at the beginning of section (6.4), shows that ϕ_k is stable relative to E. If therefore ϕ_k^* is the homomorphism to which ϕ_k gives rise on forming fractions, then

$$\operatorname{rank}_R (\phi_k, E) = \operatorname{rank}_{R_S} (\phi_k^*, E_S),$$

and hence $\mathfrak{A}(\phi_k^*, E_S) = \mathfrak{A}(\phi_k, E) R_S$. Exercise 10 of Chapter 5 now shows that

$$\mathrm{Gr}_{R_S}\{\mathfrak{A}(\phi_k^*, E_S); E_S\} \geqslant \mathrm{Gr}_R\{\mathfrak{A}(\phi_k, E); E\}$$

and with this the lemma follows.

We need one further observation before leaving this preliminary section. Suppose that $f : U \to V$ is a homomorphism of R-modules and that E is an R-module. In these circumstances, we have an isomorphism $U_S \otimes_{R_S} E_S \approx (U \otimes_R E)_S$ of R_S-modules in which the element $(u/s_1) \otimes (e/s_2)$ of the former corresponds to the element $(u \otimes e)/s_1 s_2$ of the latter. Similarly we have an isomorphism $V_S \otimes_{R_S} E_S \approx (V \otimes_R E)_S$. Furthermore these isomorphisms are compatible with the homomorphisms

$$f_S \otimes E_S : U_S \otimes_{R_S} E_S \to V_S \otimes_{R_S} E_S$$

and $\qquad (f \otimes E)_S : (U \otimes_R E)_S \to (V \otimes_R E)_S.$

It follows that if $\mathbb{Q}$ is the complex (B.1.11), then *we can identify* $\mathbb{Q}_S \otimes_{R_S} E_S$ *with* $(\mathbb{Q} \otimes_R E)_S$.

LEMMA 6. *Let the notation be as above and suppose that* $1 \leqslant k \leqslant m$. *Then* $H_k(\mathbb{Q} \otimes_R E) = 0$ *if and only if*

$$H_k(\mathbb{Q}_M \otimes_{R_M} E_M) = 0$$

for every maximal ideal M *of* R. *In particular* $\mathbb{Q} \otimes_R E$ *is exact if and only if* $\mathbb{Q}_M \otimes_{R_M} E_M$ *is exact for every maximal ideal* M.

Proof. Let M denote a typical maximal ideal. By Lemma 4, $H_k(\mathbb{Q} \otimes_R E) = 0$ if and only if $H_k((\mathbb{Q} \otimes_R E)_M) = 0$ for all M. The lemma follows because $(\mathbb{Q} \otimes_R E)_M$ can be identified with $\mathbb{Q}_M \otimes_{R_M} E_M$.

B.2 Conditions for exactness

Suppose, for the moment, that R is a quasi-local ring, that $F \xrightarrow{\phi} G \xrightarrow{\psi} H$ is a short complex of finite free R-modules, and that $\mathfrak{A}(\phi) = \mathfrak{A}(\psi) = R$. By Chapter 6 Exercise 11, Coker ϕ and Coker ψ are finite free R-modules. Now on this occasion all projective modules are free. Consequently Ker ϕ, Im ϕ, Ker ψ,

Im ψ are free modules. Moreover $\mathrm{Ker}\,\phi$ and $\mathrm{Ker}\,\psi$ are direct summands of F and G respectively whereas $\mathrm{Im}\,\phi$ and $\mathrm{Im}\,\psi$ are direct summands of G and H. Accordingly the four modules are not only free but have finite bases and, it should be noted, a base of $\mathrm{Im}\,\phi$ can be enlarged to give a base of G. From this latter observation we see that if $\phi': F \to \mathrm{Im}\,\phi$ is the epimorphism induced by ϕ, then

$$\mathrm{rank}_R(\phi) = \mathrm{rank}_R(\phi') = \mathrm{rank}_R(\mathrm{Im}\,\phi). \qquad (B.2.1)$$

(Here we have used Exercise 3 of Chapter 4.) Since $\mathrm{Im}\,\phi$ is a direct summand of G and $\mathrm{Im}\,\phi \subseteq \mathrm{Ker}\,\psi \subseteq G$, we have

$$\mathrm{Ker}\,\psi = \mathrm{Im}\,\phi \oplus U$$

for some R-module U and, because $\mathrm{Ker}\,\psi$ is a finite free module, U too is a finite free module.

Next the exact sequence

$$0 \to \mathrm{Ker}\,\psi \to G \to \mathrm{Im}\,\psi \to 0 \qquad (B.2.2)$$

of finite free modules yields

$$\begin{aligned}
\mathrm{rank}_R(G) &= \mathrm{rank}_R(\mathrm{Ker}\,\psi) + \mathrm{rank}_R(\mathrm{Im}\,\psi) \\
&= \mathrm{rank}_R(\mathrm{Im}\,\phi) + \mathrm{rank}_R(U) + \mathrm{rank}_R(\mathrm{Im}\,\psi) \\
&= \mathrm{rank}_R(\phi) + \mathrm{rank}_R(\psi) + \mathrm{rank}_R(U)
\end{aligned}$$

by (B.2.1) and its analogue for ψ. Consequently $U = 0$ *if and only if* $\mathrm{rank}_R(G) = \mathrm{rank}_R(\phi) + \mathrm{rank}_R(\psi)$.

Let $E \neq 0$ be an R-module. Since $\mathrm{Im}\,\phi$ is a direct summand of G, $\mathrm{Im}\,\phi \otimes_R E$ is a direct summand of $G \otimes_R E$ and as such it is just $\mathrm{Im}\,(\phi \otimes E)$. Similarly $\mathrm{Im}\,\psi \otimes_R E$ may be identified with $\mathrm{Im}\,(\psi \otimes E)$. But $\mathrm{Ker}\,\psi$ is a direct summand of G, hence from (B.2.2) we obtain the exact sequence

$$0 \to \mathrm{Ker}\,\psi \otimes_R E \to G \otimes_R E \to \mathrm{Im}\,(\psi \otimes E) \to 0$$

which shows that $\mathrm{Ker}\,\psi \otimes_R E = \mathrm{Ker}\,(\psi \otimes E)$. Accordingly

$$F \otimes_R E \to G \otimes_R E \to H \otimes_R E \qquad (B.2.3)$$

is exact if and only if $\mathrm{Ker}\,\psi \otimes_R E = \mathrm{Im}\,\phi \otimes_R E$. But

$$\mathrm{Ker}\,\psi \otimes_R E = (\mathrm{Im}\,\phi \otimes_R E) \oplus (U \otimes_R E).$$

Consequently (B.2.3) is exact if and only if $U \otimes_R E = 0$. But U is a finite free module and $E \neq 0$. Accordingly $U \otimes_R E = 0$ if and only if $U = 0$ that is if and only if

$$\mathrm{rank}_R(G) = \mathrm{rank}_R(\phi) + \mathrm{rank}_R(\psi).$$

We summarize these observations in

LEMMA 7. *Let R be a quasi-local ring, $F \xrightarrow{\phi} G \xrightarrow{\psi} H$ a short complex of finite free R-modules, and suppose that*

$$\mathfrak{A}(\phi) = R = \mathfrak{A}(\psi).$$

If now $E \neq 0$ is an R-module, then the following statements are equivalent:

(a) $F \otimes_R E \to G \otimes_R E \to H \otimes_R E$ *is exact;*

(b) $\mathrm{rank}_R(\phi) + \mathrm{rank}_R(\psi) = \mathrm{rank}_R(G).$

Lemma 7 will now be generalized. In particular we shall remove the condition that R be quasi-local.

THEOREM 1. *Let $F \xrightarrow{\phi} G \xrightarrow{\psi} H$ be a short complex of finite free R-modules, let $E \neq 0$ be an R-module, and suppose that*

$$\mathfrak{A}(\phi, E) = R = \mathfrak{A}(\psi, E).$$

Then the following statements are equivalent:

(a) $F \otimes_R E \to G \otimes_R E \to H \otimes_R E$ *is exact;*

(b) $\mathrm{rank}_R(\phi, E) + \mathrm{rank}_R(\psi, E) = \mathrm{rank}_R(G).$

Proof. Let M be a maximal ideal for which $E_M \neq 0$. Since $\mathfrak{A}(\phi, E) = R$, ϕ is stable relative to E. Hence

$$\mathrm{rank}_{R_M}(\phi_M, E_M) = \mathrm{rank}_R(\phi, E)$$

and $\qquad \mathfrak{A}(\phi_M, E_M) = \mathfrak{A}(\phi, E) R_M = R_M.$

Likewise $\mathrm{rank}_{R_M}(\psi_M, E_M) = \mathrm{rank}_R(\psi, E)$ and $\mathfrak{A}(\psi_M, E_M) = R_M$. Lemma 6 applied to the complex

$$\mathbb{C}: 0 \to F \xrightarrow{\phi} G \xrightarrow{\psi} H$$

now shows that it suffices to prove the theorem when R is a quasi-local ring.

Assuming that R is quasi-local, put $\bar{R} = R/\mathrm{Ann}_R(E)$. If now $\bar{\mathbb{C}} = \mathbb{C} \otimes_R \bar{R}$ is the complex

$$0 \to \bar{F} \xrightarrow{\bar{\phi}} \bar{G} \xrightarrow{\bar{\psi}} \bar{H}$$

then, by (B.1.5) and (B.1.6),

$$\mathfrak{A}(\bar{\phi}, E) = \bar{R} = \mathfrak{A}(\bar{\psi}, E), \quad \mathrm{rank}_{\bar{R}}(\bar{\phi}, E) = \mathrm{rank}_R(\phi, E),$$

and $\qquad \mathrm{rank}_{\bar{R}}(\bar{\psi}, E) = \mathrm{rank}_R(\psi, E).$

Accordingly Lemma 2 allows us to replace R by $\bar{R}$ and $F \to G \to H$ by $\bar{F} \to \bar{G} \to \bar{H}$. But $\bar{R}$ is quasi-local and $\mathrm{Ann}_{\bar{R}}(E) = 0$. Consequently we are justified in adding to the original hypotheses the conditions that R be quasi-local and $\mathrm{Ann}_R(E) = 0$.

By Chapter 7 Exercise 1, $\mathrm{rank}_R(\phi, E) = \mathrm{rank}_R(\phi)$ and therefore $\mathfrak{A}(\phi) = \mathfrak{A}(\phi, E) = R$. Likewise $\mathrm{rank}_R(\psi, E) = \mathrm{rank}_R(\psi)$ and $\mathfrak{A}(\psi) = \mathfrak{A}(\psi, E) = R$. The desired conclusion therefore follows from Lemma 7.

LEMMA 8. *Let R be a quasi-local ring with maximal ideal P and let K be an R-module. Suppose that for each prime ideal $Q \neq P$ we have $K_Q = 0$. Then either* (i) $K = 0$ *or* (ii) $K \neq 0$ *and*

$$\mathrm{Gr}_R\{P; K\} = 0.$$

Proof. Suppose that $K \neq 0$ and let $U \neq 0$ be a finitely generated submodule of K. By Chapter 5 Lemma 10, it will suffice to prove that $\mathrm{Gr}_R\{P; U\} = 0$. Let Π be a prime ideal containing $\mathrm{Ann}_R(U)$. Then $U_\Pi \neq 0$ so $K_\Pi \neq 0$ and therefore $\Pi = P$. This shows that $P = \mathrm{Rad}(\mathrm{Ann}_R(U))$. Let $\mathfrak{A}$ be a finitely generated ideal contained in P. If m is a large enough positive integer, then $\mathfrak{A}^m U = 0$ whence $0 :_U \mathfrak{A} \neq 0$. Accordingly $\mathrm{Gr}_R\{\mathfrak{A}; U\} = 0$ and now $\mathrm{Gr}_R\{P; U\} = 0$ by Chapter 5 Theorem 11.

At this point we need to introduce the notion of the *dimension*, $\mathrm{Dim}\, R$, of the ring R. This is defined as the supremum of the set of integers q for which there exists a strictly decreasing chain

$$P_0 \supset P_1 \supset P_2 \supset \dots \supset P_q$$

of prime ideals.

Once again we suppose that a complex

$$\mathbb{C}: 0 \to F_n \xrightarrow{\phi_n} F_{n-1} \xrightarrow{\phi_{n-1}} F_{n-2} \to \dots \to F_1 \xrightarrow{\phi_1} F_0 \quad \text{(B.2.4)}$$

of finite free R-modules has been given.

LEMMA 9. *Suppose that $\mathrm{Dim}\, R < \infty$ and let $E \neq 0$ be an R-module. If now conditions*

(a) $\mathrm{Gr}_R\{\mathfrak{A}(\phi_k, E); E\} \geqslant k$ *for* $k = 1, 2, \dots, n$;

(b) $\mathrm{rank}_R(\phi_n, E) = \mathrm{rank}_R(F_n)$;

(c) $\mathrm{rank}_R(\phi_{m+1}, E) + \mathrm{rank}_R(\phi_m, E) = \mathrm{rank}_R(F_m)$ *for* $1 \leqslant m < n$;

are all satisfied, then $\mathbb{C} \otimes_R E$ is exact.

Proof. Put Dim $R = s$. We use induction on s.

First suppose that $s = 0$. In view of Lemmas 5 and 6 we may suppose that R has a single maximal ideal P say. Then P is the one and only prime ideal and therefore $P = \mathrm{Rad}\,(0)$. It follows, by Chapter 5 Theorem 12, that $\mathrm{Gr}_R\{P; E\} = 0$. Consequently, by condition (a), $\mathfrak{A}(\phi_k, E) = R$ for $k = 1, 2, \ldots, n$. Theorem 1 now shows that each sequence

$$F_{k+1} \otimes_R E \to F_k \otimes_R E \to F_{k-1} \otimes_R E$$

is exact and, by the same theorem,

$$0 \to F_n \otimes_R E \to F_{n-1} \otimes_R E$$

is exact as well. This disposes of the case $s = 0$.

We next assume that $s > 0$ and that Lemma 9 has been proved for rings of smaller dimension. In deducing that $\mathbb{C} \otimes_R E$ is exact we may again suppose that R has a single maximal ideal P. We may also suppose (see Lemmas 1 and 2) that $\mathrm{Ann}_R(E) = 0$. Note that, as a consequence of the latter assumption and Lemma 5,†

$$\mathrm{rank}_R(\phi_k, E) = \mathrm{rank}_R(\phi_k) \qquad \text{(B.2.5)}$$

and therefore $\qquad \mathfrak{A}(\phi_k, E) = \mathfrak{A}(\phi_k) \qquad \text{(B.2.6)}$
for $1 \leqslant k \leqslant n$.

Let $Q \neq P$ be a prime ideal of R. If $E_Q = 0$, then certainly $\mathbb{C}_Q \otimes_{R_Q} E_Q$ is exact. On the other hand if $E_Q \neq 0$, then conditions (a), (b), (c) hold after localization at Q and therefore $\mathbb{C}_Q \otimes_{R_Q} E_Q$ is exact by virtue of the induction hypothesis. It follows, from Lemma 3, that in any case $[H_k(\mathbb{C} \otimes_R E)]_Q = 0$. Hence if $1 \leqslant k \leqslant n$ we have, by Lemma 8, either (i) $H_k(\mathbb{C} \otimes_R E) = 0$ or (ii) $H_k(\mathbb{C} \otimes_R E) \neq 0$ and $\mathrm{Gr}_R\{P; H_k(\mathbb{C} \otimes_R E)\} = 0$.

Our aim, of course, is to show that $\mathbb{C} \otimes_R E$ is exact and at this stage we use induction on n which we regard as a measure of the length of $\mathbb{C}$. By Lemma 5, ϕ_n is stable relative to E. Consequently, since $\mathrm{rank}_R(\phi_n, E) = \mathrm{rank}_R(F_n)$ as well, it follows, by Chapter 4 Theorem 4, that $\qquad 0 \to F_n \otimes_R E \to F_{n-1} \otimes_R E$

is exact. In particular we see that $\mathbb{C} \otimes_R E$ is exact in the special case $n = 1$.

† Here we use Chapter 7 Exercise 1.

We now use different arguments depending on whether $\mathfrak{A}(\phi_n)$ is or is not the whole of R.

First suppose that $\mathfrak{A}(\phi_n) \neq R$. Then, by (B.2.6),

$$\mathfrak{A}(\phi_n, E) = \mathfrak{A}(\phi_n) \subseteq P$$

and therefore

$$\mathrm{Gr}_R\{P; E\} \geqslant \mathrm{Gr}_R\{\mathfrak{A}(\phi_n, E); E\} \geqslant n.$$

Suppose that $1 \leqslant k \leqslant n$. Since $F_k \otimes_R E$ is a direct sum of a finite number of copies of E,

$$\mathrm{Gr}_R\{P; F_k \otimes_R E\} \geqslant \mathrm{Gr}_R\{P; E\} \geqslant n \geqslant k.$$

But we already know that either $H_k(\mathbb{C} \otimes_R E) = 0$ or

$$\mathrm{Gr}_R\{P; H_k(\mathbb{C} \otimes_R E)\} = 0.$$

Consequently, by Chapter 5 Theorem 21, $\mathbb{C} \otimes_R E$ is exact.

Finally assume that $\mathfrak{A}(\phi_n) = R$. Then ϕ_n is stable,

$$\mathrm{rank}_R(\phi_n) = \mathrm{rank}_R(\phi_n, E) = \mathrm{rank}_R(F_n),$$

and, by Chapter 6 Exercise 11, $\mathrm{Coker}\,\phi_n = F'_{n-1}$ (say) is a finite free module. Next, by Chapter 4 Theorem 4,

$$0 \to F_n \xrightarrow{\phi_n} F_{n-1} \to F'_{n-1} \to 0 \qquad (\text{B}.2.7)$$

is a split exact sequence and therefore

$$0 \to F_n \otimes_R E \to F_{n-1} \otimes_R E \to F'_{n-1} \otimes_R E \to 0$$

is exact as well.

Since ϕ_{n-1} vanishes on $\phi_n(F_n)$, it induces a homomorphism

$$\phi'_{n-1}\colon F'_{n-1} \to F_{n-2}$$

which is such that $\phi_{n-1}(F_{n-1}) = \phi'_{n-1}(F'_{n-1})$. We now have a new complex

$$\mathbb{C}'\colon 0 \to F'_{n-1} \xrightarrow{\phi'_{n-1}} F_{n-2} \xrightarrow{\phi_{n-2}} F_{n-3} \to \dots \to F_1 \xrightarrow{\phi_1} F_0$$

of finite free modules. Note that the theorem will follow if we show that $\mathbb{C}' \otimes_R E$ is exact.

By Chapter 6 Exercise 13, $\mathrm{rank}_R(\phi_{n-1}) = \mathrm{rank}_R(\phi'_{n-1})$ and therefore

$$\mathrm{rank}_R(\phi_{n-1}, E) = \mathrm{rank}_R(\phi'_{n-1}, E) \qquad (\text{B}.2.8)$$

because $\operatorname{Ann}_R(E) = 0$. Next (B. 2.7) yields

$$
\begin{aligned}
\operatorname{rank}_R(F'_{n-1}) &= \operatorname{rank}_R(F_{n-1}) - \operatorname{rank}_R(F_n) \\
&= \operatorname{rank}_R(F_{n-1}) - \operatorname{rank}_R(\phi_n, E) \\
&= \operatorname{rank}_R(\phi_{n-1}, E) \\
&= \operatorname{rank}_R(\phi'_{n-1}, E)
\end{aligned}
$$

by (B. 2.8). Now from Chapter 6 Exercise 13 and (B. 2.8) it follows that $\mathfrak{A}(\phi'_{n-1}, E) = \mathfrak{A}(\phi_{n-1}, E)$; consequently

$$
\operatorname{Gr}_R\{\mathfrak{A}(\phi'_{n-1}, E); E\} \geqslant n-1.
$$

Thus $\mathbb{C}'$ and E satisfy conditions analogous to (a), (b), (c) but the length of $\mathbb{C}'$ is smaller by one than the length of $\mathbb{C}$. Induction now ensures that $\mathbb{C}' \otimes_R E$ is exact and with this the lemma is proved.

We are now ready to prove the main theorem of the appendix. $\mathbb{C}$ continues to denote a complex of finite free R-modules as in (B. 2.4).

THEOREM 2. *Let $E \neq 0$ be an R-module. Then $\mathbb{C} \otimes_R E$ is exact if and only if the following three conditions are satisfied:*

(a) $\operatorname{Gr}_R\{\mathfrak{A}(\phi_k, E); E\} \geqslant k$ *for* $k = 1, 2, \ldots, n$;

(b) $\operatorname{rank}_R(\phi_n, E) = \operatorname{rank}_R(F_n)$;

(c) $\operatorname{rank}_R(\phi_{m+1}, E) + \operatorname{rank}_R(\phi_m, E) = \operatorname{rank}_R(F_m)$ *for* $1 \leqslant m < n$.

Proof. We shall assume that (a), (b), (c) hold and seek to deduce that $\mathbb{C} \otimes_R E$ is exact. This is sufficient in view of Chapter 6 Theorem 14.

Choose a base for each of $F_0, F_1, \ldots, F_n$ and with respect to these bases let the matrices of $\phi_1, \phi_2, \ldots, \phi_n$ be $A_1, A_2, \ldots, A_n$ respectively. Denote by S the smallest subring of R containing 1_R and all the entries in $A_1, A_2, \ldots, A_n$. Then $\operatorname{Dim} S < \infty$ because S is a homomorphic image of a ring of polynomials with integer coefficients.[†]

We can regard F_k as an S-module. Let G_k be the free S-submodule generated by the chosen base of the R-module F_k, and define the S-homomorphism

$$
\psi_k \colon G_k \to G_{k-1}
$$

[†] If $\mathbb{Z}$ denotes the integers and $X_1, X_2, \ldots, X_n$ are indeterminates, then $\operatorname{Dim}\{\mathbb{Z}[X_1, X_2, \ldots, X_n]\} = n+1$. See [(37) Theorem 16, p. 269].

so that its matrix is A_k. Then

$$\mathbb{C}': 0 \to G_n \xrightarrow{\psi_n} G_{n-1} \xrightarrow{\psi_{n-1}} G_{n-2} \to \dots \to G_1 \xrightarrow{\psi_1} G_0$$

is a *complex* of finite free S-modules. Also

$$\mathrm{rank}_S(G_k) = \mathrm{rank}_R(F_k),$$

$$\mathfrak{A}_\nu(\psi_k)\,R = \mathfrak{A}_\nu(\phi_k),$$

and, bearing in mind that E is an S-module,

$$\mathfrak{A}_\nu(\psi_k)\,E = \mathfrak{A}_\nu(\phi_k)\,E$$

for all $\nu \geqslant 0$. In particular we see that

$$\mathrm{rank}_S(\psi_k, E) = \mathrm{rank}_R(\phi_k, E)$$

and therefore $\quad\quad \mathfrak{A}(\psi_k, E)\,R = \mathfrak{A}(\phi_k, E).$

It now follows, from Chapter 5 Theorem 19, that

$$\mathrm{Gr}_S\{\mathfrak{A}(\psi_k, E);\, E\} = \mathrm{Gr}_R\{\mathfrak{A}(\phi_k, E);\, E\} \geqslant k.$$

Hence, by Lemma 9, $\mathbb{C}' \otimes_S E$ is exact. But $\mathbb{C}$ can be identified with $\mathbb{C}' \otimes_S R$ and so it follows, from Lemma 2, that $\mathbb{C} \otimes_R E$ is exact. This completes the proof.

Appendix C:
Some special free complexes

General remarks

The main text has been concerned with the principal properties of finite free resolutions, but in fact the actual examples that we have encountered have been relatively trivial. In the present appendix we shall seek to remedy this by considering in detail some special free complexes. The best known of these is the *Koszul Complex*.† This remarkable complex has been generalized by J. A. Eagon and D. G. Northcott (**17**) and by D. A. Buchsbaum and D. S. Rim (**8, 10, 11**). Here we shall follow very closely the unified treatment of D. Kirby (**30**) which yields an infinite sequence of interesting complexes.

As usual R denotes a commutative ring with a non-zero identity element. The discussion assumes a familiarity with the more elementary parts of the theory of exterior algebras.

C. 1 Differentiations on an exterior algebra

Let $y_1, y_2, ..., y_n$ be indeterminates and $a_1, a_2, ..., a_n$ elements of R. Denote by Λ the exterior R-algebra on $y_1, y_2, ..., y_n$ so that

$$\Lambda = \sum_{-\infty}^{\infty} \Lambda_k \quad \text{(direct sum)},$$

where Λ_k is the R-module consisting of all exterior forms of degree k. Accordingly $\Lambda_k = 0$ if either $k < 0$ or $k > n$; also, for $0 \leqslant k \leqslant n$, Λ_k is a free R-module with a base consisting of all exterior products $y_{j_1} y_{j_2} ... y_{j_k}$, where $1 \leqslant j_1 < j_2 < ... < j_k \leqslant n$.

† See for example [(**37**) Chapter 8] and also (**3**).

It is now easy to verify that there exists an R-homomorphism $\delta\colon \Lambda \to \Lambda$ which is such that

$$\delta(y_{j_1}y_{j_2}\cdots y_{j_s}) = \sum_{\mu=0}^{s} (-1)^{\mu+1} a_{j_\mu} y_{j_1} \cdots \hat{y}_{j_\mu} \cdots y_{j_s}, \quad \text{(C. 1.1)}$$

where the $\hat{}$ over y_{j_μ} indicates that this term is to be omitted from the product. It should be emphasized that, in (C. 1.1), we only require the integers $j_1, j_2, \ldots, j_s$ to lie between 1 and n; it is not postulated that they form an increasing sequence nor are repetitions excluded. Note that $\delta(y_j) = a_j$ and that δ lowers degrees by one, that is to say

$$\delta(\Lambda_k) \subseteq \Lambda_{k-1}. \quad \text{(C. 1.2)}$$

The main properties of δ are set out in

LEMMA 1. *If $\delta\colon \Lambda \to \Lambda$ is defined as above, then $\delta^2 = 0$. Also when $f\in\Lambda_k$ and g is an arbitrary element of Λ we have*

$$\delta(fg) = (\delta f)g + (-1)^k f(\delta g). \quad \text{(C. 1.3)}$$

As the verification of these statements is straightforward we shall omit the details. In view of (C. 1.3), δ is sometimes called a *differentiation homomorphism*. Note that δ gives rise to a complex

$$\cdots \to 0 \to 0 \to \Lambda_n \to \Lambda_{n-1} \to \cdots \to \Lambda_1 \to \Lambda_0 \to 0 \to 0 \to \cdots$$

of finite free R-modules. This is the *Koszul complex* derived from the given sequence $a_1, a_2, \ldots, a_n$.

Now assume that in addition to $a_1, a_2, \ldots, a_n$ we have a second sequence $a_1', a_2', \ldots, a_n'$ also composed of n elements of R. An easy verification shows that if $\delta'\colon \Lambda \to \Lambda$ is the differentiation homomorphism to which the latter sequence gives rise, then

$$\delta\delta' + \delta'\delta = 0. \quad \text{(C. 1.4)}$$

After these preliminaries let $A = \|a_{ij}\|$ be an $m \times n$ matrix with entries in R and denote by δ_i the differentiation homomorphism $\delta_i\colon \Lambda \to \Lambda$ associated with the ith row of A. Then, by Lemma 1,

$$\delta_i \delta_i = 0; \quad \text{(C. 1.5)}$$

furthermore, by (C. 1.4), we have

$$\delta_i \delta_j + \delta_j \delta_i = 0 \quad \text{(C. 1.6)}$$

when $i \neq j$.

LEMMA 2. *Let* $i_1, i_2, \ldots, i_k$ *lie between* 1 *and* m *and* $j_1, j_2, \ldots, j_k$ *between* 1 *and* n. *Then*

$$\delta_{i_k}\delta_{i_{k-1}}\ldots\delta_{i_1}(y_{j_1}y_{j_2}\ldots y_{j_k}) = \begin{vmatrix} a_{i_1 j_1} & a_{i_1 j_2} & \cdots & a_{i_1 j_k} \\ a_{i_2 j_1} & a_{i_2 j_2} & \cdots & a_{i_2 j_k} \\ \cdot & \cdot & \cdots & \cdot \\ a_{i_k j_1} & a_{i_k j_2} & \cdots & a_{i_k j_k} \end{vmatrix}.$$

The proof (which we omit) is a simple argument using induction on k. Note that repetitions are allowed among $i_1, i_2, \ldots, i_k$ and also among $j_1, j_2, \ldots, j_k$.

Now assume that $1 \leqslant i \leqslant m$ and that $f \in \Lambda_k$. By Lemma 1,

$$\delta_i((\delta_{i-1}\ldots\delta_1 f)\,g) = (\delta_i\delta_{i-1}\ldots\delta_1 f)\,g + (-1)^{k-i+1}(\delta_{i-1}\ldots\delta_1 f)(\delta_i g),$$

whence

$$(-1)^{k-i}\delta_m\ldots\delta_{i+1}((\delta_{i-1}\ldots\delta_1 f)(\delta_i g))$$
$$= \delta_m\ldots\delta_{i+1}((\delta_i\ldots\delta_1 f)\,g) - \delta_m\ldots\delta_i((\delta_{i-1}\ldots\delta_1 f)\,g)$$
$$= \xi_i - \xi_{i-1} \quad \text{(say)}.$$

This yields

LEMMA 3. *Suppose that* $1 \leqslant r \leqslant m+1$ *and* $f \in \Lambda_k$. *Then*

$$\sum_{i=r}^{m} (-1)^{k-i}\delta_m\ldots\delta_{i+1}((\delta_{i-1}\ldots\delta_1 f)(\delta_i g))$$

$$= (\delta_m\ldots\delta_1 f)\,g - \delta_m\ldots\delta_r((\delta_{r-1}\ldots\delta_1 f)\,g).$$

Note that the extreme case $r = m+1$ is included.

C. 2   Generalized Koszul complexes

Throughout section (C.2), $A = \|a_{ij}\|$ will denote an $m \times n$ R-matrix and $x_1, x_2, \ldots, x_m$ and $y_1, y_2, \ldots, y_n$ two sets of indeterminates having m and n members respectively. As before Λ will denote the exterior R-algebra on $y_1, y_2, \ldots, y_n$ and $\delta_i\colon \Lambda \to \Lambda$ the differentiation homomorphism determined by the ith row of A. The complex we are now going to construct depends only on the ring R, the matrix A, and an integer t. No restriction is placed on t and in particular it may be negative.

Let $S = R[x_1, x_2, ..., x_m]$ be the (commutative) polynomial ring in the indeterminates $x_1, x_2, ..., x_m$. Then

$$S = R[x_1, ..., x_m] = \sum_{-\infty}^{\infty} S_k \quad \text{(direct sum)} \qquad \text{(C.2.1)}$$

where S_k is the finite free R-module consisting of all homogeneous polynomials of degree k. It will be convenient to use x_i to denote not only the ith indeterminate but also the R-homomorphism

$$x_i \colon S \to S \qquad \text{(C.2.2)}$$

which results from multiplication by x_i. Closely associated with this homomorphism is a second one which will be denoted by

$$x_i^{-1} \colon S \to S. \qquad \text{(C.2.3)}$$

This is defined so that if $\mu_1, \mu_2, ..., \mu_m$ are non-negative integers, then

$$x_i^{-1}(x_1^{\mu_1} x_2^{\mu_2} ... x_m^{\mu_m}) = \begin{cases} x_1^{\mu_1} ... x_i^{\mu_i - 1} ... x_m^{\mu_m} & \text{if} \quad \mu_i \geqslant 1, \\ 0 & \text{if} \quad \mu_i = 0. \end{cases} \qquad \text{(C.2.4)}$$

Observe that $S_k = 0$ when $k < 0$, $x_i(S_k) \subseteq S_{k+1}$, and $x_i^{-1}(S_k) \subseteq S_{k-1}$.

We are now ready to define a complex

$$... \to K_{h+1} \xrightarrow{d_{h+1}} K_h \xrightarrow{d_h} K_{h-1} \xrightarrow{d_{h-1}} K_{h-2} \to ... \qquad \text{(C.2.5)}$$

which will be denoted by $K_R(A, t)$. To this end put†

$$K_h = \begin{cases} \Lambda_{m+h-1} \otimes_R S_{h-t-1} & \text{for} \quad h > t, \\ \Lambda_h \otimes_R S_{t-h} & \text{for} \quad h \leqslant t. \end{cases} \qquad \text{(C.2.6)}$$

Then K_h is a finite free R-module and there are only finitely many values of h for which $K_h \neq 0$. It should be noted that the entries in A play no part in the definition of these modules. Their role has solely to do with the homomorphisms $d_h \colon K_h \to K_{h-1}$ ($h = 0, \pm 1, \pm 2, ...$) which are defined as follows:

$$d_h(f \otimes \alpha) = \begin{cases} \sum_{i=1}^{m} (\delta_i f \otimes x_i^{-1} \alpha) & \text{for} \quad h > t+1, \\ \delta_m \delta_{m-1} ... \delta_1 f \otimes \alpha & \text{for} \quad h = t+1, \\ \sum_{i=1}^{m} (\delta_i f \otimes x_i \alpha) & \text{for} \quad h \leqslant t. \end{cases} \qquad \text{(C.2.7)}$$

---

† It is sometimes helpful to use $K_h(A, t)$ as an alternative.

Using the relations $\delta_i \delta_i = 0$ and $\delta_i \delta_j + \delta_j \delta_i = 0$ $(i \neq j)$, it is easily verified that (C. 2.5) really is a complex of finite free modules. Also when $m = 1$ it is not difficult to see that, regardless of the value of t, $K_R(A, t)$ is essentially the Koszul complex associated with the (single) row of the matrix A. This provides the justification for regarding $K_R(A, t)$ as a generalized Koszul complex.

Let z be a new indeterminate. On adjoining z to $K_R(A, t)$ we obtain a complex

$$\ldots \to K_{h+1}[z] \to K_h[z] \to K_{h-1}[z] \to K_{h-2}[z] \to \ldots$$

of finite free $R[z]$-modules. Let us denote this complex by $K_R(A, t)[z]$. Then a little reflection makes it clear that

$$K_R(A, t)[z] = K_{R[z]}(A, t). \qquad (C. 2.8)$$

This observation will be needed later.

In preparation for Lemma 4 put

$$D = D(j_1, j_2, \ldots, j_m) = \begin{vmatrix} a_{1j_1} & a_{1j_2} & \ldots & a_{1j_m} \\ a_{2j_1} & a_{2j_2} & \ldots & a_{2j_m} \\ . & . & \ldots & . \\ . & . & \ldots & . \\ a_{mj_1} & a_{mj_2} & \ldots & a_{mj_m} \end{vmatrix}, \quad (C. 2.9)$$

where $j_1, j_2, \ldots, j_m$ denote integers between 1 and n. These integers, which are to be kept fixed until Lemma 4 has been established, will be used to define, for each p, a homomorphism $\sigma_p : K_p \to K_{p+1}$. The importance of these new homomorphisms resides in the fact that

$$(d_{p+1}\sigma_p + \sigma_{p+1}d_p): K_p \to K_p$$

is just multiplication by D. The definition of σ_p is a little complicated. For convenience we consider separately four different cases.

Case (1) $p > t+1$. Here

$$K_p = \Lambda_{m+p-1} \otimes S_{p-t-1} \quad \text{and} \quad K_{p+1} = \Lambda_{m+p} \otimes S_{p-t},$$

where we have used plain $\otimes$ in place of $\otimes_R$. In this situation

$$\sigma_p(f \otimes x_1^{\mu_1} x_2^{\mu_2} \ldots x_m^{\mu_m})$$
$$= \sum_{i=q}^{m} (\delta_{i-1} \ldots \delta_1 y_{j_1} \ldots y_{j_m})(\delta_m \ldots \delta_{i+1} f) \otimes x_i(x_1^{\mu_1} x_2^{\mu_2} \ldots x_m^{\mu_m}),$$

where q is the largest integer for which $\mu_q \neq 0$. To remove an ambiguity we adopt the convention that in this formula, and in similar situations which occur later, $\delta_{i-1} \ldots \delta_1 y_{j_1} \ldots y_{j_m}$ stands for $\delta_{i-1} \ldots \delta_1 (y_{j_1} \ldots y_{j_m})$.

Case (2) $p = t+1$. This time
$$K_p = \Lambda_{m+t} \otimes S_0, K_{p+1} = \Lambda_{m+t+1} \otimes S_1,$$
and
$$\sigma_p(f \otimes 1) = \sum_{i=1}^{m} (\delta_{i-1} \ldots \delta_1 y_{j_1} \ldots y_{j_m})(\delta_m \ldots \delta_{i+1} f) \otimes x_i.$$

Case (3) $p = t$. In this case $K_p = \Lambda_t \otimes S_0, K_{p+1} = \Lambda_{m+t} \otimes S_0$ and σ_p is defined by
$$\sigma_p(f \otimes 1) = y_{j_1} \ldots y_{j_m} f \otimes 1.$$

Case (4) $p < t$. This is the final case. We have $K_p = \Lambda_p \otimes S_{t-p}$, $K_{p+1} = \Lambda_{p+1} \otimes S_{t-p-1}$, and we put
$$\sigma_p(f \otimes x_1^{\mu_1} x_2^{\mu_2} \ldots x_m^{\mu_m})$$
$$= (-1)^{m-q} \delta_m \ldots \delta_{q+1}((\delta_{q-1} \ldots \delta_1 y_{j_1} \ldots y_{j_m})f) \otimes x_q^{-1}(x_1^{\mu_1} x_2^{\mu_2} \ldots x_m^{\mu_m}),$$
where, once again, q is the largest integer for which $\mu_q \neq 0$.

We are now ready to prove

LEMMA 4. *With the above notation, the homomorphism* $(d_{p+1}\sigma_p + \sigma_{p-1}d_p): K_p \to K_p$ *consists in multiplication by* D.

Proof. The cases $p > t+1$, $p = t+1$, $p = t$ and $p < t$ require separate consideration.

First suppose that $p > t+1$. Let $\xi = f \otimes x_1^{\mu_1} x_2^{\mu_2} \ldots x_m^{\mu_m}$ belong to K_p and let q be the largest integer for which $\mu_q \neq 0$. Then
$$d_{p+1}\sigma_p(\xi)$$
$$= d_{p+1}\left[\sum_{k=q}^{m} (\delta_{k-1} \ldots \delta_1 y_{j_1} \ldots y_{j_m})(\delta_m \ldots \delta_{k+1}f) \otimes x_k(x_1^{\mu_1} \ldots x_m^{\mu_m}) \right]$$
$$= \sum_{i=1}^{m} \sum_{k=q}^{m} \delta_i((\delta_{k-1} \ldots \delta_1 y_{j_1} \ldots y_{j_m})(\delta_m \ldots \delta_{k+1}f)) \otimes x_i^{-1} x_k(x_1^{\mu_1} \ldots x_m^{\mu_m}).$$

But, by Lemma 1,
$$\delta_i((\delta_{k-1} \ldots \delta_1 y_{j_1} \ldots y_{j_m})(\delta_m \ldots \delta_{k+1}f))$$
$$= (\delta_i \delta_{k-1} \ldots \delta_1 y_{j_1} \ldots y_{j_m})(\delta_m \ldots \delta_{k+1}f)$$
$$- (\delta_{k-1} \ldots {}_1 \delta_1 y_j \ldots y_{j_m})(\delta_m \ldots \delta_{k+1} \delta_i f).$$

Now if $i < k$ then, from (C.1.5) and (C.1.6), $\delta_i \delta_{k-1} \ldots \delta_1 = 0$

whereas if $i > k \geqslant q$ then $x_i^{-1} x_k (x_1^{\mu_1} \ldots x_m^{\mu_m}) = 0$. Accordingly

$$d_{p+1}\sigma_p(\xi) = \sum_{i=q}^{m} (\delta_i \ldots \delta_1 y_{j_1} \ldots y_{j_m})(\delta_m \ldots \delta_{i+1}f) \otimes x_1^{\mu_1} \ldots x_m^{\mu_m}$$
$$- \sum_{i=1}^{m}\sum_{k=q}^{m} (\delta_{k-1} \ldots \delta_1 y_{j_1} \ldots y_{j_m})(\delta_m \ldots \delta_{k+1}\delta_i f) \otimes x_i^{-1}x_k(x_1^{\mu_1} \ldots x_m^{\mu_m}).$$

Next we have

$$\sigma_{p-1}d_p(\xi) = \sigma_{p-1}\left[\sum_{i=1}^{m} \delta_i f \otimes x_i^{-1}(x_1^{\mu_1} \ldots x_m^{\mu_m})\right]$$
$$= \sum_{\substack{i=1\\i\neq q}}^{m}\sum_{k=q}^{m} (\delta_{k-1} \ldots \delta_1 y_{j_1} \ldots y_{j_m})(\delta_m \ldots \delta_{k+1}\delta_i f) \otimes x_k x_i^{-1}(x_1^{\mu_1} \ldots x_m^{\mu_m})$$
$$+ \sum_{k=\nu}^{m} (\delta_{k-1} \ldots \delta_1 y_{j_1} \ldots y_{j_m})(\delta_m \ldots \delta_{k+1}\delta_q f) \otimes x_k x_q^{-1}(x_1^{\mu_1} \ldots x_m^{\mu_m}),$$

where in the latter sum ν is an integer satisfying $\nu \leqslant q$. However if $k < q$, then $\delta_m \ldots \delta_{k+1}\delta_q = 0$ and thus we see that
$$\sigma_{p-1}d_p(\xi)$$
$$= \sum_{i=1}^{m}\sum_{k=q}^{m} (\delta_{k-1} \ldots \delta_1 y_{j_1} \ldots y_{j_m})(\delta_m \ldots \delta_{k+1}\delta_i f) \otimes x_k x_i^{-1}(x_1^{\mu_1} \ldots x_m^{\mu_m}).$$
It now follows that

$$(d_{p+1}\sigma_p + \sigma_{p-1}d_p)(\xi)$$
$$= \sum_{i=q}^{m} (\delta_i \ldots \delta_1 y_{j_1} \ldots y_{j_m})(\delta_m \ldots \delta_{i+1}f) \otimes x_1^{\mu_1} \ldots x_m^{\mu_m}$$
$$- \sum_{i=1}^{m}\sum_{k=q}^{m} (\delta_{k-1} \ldots \delta_1 y_{j_1} \ldots y_{j_m})(\delta_m \ldots \delta_{k+1}\delta_i f)$$
$$\otimes (x_i^{-1}x_k - x_k x_i^{-1})(x_1^{\mu_1} \ldots x_m^{\mu_m}).$$

But $(x_i^{-1}x_k - x_k x_i^{-1})(x_1^{\mu_1} \ldots x_m^{\mu_m}) = 0$ if either $i \neq k$ or $i = k = q$, whereas if $i = k > q$, then $(x_i^{-1}x_k - x_k x_i^{-1})(x_1^{\mu_1} \ldots x_m^{\mu_m})$ is just $x_1^{\mu_1}x_2^{\mu_2} \ldots x_m^{\mu_m}$. In view of this, the double sum in the expression for $(d_{p+1}\sigma_p + \sigma_{p-1}d_p)(\xi)$ reduces to

$$\sum_{i=q+1}^{m} (\delta_{i-1} \ldots \delta_1 y_{j_1} \ldots y_{j_m})(\delta_m \ldots \delta_i f) \otimes x_1^{\mu_1}x_2^{\mu_2} \ldots x_m^{\mu_m}$$

and thus we arrive at

$$(d_{p+1}\sigma_p + \sigma_{p-1}d_p)(\xi) = (\delta_m \ldots \delta_1 y_{j_1} \ldots y_{j_m})f \otimes x_1^{\mu_1}x_2^{\mu_2} \ldots x_m^{\mu_m}$$
$$= Df \otimes x_1^{\mu_1}x_2^{\mu_2} \ldots x_m^{\mu_m}$$

by Lemma 2. This, in effect, is what we were aiming to prove.

The next case for consideration is that where $p = t+1$. However this is entirely straightforward so we shall omit the details and pass on to deal with the situation where $p = t$. This time $K_p = \Lambda_t \otimes S_0$ and

$$d_{p+1}\sigma_p(f \otimes 1)$$
$$= d_{t+1}(y_{j_1} \ldots y_{j_m} f \otimes 1) = (\delta_m \ldots \delta_1(y_{j_1} \ldots y_{j_m} f)) \otimes 1.$$

But, by Lemma 3 (with $r = 1$),

$$\delta_m \ldots \delta_1(y_{j_1} \ldots y_{j_m} f)$$
$$= (\delta_m \ldots \delta_1 y_{j_1} \ldots y_{j_m})f$$
$$- \sum_{i=1}^{m} (-1)^{m-i} \delta_m \ldots \delta_{i+1}((\delta_{i-1} \ldots \delta_1 y_{j_1} \ldots y_{j_m})(\delta_i f))$$

and therefore

$$d_{p+1}\sigma_p(f \otimes 1) = Df \otimes 1$$
$$- \sum_{i=1}^{m} (-1)^{m-i} \delta_m \ldots \delta_{i+1}((\delta_{i-1} \ldots \delta_1 y_{j_1} \ldots y_{j_m})(\delta_i f)) \otimes 1.$$

Next

$$\sigma_{p-1} d_p(f \otimes 1) = \sigma_{t-1}\left[\sum_{i=1}^{m} (\delta_i f \otimes x_i)\right]$$
$$= \sum_{i=1}^{m} (-1)^{m-i} \delta_m \ldots \delta_{i+1}((\delta_{i-1} \ldots \delta_1 y_{j_1} \ldots y_{j_m})(\delta_i f)) \otimes 1.$$

Accordingly $(d_{p+1}\sigma_p + \sigma_{p-1}d_p)(f \otimes 1) = Df \otimes 1$ and now we have disposed of this case as well.

Finally assume that $p < t$. Suppose that $\xi = f \otimes x_1^{\mu_1} \ldots x_m^{\mu_m}$ belongs to $K_p = \Lambda_p \otimes S_{t-p}$ and let q be the largest integer for which $\mu_q \neq 0$. Then

$$d_{p+1}\sigma_p(\xi) = (-1)^{m-q} \sum_{i=1}^{m} \delta_i \delta_m \ldots \delta_{q+1}((\delta_{q-1} \ldots \delta_1 y_{j_1} \ldots y_{j_m})f)$$
$$\otimes x_i x_q^{-1}(x_1^{\mu_1} \ldots x_m^{\mu_m})$$
$$= \sum_{i=1}^{q} \delta_m \ldots \delta_{q+1} \delta_i((\delta_{q-1} \ldots \delta_1 y_{j_1} \ldots y_{j_m})f)$$
$$\otimes x_i x_q^{-1}(x_1^{\mu_1} \ldots x_m^{\mu_m})$$
$$= \delta_m \ldots \delta_{q+1}((\delta_q \delta_{q-1} \ldots \delta_1 y_{j_1} \ldots y_{j_m})f) \otimes x_1^{\mu_1} \ldots x_m^{\mu_m}$$
$$+ (-1)^{m-q+1} \sum_{i=1}^{q} \delta_m \ldots \delta_{q+1}((\delta_{q-1} \ldots \delta_1 y_{j_1} \ldots y_{j_m})(\delta_i f))$$
$$\otimes x_i x_q^{-1}(x_1^{\mu_1} \ldots x_m^{\mu_m}).$$

On the other hand

$$\sigma_{p-1}d_p(\xi) = \sigma_{p-1}\Bigg[\sum_{i=1}^{q} \delta_i f \otimes x_i(x_1^{\mu_1}\dots x_m^{\mu_m})$$
$$+ \sum_{i=q+1}^{m} \delta_i f \otimes x_i(x_1^{\mu_1}\dots x_m^{\mu_m})\Bigg]$$
$$= (-1)^{m-q}\sum_{i=1}^{q} \delta_m\dots\delta_{q+1}((\delta_{q-1}\dots\delta_1 y_{j_1}\dots y_{j_m})(\delta_i f))$$
$$\otimes x_q^{-1}x_i(x_1^{\mu_1}\dots x_m^{\mu_m})$$
$$+ \sum_{i=q+1}^{m} (-1)^{m-i}\delta_m\dots\delta_{i+1}((\delta_{i-1}\dots\delta_1 y_{j_1}\dots y_{j_m})(\delta_i f))$$
$$\otimes x_1^{\mu_1}\dots x_m^{\mu_m}.$$

It follows that

$$(d_{p+1}\sigma_p + \sigma_{p-1}d_p)(\xi)$$
$$= \delta_m\dots\delta_{q+1}((\delta_q\dots\delta_1 y_{j_1}\dots y_{j_m})f) \otimes x_1^{\mu_1}\dots x_m^{\mu_m}$$
$$+ \sum_{i=q+1}^{m} (-1)^{m-i}\delta_m\dots\delta_{i+1}((\delta_{i-1}\dots\delta_1 y_{j_1}\dots y_{j_m})(\delta_i f))$$
$$\otimes x_1^{\mu_1}\dots x_m^{\mu_m}$$
$$= \delta_m\dots\delta_{q+1}((\delta_q\dots\delta_1 y_{j_1}\dots y_{j_m})f) \otimes x_1^{\mu_1}\dots x_m^{\mu_m}$$
$$+ (\delta_m\dots\delta_1 y_{j_1}\dots y_{j_m})f \otimes x_1^{\mu_1}\dots x_m^{\mu_m}$$
$$- \delta_m\dots\delta_{q+1}((\delta_q\dots\delta_1 y_{j_1}\dots y_{j_m})f) \otimes x_1^{\mu_1}\dots x_m^{\mu_m}$$

by Lemma 3. This reduces to

$$(d_{p+1}\sigma_p + \sigma_{p-1}d_p)(\xi) = Df \otimes x_1^{\mu_1}\dots x_m^{\mu_m}$$

and with this Lemma 4 is finally established.

Let E be an R-module. Put

$$K_R(A,t,E) = K_R(A,t) \otimes_R E. \qquad (\text{C.}2.10)$$

THEOREM 1. *For every R-module E and for every integer p, the pth homology module $H_p K_R(A,t,E)$ of the complex $K_R(A,t,E)$ is annihilated by the ideal $\mathfrak{A}_m(A)$.*

Remark. Since $\mathfrak{A}_m(A) = 0$ if $m > n$, only the case $m \leqslant n$ is significant.

Proof. Let $j_1, j_2, \dots, j_m$ be integers between 1 and n, let D be as in (C.2.9), and define homomorphisms $\sigma_p\colon K_p \to K_{p+1}$ as for

Lemma 4. Put $d_p^* = d_p \otimes E$ and $\sigma_p^* = \sigma_p \otimes E$. Since D is a typical generator of $\mathfrak{A}_m(A)$, it will suffice to show that

$$D(\operatorname{Ker} d_p^*) \subseteq \operatorname{Im} d_{p+1}^*.$$

By Lemma 4, the endomorphism $d_{p+1}^* \sigma_p^* + \sigma_{p-1}^* d_p^*$ of $K_p \otimes E$ consists in multiplication by D. Hence if $\xi \in \operatorname{Ker} d_p^*$, then

$$D\xi = (d_{p+1}^* \sigma_p^* + \sigma_{p-1}^* d_p^*)\,(\xi) = d_{p+1}^* \sigma_p^*(\xi) \in \operatorname{Im} d_{p+1}^*.$$

Thus the theorem is proved.

In connection with the next theorem it should be noted that if $t \leqslant n-m$, then $K_h = 0$ when $h > n-m+1$ whereas

$$K_{n-m+1} = \Lambda_n \otimes S_{n-m-t}$$

is non-zero.

THEOREM 2. *Suppose that $t \leqslant n-m$ and let E be an R-module. Assume that*
$$0 \leqslant n-m-q+1 \leqslant \operatorname{Gr}_R\{\mathfrak{A}_m(A); E\}.$$

Then the truncated complex

$$\ldots \to 0 \to 0 \to K_{n-m+1}(A,t,E) \to K_{n-m}(A,t,E) \to \ldots \to K_q(A,t,E)$$

is exact.

Proof. By Theorem 1, $\mathfrak{A}_m(A)$ annihilates all the homology modules of the complex. Also

$$K_p(A,t,E) = K_p(A,t) \otimes_R E$$

is a direct sum of a finite number (possibly zero) of copies of E. Accordingly, by Chapter 5 Theorem 18,

$$\operatorname{Gr}_R\{\mathfrak{A}_m(A); K_p(A,t,E)\} \geqslant \operatorname{Gr}_R\{\mathfrak{A}_m(A); E\}.$$

The desired result therefore follows from Chapter 5 Theorem 22.

THEOREM 3. *Suppose that $t \leqslant n-m$ and that E is an R-module. If now $H_{n-m+1}K_R(A,t,E) = 0$, then $0 :_E \mathfrak{A}_m(A) = 0$.*

Note that we do not assume that $m \leqslant n$.

Proof. By hypothesis the sequence

$$0 \to K_{n-m+1}^* \xrightarrow{d_{n-m+1}^*} K_{n-m}^* \qquad \text{(C.2.11)}$$

is exact, where $K_p^* = K_p \otimes E$ and $d_p^* = d_p \otimes E$. Note that

$$K_{n-m+1}^* = \Lambda_n \otimes S_{n-m-t} \otimes E.$$

Following D. Kirby (**30**) we shall treat separately the cases $t = n - m$ and $t < n - m$.

First suppose that $t = n - m$. Let e belong to $0:_E \mathfrak{A}_m(A)$. Then $y_1 y_2 \cdots y_n \otimes 1 \otimes e$ is in K^*_{n-m+1} and

$$d^*_{n-m+1}(y_1 y_2 \cdots y_n \otimes 1 \otimes e) = \delta_m \cdots \delta_1 (y_1 \cdots y_n) \otimes 1 \otimes e.$$

But $\delta_m \cdots \delta_1 (y_1 \cdots y_n)$ can be expressed as a sum of the form

$$\Sigma \pm D(j_1, \ldots, j_m) y_{k_1} y_{k_2} \cdots y_{k_{n-m}},$$

where $D(j_1, \ldots, j_m)$ is defined as in (C. 2.9) and $j_1, \ldots, j_m, k_1, \ldots, k_{n-m}$ is a permutation of $1, 2, \ldots, n$. (If $m > n$, then, of course, $\delta_m \cdots \delta_1 (y_1 \cdots y_n) = 0$.) But $D(j_1, \ldots, j_m) e = 0$ and so we see that

$$d^*_{n-m+1}(y_1 y_2 \cdots y_n \otimes 1 \otimes e) = 0.$$

Accordingly $y_1 y_2 \cdots y_n \otimes 1 \otimes e = 0$ and therefore $e = 0$. Thus $0:_E \mathfrak{A}_m(A) = 0$ as required.

The case $t < n - m$ is more complicated and to facilitate the discussion we introduce some temporary notation.

Suppose that $0 \leqslant s \leqslant m$, that $j_1, j_2, \ldots, j_s$ lie between 1 and n, and that $1 \leqslant i \leqslant s + 1$. We shall use $D(j_1, j_2, \ldots, j_s | i)$ to denote the $s \times s$ minor of A obtained from the columns $j_1, j_2, \ldots, j_s$ (in that order) together with rows† $1, \ldots, i - 1, i + 1, \ldots, s + 1$. For $s = 0$ we have $D(\ | 1) = 1_R$. Note that the elements $D(j_1, \ldots, j_m | m + 1)$ generate $\mathfrak{A}_m(A)$ and that

$$D(j_1, j_2, \ldots, j_s | s + 1) = \sum_{k=1}^{s} (-1)^{k+1} a_{kj_1} D(j_2, \ldots, j_s | k) \quad \text{(C. 2.12)}$$

when $s \geqslant 1$. The latter relation is obtained by expanding the minor on the left hand side by its first column.

For a given s let $\mathfrak{B}_s$ be the ideal generated by all the elements $D(j_1, j_2, \ldots, j_s | s + 1)$. Thus $\mathfrak{B}_0 = R$ and $\mathfrak{B}_m = \mathfrak{A}_m(A)$. Still assuming that $t < n - m$ and that $H_{n-m+1} K_R(A, t, E) = 0$, we now prove the following lemma which suffices to complete the proof of Theorem 3.

LEMMA 5. *The relation* $0:_E \mathfrak{B}_s = 0$ *holds for* $s = 0, 1, \ldots, m$.

† $D(j_1, j_2, \ldots, j_m | k)$ is only defined for $k = m + 1$.

Proof. We use induction on s. The case $s = 0$ is trivial. We therefore assume that $0 < s \leqslant m$ and that we already know that $0 :_E \mathfrak{B}_{s-1} = 0$.

Suppose that e belongs to $0 :_E \mathfrak{B}_s$, i.e. that

$$D(j_1, ..., j_s | s + 1) e = 0$$

for all choices of $j_1, j_2, ..., j_s$. The lemma and the theorem will follow if we can show that $e = 0$.

In what follows $\mu = (\mu_1, \mu_2, ..., \mu_s)$ stands for a typical sequence of s non-negative integers such that $\mu_1 + \mu_2 + ... + \mu_s = n - m - t$. For a given μ and sequence $j_2, j_3, ..., , j_s$ put

$$\Delta_\mu(j_2, ..., j_s) = \prod_{i=1}^{s} [(-1)^{i+1} D(j_2, ..., j_s | i)]^{\mu_i}. \quad \text{(C.2.13)}$$

For the moment we propose to keep $j_2, j_3, ..., j_s$ fixed. We take advantage of this fact to write Δ_μ in place of $\Delta_\mu(j_2, ..., j_s)$.

After these preliminaries put

$$\eta = \sum_\mu x_1^{\mu_1} ... x_s^{\mu_s} \otimes \Delta_\mu e \quad \text{(C.2.14)}$$

and consider the element $y_1 y_2 ... y_n \otimes \eta$ of

$$K_{n-m+1}^* = \Lambda_n \otimes S_{n-m-t} \otimes E.$$

We shall show that $y_1 y_2 ... y_n \otimes \eta = 0$ by establishing that it is annihilated by the monomorphism† d_{n-m+1}^*. In fact

$$d_{n-m+1}^*(y_1 y_2 ... y_n \otimes \eta)$$
$$= \sum_\mu \sum_i \delta_i(y_1 ... y_n) \otimes x_i^{-1}(x_1^{\mu_1} ... x_s^{\mu_s}) \otimes \Delta_\mu e$$
$$= \sum_\mu \sum_i \sum_k (-1)^{k+1} a_{ik} y_1 ... \hat{y}_k ... y_n \otimes x_i^{-1}(x_1^{\mu_1} ... x_s^{\mu_s}) \otimes \Delta_\mu e$$
$$= \sum_\mu \sum_i \sum_k (-1)^{k+1} a_{ik} \Delta_\mu(y_1 ... \hat{y}_k ... y_n \otimes x_i^{-1}(x_1^{\mu_1} ... x_s^{\mu_s}) \otimes e).$$

Now assume that $\nu = (\nu_1, \nu_2, ..., \nu_s)$ is a sequence of s non-negative integers satisfying $\nu_1 + \nu_2 + ... + \nu_s = n - m - t - 1$ and let us pick out the terms involving $y_1 ... \hat{y}_k ... y_n \otimes x_1^{\nu_1} ... x_s^{\nu_s} \otimes e$ in the last multiple sum. These arise when μ has the form

† See (C. 2.11).

$(\nu_1, ..., \nu_h + 1, ..., \nu_s)$. Accordingly the coefficient of the expression in question is

$$\sum_{h=1}^{s} (-1)^{k+1} a_{hk} \Delta_{(\nu_1, ..., \nu_h+1, ..., \nu_s)}$$

$$= (-1)^{k+1} \Delta_\nu \sum_{h=1}^{s} (-1)^{h+1} a_{hk} D(j_2, ..., j_s|h)$$

$$= (-1)^{k+1} \Delta_\nu D(k, j_2, ..., j_s|s+1)$$

by virtue of (C.2.12). But $D(k, j_2, ..., j_s|s+1) e = 0$ and thus we see that $d^*_{n-m+1}(y_1 y_2 ... y_n \otimes \eta) = 0$ and hence that

$$y_1 y_2 ... y_n \otimes \eta = 0.$$

Accordingly $\quad y_1 ... y_n \otimes \sum_\mu (x_1^{\mu_1} ... x_s^{\mu_s} \otimes \Delta_\mu e) = 0$

and therefore $\Delta_\mu e = 0$ for all μ. In other terms, for all choices of μ, e is annihilated by

$$\prod_{i=1}^{s} [(-1)^{i+1} D(j_2, ..., j_s|i)]^{\mu_i}$$

and, in particular, $(D(j_2, ..., j_s|s))^{n-m-t} e = 0$. As this is valid for all sequences $j_2, j_3, ..., j_s$ it follows that if p is a large enough positive integer, then $\mathfrak{B}^p_{s-1} e = 0$. But this implies that $e = 0$ because, under our present assumptions, we know that

$$0 :_E \mathfrak{B}_{s-1} = 0.$$

The lemma and Theorem 3 now follow for reasons already explained.

Theorem 3 provides the means whereby we can sharpen Theorem 2 when the grade of $\mathfrak{A}_m(E)$ on E is finite.

THEOREM 4. *Suppose that* $t \leqslant n-m$, *that* E *is an* R-*module, and* $\mathrm{Gr}_R\{\mathfrak{A}_m(A); E\} = \mu$, *where* μ *is finite. Then* $H_p K_R(A, t, E)$ *is zero when* $p > n-m-\mu+1$ *and is non-zero when* $p = n-m-\mu+1$.

Proof. The first assertion follows from Theorem 2 so we have only to show that $H_{n-m-\mu+1} K_R(A, t, E) \neq 0$.

This will be established by an argument using induction on μ. When $\mu = 0$ our hypotheses ensure that $0 :_E \mathfrak{A}_m(A) \neq 0$ and therefore the desired result is a consequence of Theorem 3. From here on it will be assumed that $\mu > 0$ and that the statement under consideration holds for modules on which the grade of $\mathfrak{A}_m(A)$ is less than μ.

First suppose that $\mathfrak{A}_m(A)$ contains an element α which is not a zerodivisor on E. Then

$$\mathrm{Gr}_R\{\mathfrak{A}_m(A); E/\alpha E\} = \mu - 1$$

and therefore

$$H_{n-m-\mu+2}K_R(A, t, E/\alpha E) \neq 0$$

by the induction hypothesis. Next the exact sequence

$$0 \to E \xrightarrow{\ \alpha\ } E \to E/\alpha E \to 0$$

gives rise to an exact sequence

$$0 \to K_R(A, t) \otimes E \to K_R(A, t) \otimes E \to K_R(A, t) \otimes E/\alpha E \to 0$$

of complexes and this in turn gives rise to an exact sequence involving the homology modules of the complexes.† In particular we have an exact sequence

$$H_{n-m-\mu+2}K_R(A, t, E) \to H_{n-m-\mu+2}K_R(A, t, E/\alpha E)$$
$$\to H_{n-m-\mu+1}K_R(A, t, E) \to H_{n-m-\mu+1}K_R(A, t, E).$$

The opening remarks of the proof show that

$$H_{n-m-\mu+2}K_R(A, t, E) = 0.$$

Moreover the final homomorphism in this exact sequence consists in multiplication by the element α of $\mathfrak{A}_m(A)$ and therefore, in view of Theorem 1, it is a null homomorphism. It follows that $H_{n-m-\mu+2}K_R(A, t, E/\alpha E)$ and $H_{n-m-\mu+1}K_R(A, t, E)$ are isomorphic and therefore the latter is not zero.

It remains for us to remove the assumption that $\mathfrak{A}_m(A)$ contains a non-zerodivisor on E. To this end let z be an indeterminate. Since $\mu > 0$, $\mathfrak{A}_m(A)$ contains a latent non-zerodivisor on E and therefore $\mathfrak{A}_m(A)R[z]$ contains a non-zerodivisor on $E[z]$. The results of the last paragraph now show that

$$H_{n-m-\mu+1}\{K_{R[z]}(A, t) \otimes_{R[z]} E[z]\} \neq 0.$$

But, by (C.2.8), $K_{R[z]}(A, t)$ is the complex of $R[z]$-modules that

---

† See for example [(**37**) Theorem 1, p. 355].

one obtains by adjoining z to $K_R(A,t)$. Accordingly the sequence

$$K_{n-m-\mu+2} \otimes_R E \to K_{n-m-\mu+1} \otimes_R E \to K_{n-m-\mu} \otimes_R E$$

does not become exact when z is adjoined† and therefore it cannot itself be exact. Since this means that

$$H_{n-m-\mu+1} K_R(A,t,E) \neq 0$$

the proof is now complete.

† See the discussion immediately following Exercise 12 in Chapter 6.

References

1. AUSLANDER M. and BUCHSBAUM D. A. Homological dimension in local rings. *Trans. Amer. Math. Soc.* **85** (1957), 390–405.
2. AUSLANDER M. and BUCHSBAUM D. A. Homological dimension in Noetherian rings II. *Trans. Amer. Math. Soc.* **88** (1958), 194–206.
3. AUSLANDER M. and BUCHSBAUM D. A. Codimension and multiplicity. *Ann. Math.* **68** (1958), 625–57.
4. BARGER S. F. A theory of grade for commutative rings. *Proc. Amer. Math. Soc.* **36** (1972), 365–8.
5. BOURBAKI N. *Éléments de Mathematique, Livre II*, Chap. III, Algèbre Multilinéaire. Hermann (Paris), 1948.
6. BOURBAKI N. *Éléments de Mathematique, Livre II*, Chap. VII, Modules sur les anneaux principaux. Hermann (Paris), 1952.
7. BOURBAKI N. *Éléments de Mathematique, Algèbre Commutative*, Chaps. I, II. Hermann (Paris), 1961.
8. BUCHSBAUM D. A. A generalized Koszul complex, I. *Trans. Amer. Math. Soc.* **111** (1964), 183–96.
9. BUCHSBAUM D. A. Complexes associated with the minors of a matrix. *Symposia Mathematica, dell' Istituto Nazionale di Alta Matematica, Roma* **14** (1970), 255–83.
10. BUCHSBAUM D.A. and RIM D. S. A generalized Koszul complex, II. Depth and multiplicity. *Trans. Amer. Math. Soc.* **111** (1964), 197–224.
11. BUCHSBAUM D. A. and RIM D. S. A generalized Koszul complex, III. A remark on generic acyclicity. *Proc. Amer. Math. Soc.* **16** (1965), 555–8.
12. BUCHSBAUM D. A. and EISENBUD D. Lifting modules and a theorem on finite free resolutions. *Ring Theory*, 63–74. Academic Press, 1972.
13. BUCHSBAUM D. A. and EISENBUD D. What makes a complex exact? *Journ. of Algebra* **25** (1973), 259–68.
14. BUCHSBAUM D. A. and EISENBUD D. Some structure theorems for finite free resolutions. Preprint (Brandeis Univ.).
15. BURCH L. On ideals of finite homological dimension in local rings. *Proc. Cambridge Phil. Soc.* **64** (1968), 941–6.
16. CHEVALLEY C. *Fundamental concepts of algebra*. Monographs in Pure and Applied Mathematics, No. 7. Academic Press, 1956.
17. EAGON J. A. and NORTHCOTT D. G. Ideals defined by matrices and a certain complex associated with them. *Proc. Roy. Soc.* A **269** (1962), 188–204.

18. EAGON J. A. and NORTHCOTT D. G. Generically acyclic complexes and generically perfect ideals. *Proc. Roy. Soc.* A **299** (1967), 147–72.

19. EAGON J. A. and NORTHCOTT D. G. On the Buchsbaum–Eisenbud theory of finite free resolutions. *Journ. für die reine und angewandte Mathematik.* **262/263** (1973), 205–19.

20. FITTING H. Die Determinantenideale eines Moduls. *Jahresbericht. Deutsch Math.-Verein* **46** (1936), 195–228.

21. FLANDERS H. Tensor and exterior powers. *Journ. of Algebra* **7** (1967), 1–24.

22. GRÖBNER W. *Moderne algebraische Geometrie.* Springer, Berlin, 1949.

23. HILBERT D. Über die Theorie der algebraischen Formen. *Math. Ann.* **36** (1890), 473–534.

24. HILTON P. J. and WYLIE, S. *Homology theory, an introduction to algebraic topology.* Cambridge Univ. Press, 1960.

25. HOCHSTER M. Grade-sensitive modules and perfect modules. Preprint (Univ. of Minnesota).

26. HOCHSTER M. and EAGON J. A. A class of perfect determinantal ideals. *Bull. Amer. Math. Soc.* **76** (1970), 1026–9. Erratum. *Bull. Amer. Math. Soc.* **77** (1971), 1120.

27. HOCHSTER M. and EAGON J. A. Cohen–Macaulay rings, invariant theory, and the generic perfection of determinantal loci. *Amer. Journ. Math.* **93** (1971), 1020–58.

28. KAPLANSKY I. Projective modules. *Ann. Math.* **68** (1958), 372–7.

29. KAPLANSKY I. *Commutative rings.* Allyn and Bacon, Boston, Mass., 1970.

30. KIRBY D. A sequence of complexes associated with a matrix. *J. London Math. Soc.* (2) **7** (1973), 523–30.

31. KIRBY D. Relations between generalized Koszul complexes. *Annali di Mathematica* (4) **98** (1974), 201–20.

32. MACRAE R. E. On an application of Fitting invariants. *Journ. of Algebra* **2** (1965), 153–69.

33. MCCOY N. H. *Rings and ideals.* Carus Mathematical Monographs No. 8. Math. Asssoc. of America, 1948.

34. NAGATA M. *Local rings.* Interscience Tracts in Pure and Applied Mathematics, No. 13. J. Wiley and Sons, 1962.

35. NORTHCOTT D. G. *An introduction to homological algebra.* Cambridge Univ. Press, 1960.

36. NORTHCOTT D. G. Some remarks on the theory of ideals defined by matrices. *Quart. Journ. Math. Oxford* (2) **14** (1963), 193–204.

37. NORTHCOTT D. G. *Lessons on rings, modules and multiplicities.* Cambridge Univ. Press, 1968.

38. NORTHCOTT D. G. *A first course of homological algebra.* Cambridge Univ. Press, 1973.

39. PESKINE L. and SZPIRO C. Dimension projective finie et cohomologie locale. Publ. Math., Inst. des Hautes Etudes Sci., Paris, No. 42, 1973.

40. REES D. The grade of an ideal or module. *Proc. Cambridge Phil. Soc.* **53** (1957), 28–42.

41. SHARPE D. W. On certain ideals defined by matrices. *Quart. Journ. Math. Oxford* (2) **15** (1964), 155–75.
42. SHARPE D. W. The syzygies and semi-regularity of certain ideals defined by matrices. *Proc. London Math. Soc.* (3) **15** (1965), 645–79.
43. VASCONCELOS W. V. On finitely generated flat modules. *Trans. Amer. Math. Soc.* **138** (1969), 505–12.
44. VASCONCELOS W. V. Annihilators of modules with a finite free resolution. *Proc. Amer. Math. Soc.* **29** (1971), 440–2.

General Index

The numbers refer to pages

Index of special symbols